Short Course in Biochemistry

Short Course in Biochemistry

ALBERT L. LEHNINGER

THE JOHNS HOPKINS UNIVERSITY

SCHOOL OF MEDICINE

WORTH PUBLISHERS, INC.

SHORT COURSE IN BIOCHEMISTRY

by Albert L. Lehninger

Library of Congress Catalog Card No. 72-93199

ISBN: 0-87901-024-X

Designed by Malcolm Grear Designers, Inc.

Second printing January 1974

Worth Publishers, Inc.

444 Park Avenue South

New York, New York 10016

PREFACE

Biochemistry was once little more than a vast collection of facts about the biological occurrence of a large number of organic compounds and their functions in living organisms. But in the last ten years it has acquired a set of organizing principles which has made it a much simpler science to study, and a far more coherent way of analyzing many important problems in the life sciences. I believe that students who study biochemistry should develop an appreciation of the simplicity of these central principles.

This belief led me to write my earlier and more comprehensive textbook, *Biochemistry* (Worth Publishers, Inc., 1970). The warm reception the larger book has received, from both students and teachers in many colleges and universities, has convinced me that a shorter and more elementary version would be useful to students taking one-quarter and one-semester courses in biochemistry as part of their preparation in biology, chemistry, or the applied life and health sciences.

The central motif and the general organization of *Short Course in Biochemistry* follows that developed in *Biochemistry*. However, this is not a scissors-and-paste collection of excerpts from the larger book. It is totally rewritten in language, content, and depth of penetration. I have emphasized the main story-line of biochemistry, rather than masses of biochemical detail, yet I have tried to be rigorous in the fundamentals.

The book has four major sections:

Part I   Biomolecules
Part II   Energy-yielding processes
Part III Energy-requiring processes
Part IV Transfer of genetic information

Each section consists of a logical progression of chapters and each chapter is the equivalent of one lecture or discussion period. Some teachers may prefer to teach the structure and properties of the biomolecules in close context with their metabolism; this can be readily accomplished using the book, although I have chosen to collect the material on biomolecules and their properties into the first section for convenience of reference.

An important feature of the book is a chapter on the special biochemistry of the mammalian organism, Chapter 20. It includes the biochemical aspects of digestion; the transport of nutrients, oxygen, and carbon dioxide via the blood; and the hormonal regulation of metabolism. It provides a unique summary of the metabolic relationships among the major organs and also describes the use of metabolic energy for muscular work and for the formation of body fluids. This chapter should make the book particularly useful for students of the health sciences.

The book also contains a number of study aids for the student. Each chapter has a summary—useful for review—as well as a short list of readable references. Most chapters have a set of problems, some emphasizing quantitative aspects of biochemistry and others emphasizing important relationships in intermediary metabolism. Carefully checked answers and solutions are provided. Another important study aid is an extensive glossary of biochemical terms and definitions.

In offering this book, I shall at all times welcome comments, suggestions, and criticisms from students and teachers alike.

Acknowledgments

It is a pleasure to thank those who have helped me to prepare this book. The entire manuscript was reviewed for content, level and style by teachers of short courses in many junior colleges, colleges, and universities. Their criticisms and advice, for which I am very grateful, have been most helpful. I owe much to Linda Hansford, who gave me indispensable help in selection of material, proofreading, indexing, preparation of problems, and the planning of page layouts. Dr. Robert Egan also provided much help by reading the entire manuscript and checking the problems. I am also very grateful to my secretary, Peggy Ford, who marshalled my time and attention in response to the competing demands of teaching, research, departmental administration, and book-writing. I also wish to thank the entire staff of Worth Publishers for their encouragement and help in the writing and preparation of the book.

Finally, I must acknowledge with deep appreciation the indispensable help of my wife, who not only typed the entire manuscript through at least three drafts, but also served as my keenest critic of style.

ALBERT L. LEHNINGER

Sparks, Maryland
November, 1972

CONTENTS

Contents

INTRODUCTION   **THE MOLECULAR LOGIC OF LIVING ORGANISMS**

THE MOLECULAR LOGIC OF LIVING ORGANISMS

Living things are composed of lifeless molecules. These molecules, when isolated and examined individually, conform to all the physical and chemical laws that describe the behavior of inanimate matter. Yet living organisms possess extraordinary attributes not shown by collections of inanimate matter. If we examine some of these special properties, we can approach the study of biochemistry with a better understanding of the fundamental questions it seeks to answer.

The Identifying Characteristics of Living Matter

Perhaps the most conspicuous attribute of living organisms is that they are complicated and highly organized. They possess intricate internal structures, containing many kinds of complex molecules. Furthermore, they occur in an enormous number of different species. In contrast, the inanimate matter in our environment, as represented by soil, water, and rocks, usually consists of random mixtures of simple chemical compounds, with comparatively little structural organization.

Second, each component part of a living organism appears to have a specific purpose or function. This is true not only of intracellular structures, such as the nucleus and the cell membrane, but also of individual chemical compounds in the cell, such as lipids and proteins and nucleic acids. In living organisms it is quite legitimate to ask what the function of a given molecule is. However, to ask such questions about molecules in collections of inanimate matter is irrelevant and meaningless.

Third, living organisms have the capacity to extract and transform energy from their environment, which they use together with simple raw materials to build and maintain their own intricate structures. They can also carry out other forms of purposeful work such as the mechanical work of locomotion. Inanimate matter does not have this capacity to utilize external energy to maintain its own structural organization. In fact, inanimate matter usually decays to a more random state when it absorbs external energy such as heat or light.

But the most extraordinary attribute of living organisms is their capacity for precise self-replication, a property which can be regarded as the very quintessence of the living state. Collections of inanimate matter with which we are familiar show no apparent capacity to reproduce themselves in forms identical in mass, shape, and internal structure, through "generation" after "generation."

Biochemistry and the Living State

We may now ask: If living organisms are composed of molecules that are intrinsically inanimate, why is it that living matter differs so radically from nonliving matter, which also consists of intrinsically inanimate molecules? Why does the living organism appear to be more than the sum of its inanimate parts? The medieval philosopher would have answered that living organisms are endowed with a mysterious and divine life-force. But this doctrine, called vitalism, has been rejected by modern science. Today it is the basic goal of biochemistry to determine how the collections of inanimate molecules that constitute living organisms interact with each other to maintain and perpetuate the living state.

The molecules comprising living organisms conform to all the familiar laws of chemistry. In addition, however, they interact with each other in accordance with another set of principles, which we shall refer to collectively as the _molecular logic of the living state_. These principles do not necessarily involve any new or as yet undiscovered physical laws or forces. Rather they should be regarded as a unique set of "ground rules" that govern the nature, function, and interactions of the specific types of molecules found in living organisms, which we shall call biomolecules.

Now let us see if we can identify some of the important axioms in the molecular logic of the living state.

Biomolecules

Most of the chemical components of living organisms are organic compounds of carbon, in which the carbon is relatively reduced, or hydrogenated. Many organic biomolecules also contain nitrogen.

The organic compounds in living matter occur in extraordinary variety and most of them are extremely complex. For example, even the simplest and smallest cells, the bacteria, contain a very large number of different organic molecules. It is estimated that the bacterium *Escherichia coli* contains about 5000 different organic compounds, including some 3000 different kinds of proteins and 1000 different kinds of nucleic acids. Moreover, proteins and nucleic acids are very large molecules (often called *macromolecules*) and the structures of only a few of them are known. In the human organism, there may be as many as 5 million different kinds of proteins. None of the protein molecules of *E. coli* is identical with any of the proteins found in man, although some function in quite similar ways. In fact, each species of organism has its own chemically distinct set of protein molecules and nucleic acid molecules. Since there are probably over 1,200,000 species of living organisms, ranging in complexity from *E. coli* to the human organism, it may be calculated that all living species together must contain somewhere between 10^{10} and 10^{12} different kinds of protein molecules and about 10^{10} different kinds of nucleic acids.

For biochemists to attempt to isolate, identify, and synthesize all the different organic molecules present in living matter would appear to be a hopeless undertaking. Paradoxically, however, the immense diversity of organic molecules in living organisms is reducible to an almost absurd simplicity. All the macromolecules of cells are composed of simple, small, building-block molecules strung together in long chains. Proteins, for example, consist of covalently linked chains of 100 or more molecules of amino acids, small compounds of known structure. Only 20 different kinds of amino acids are found in proteins, but they are arranged in many different sequences to form many different kinds of proteins. Thus, all the 3000 or more proteins in the *E. coli* cell are built from only 20 different small molecules. Similarly, the 1000 or more nucleic acids of the *E. coli* cell, which also are long, polymeric molecules, are constructed from sets of only 4 different building blocks, the nucleotides. Moreover, the 20 different amino acids from which proteins are built and the sets of 4 different nucleotides from which nucleic acids are built are identical in all living species, suggesting that all living organisms had a common ancestor.

The building-block molecules from which all macromolecules are constructed have another striking characteristic. Each of them serves more than one function in living cells. The amino acids serve not only as building blocks of protein molecules but also as precursors of hormones, alkaloids, porphyrins, pigments, and many other biomolecules. The mononucleotides serve not only as building blocks of nucleic acids but also as coenzymes and as energy-carrying molecules. So far as we know, living organisms normally contain no functionless compounds, although there are some biomolecules whose functions are not yet understood.

Now we can identify some of the axioms in the molecular logic of the living state: *There is an underlying simplicity in the molecular organization of the cell. All living organisms appear to have had a common ancestor. The identity of each species of organism is preserved by its possession of a distinctive set of nucleic acids and proteins. Moreover, there is an underlying principle of molecular economy.* Living cells appear to contain only the simplest possible molecules in the least number of different types, just enough to endow them with the attribute of life and with species identity under the environmental conditions in which they exist.

Energy Transformations in Living Cells

Living organisms do not constitute exceptions to the laws of thermodynamics. Their high degree of molecular orderliness must be paid for in some way, since it cannot arise spontaneously from disorder. Living organisms absorb from their environment forms of energy that are useful to them under the special conditions of temperature and pressure in which they live and then return to the environment an equivalent amount of energy in some other, less useful, form. The useful form of energy that cells take in is *free energy*, which may be simply defined as that type of energy that can do work at constant temperature and pressure. The less useful type of energy that cells return to their environment consists largely of heat, which quickly becomes randomized in the environment and increases its disorder or entropy. Thus we have another axiom in the molecular logic of the living state: *Living organisms create and maintain their essential orderliness at the expense of their environment, which they cause to become more disordered and random.*

The energy-transforming machinery of living cells is built entirely of relatively fragile and unstable organic molecules that are unable to withstand high temperatures, strong electrical currents, or extremely acid or basic conditions. The living cell is also essentially isothermal; at any given time, all parts of the cell have essentially the same temperature. Furthermore, there are no significant differences in pressure from one part of the cell to another. For these reasons, cells are unable to use heat as a source of energy, since heat can do work at constant pressure only if it passes from a zone of higher temperature to a zone of lower temperature. Living cells therefore do not resemble heat engines or electrical engines, the types of engines with which we are most familiar. Instead, and this is another important axiom in the molecular logic of the living state: *living cells are chemical engines which function at constant temperature.* Cells extract chemical energy from their organic nutrients or from sunlight and then use it to carry out the chemical work involved in the biosynthesis of cell components, the osmotic work required to transport materials into the cell, and the mechanical work of contraction and locomotion.

Enzymes and the Catalysis of Chemical Reactions in Living Cells

Cells can function as chemical engines because they possess enzymes, catalysts capable of greatly enhancing the rate of specific chemical reactions. The enzymes are highly specialized protein molecules, made by cells from simple amino acids. Each type of enzyme can catalyze only one specific type of chemical reaction; well over a thousand different enzymes are known. Enzymes far exceed man-made catalysts in their reaction specificity, their catalytic efficiency, and their capacity to operate under mild conditions of temperature and hydrogen ion concentration. They can catalyze in milliseconds complex sequences of reactions that would require days, weeks, or months of work in the chemical laboratory. Enzyme-catalyzed reactions proceed with a 100 percent yield; there are no by-products. In contrast, the reactions of organic chemistry carried out in the laboratory with man-made catalysts are nearly always accompanied by the formation of one or more by-products. Because enzymes can enhance a single reaction pathway of a given molecule without enhancing its other possible reactions, living organisms can carry out simultaneously many different chemical reactions without bogging down in a morass of useless by-products.

The hundreds of enzyme-catalyzed chemical reactions are linked into many different sequences of consecutive reactions. Such sequences, which may have anywhere from 2 to 20 or more reaction steps, are in turn linked to form networks of converging or diverging patterns. This arrangement has several important biological implications. One is that such systems of consecutive reactions provide for the channeling of chemical reactions along specific routes. Another is that sequential reactions make possible transfer of chemical energy from one biomolecule to another.

The Energy Cycle in Cells

Living organisms recover and use energy largely in the form of one specific molecule—*adenosine triphosphate,* or *ATP.* This compound functions as the major carrier of chemical energy in the cells of all living species. As it transfers its energy to other molecules, it loses its terminal phosphate group and becomes *adenosine diphosphate,* or *ADP,* which can in turn accept chemical energy again by regaining a phosphate group to become ATP, at the expense of either solar energy in photosynthetic cells or chemical energy in animal cells. The ATP system is the connecting link between two large networks of enzyme-catalyzed reactions in the cell. One of these networks conserves chemical energy derived from the environment by causing the phosphorylation of the energy-poor ADP to the energy-rich ATP. The other network utilizes the energy of ATP to carry out the biosynthesis of cell components from simple precursors, with simultaneous breakdown of the ATP to ADP. Like the building-block biomolecules, these consecutively linked networks of enzyme-catalyzed reactions are essentially identical in all living species.

Growing cells can simultaneously synthesize thousands of different kinds of protein and nucleic acid molecules in the precise proportions required to constitute living, functional protoplasm. The enzyme-catalyzed reactions of metabolism are thus tightly regulated so as to make only the requisite number of each type of building block molecule and to assemble these into a certain number of molecules of each type of protein, each nucleic acid, and each type of lipid or polysaccharide. Living cells also possess the power to regulate the synthesis of their own catalysts. Thus the cell can "turn off" the synthesis of the enzymes required to make a given product from its precursors whenever that product is available, ready-made, from the environment. Such self-adjusting and self-regulating properties are fundamental in the maintenance of the steady state of the living cell and are essential to its energy-transforming efficiency. We may then define another axiom in the molecular logic of the living state: *Living cells are self-regulating engines, so tuned as to operate on the pervading principle of maximum economy*.

The Self-Replication of Living Organisms

The most remarkable of all the properties of living cells is their capacity to reproduce themselves with nearly perfect fidelity, not just once or twice, which would be remarkable enough, but for hundreds and thousands of generations. Three features immediately stand out. First, some living organisms are so immensely complex that the amount of genetic information that is transmitted seems out of all proportion to the minute size of the cells that must carry it, namely, the single sperm cell and the single egg cell. We know today that all this information is compressed into the nucleus of these cells, contained in the nucleotide sequence of a one or a few large *deoxyribonucleic acid* (DNA) molecules weighing altogether no more than 6×10^{-12} gram. We therefore come to another axiom in the molecular logic of the living state: *The symbols in which the genetic information is coded have the dimensions of parts of single DNA molecules*.

A second remarkable characteristic of the self-replicating property of living organisms is the extraordinary stability of the genetic information stored in DNA. Very few early historical records prepared by man have survived for long, even though they have been etched in copper or stone and preserved against the elements. The Dead Sea scrolls and the Rosetta stone, for example, are only a few thousand years old. But there is good reason to believe that present-day bacteria have nearly the same size, shape, internal structure, and contain the same kinds of building-block molecules and the same kinds of enzymes as those that lived billions of years ago, despite the fact that bacteria, like all organisms, have been undergoing constant evolutionary change. Genetic information is preserved, not on a copper scroll or etched in stone, but in the form of deoxyribonucleic acid (DNA), an organic molecule so fragile that when isolated in solution, it will break into many pieces if the solution is merely stirred or pipetted. It now appears certain that, even in the intact cell, DNA strands may break frequently, but they are quickly and automatically repaired. The remarkable capacity of living cells to preserve their genetic material is the result of _structural complementarity_. One DNA strand serves as the template for the enzymatic replication or repair of a structurally complementary DNA strand.

There is a third remarkable characteristic of genetic information transfer in living organisms. The genetic information is encoded in the form of a linear one-dimensional sequence of different building blocks of DNA. But living cells are three-dimensional in structure and they have three-dimensional parts or components. The one-dimensional information of DNA is translated into the three-dimensional information inherent in living organisms by translation of DNA structure into protein structure. Unlike DNA molecules, protein molecules spontaneously curl up and fold into a specific three-dimensional structures. The precise geometry of each type of protein is determined by its amino acid sequence.

We may now summarize the various axioms of the living state in the following statement: _A living cell is a self-assembling, self-adjusting, self-perpetuating isothermal system of molecules which exchanges matter and energy with its environment. This system carries out many consecutive organic reactions that are promoted by organic catalysts produced by the cell. It operates on the principle of maximum economy of parts and processes and its precise self-replication is ensured by a linear molecular code._

At no point in our examination of the molecular logic of living cells have we encountered any violation of known physical laws, nor have we needed to define new ones. The machinery of living cells functions within the same set of laws that governs the operation of man-made machines. However, the chemical reactions and processes of cells have been refined far beyond the present-day capabilities of chemical engineering.

In this orienting survey we have seen that biochemistry has an underlying system, a set of organizing principles. It is not merely a collection of unrelated chemical facts about living matter. As we now begin the study of biochemistry, these organizing principles should serve as our framework of reference. First we shall start with a description of the various classes of biomolecules (Part I). We shall then proceed to analyze the isothermal, self-adjusting, consecutively linked, enzyme-catalyzed reactions which make possible metabolism, the flow of matter and energy between the organism and the environment. Metabolism consists of two networks of reactions. That network which yields chemical energy as ATP will be the subject of Part II of this book. In Part III, we shall examine the other great network, that which utilizes ATP for the performance of the chemical work of biosynthesis. Finally, in Part IV we shall consider the molecular basis of the self-replication of cells and the translation of one-dimensional information of DNA into three-dimensional proteins.

PART I BIOMOLECULES

PART I BIOMOLECULES

Table I-1   The bioelements

Those in color are found in all organisms. The remainder are essential for only certain species.

The elements of organic matter
   O
   C
   N
   H
   P
   S

The monoatomic ions
   Na^+
   K^+
   Mg^{2+}
   Ca^{2+}
   Cl^-

The trace elements
   Mn
   Fe
   Co
   Cu
   Zn
   B
   Al
   V
   Mo
   I
   Si

Part I of this book is devoted to the structures and properties of the major classes of biomolecules, the term we use to refer to the characteristic organic components of living cells. As we begin the study of the biomolecules, we should of course examine their properties as we would those of nonbiological molecules, by the principles and approaches used in classical chemistry. But we must also examine them in the light of the hypothesis that the biomolecules are the products of evolutionary selection, that they may be the fittest possible molecules for their biological function, and that they interact with each other in the set of very specific relationships which we have called the molecular logic of the living state.

The elementary composition of living matter is very different from that of the lithosphere and atmosphere. Only 22 of the 100 chemical elements found in the earth's crust are essential components in living organisms (Table I-1), and of these, only 16 are found in all types of organisms. Moreover, the distribution of these elements in living organisms is not in proportion to their occurrence in the earth's crust. The four most abundant elements in the solid matter of living organisms, hydrogen, oxygen, carbon, and nitrogen, which make up about 99 percent of the mass of most cells, are far more abundant in living matter than in the earth's crust. We may therefore suppose that compounds of these elements possess unique molecular fitness for the processes that collectively constitute the living state.

Nearly all of the solid matter of living cells consists of organic compounds, compounds of the element carbon with the elements hydrogen, oxygen, and nitrogen. These four elements possess common properties. They readily form covalent bonds by electron-pair sharing. They also can combine with each other to fill their outer electron shells and thus form covalent bonds. Furthermore, three of these elements (C, N, and O) can share either one or two electron pairs to yield either single or double bonds, a capacity which endows them with considerable versatility of bonding.

Particularly significant is the capacity of carbon atoms to interact with each other to form stable, covalent carbon-carbon bonds, which is due to the fact that carbon atoms may either accept or donate four electrons to complete an outer octet. Thus each carbon atom can form covalent bonds with four other carbon atoms, to constitute the backbones of an immense variety of different organic molecules. Moreover, since carbon atoms readily form covalent bonds with oxygen, hydrogen, and nitrogen, as well as with sulfur, a large number of different kinds of functional groups can be introduced into the structure of organic molecules. Organic compounds of carbon have yet another distinctive feature important in biology. Because of the tetrahedral configuration of the shared electron pairs around each carbon atom, different types of organic molecules possess different three-dimensional structures. No other chemical element can form stable molecules of such widely different sizes and shapes, nor with such a variety of functional groups.

Table I-2 shows the approximate molecular composition of the bacterium *Escherichia coli*. We note first that water is the most abundant single compound in the *E. coli* cell, as it is in all types of cells and organisms. We also note that inorganic elements make up only a very small portion of the total solid matter of the cell contents. Four major classes of compounds make up nearly all of the large amounts of organic matter in cells: proteins, nucleic acids, carbohydrates, and lipids. Proteins (Greek *proteios*, first) are the most abundant organic molecules in *E. coli* cells, making up about one half of all the organic matter.

The four major classes of biomolecules have identical functions in all species of cells. The nucleic acids universally function to store and transmit genetic information. The proteins are the direct products and effectors of gene action, and into them the genetic information is incorporated. Most proteins have specific catalytic activ-

Table I-2 Molecular components of an *E. coli* cell

	Percent total weight	Number of each kind
Water	70	
Proteins	15	~3,000
Nucleic acids		
DNA	1	1
RNA	6	~1,000
Carbohydrates	3	~50
Lipids	2	~40
Building-block molecules and intermediates	2	~500
Inorganic ions	1	12

ity and function as enzymes; others serve as structural elements. Many other biological functions are served by proteins, which are the most versatile of all biomolecules. The polysaccharides have two major functions: some, such as starch, serve as storage forms of energy-yielding fuels for cell activity; and others such as cellulose, serve as extracellular structural elements. The lipids serve two chief roles: as major structural components of membranes, and as a storage form of energy-rich fuel.

We have seen [Introduction] that the immensely large number of different proteins and nucleic acids in living matter are made from a small number of different building-block molecules, which are identical in all species of living organism. All proteins are constructed from only 20 different amino acids and all nucleic acids are built from only a few nitrogenous bases, two sugars, and phosphoric acid. In fact, it has been calculated that well over 90 percent of all the solid organic matter of cells, containing many thousands of different compounds, are constructed from only about 40 simple, small organic molecules. Thus we need to know the structure and properties of only a relatively small number of different compounds in order to understand the organizing principles of biochemistry.

Several lines of evidence suggest that the first living cells originated about 4,000,000,000 years ago. They arose by the coming together of the first primordial biomolecules, which were probably formed on the primitive earth from the gases ammonia, methane, and water vapor exposed to such sources of energy as heat, ultraviolet light, lightning discharges, or shock waves, by processes that can easily be simulated in the laboratory. From many kinds of organic molecules generated in this manner, the first living cells arose, presumably by selection of certain specific organic molecules which happened to be more "fit" than others for the survival of the first primitive cells or parts of cells. We may therefore presume that each biomolecule we know today is probably the simplest and best adapted for its special role in the cell. We must regard the simple building block molecules of the cell with some awe and wonder, since an extraordinary and unique relationship exists among them: they are the alphabet of life.

The study of biochemistry begins with the study of water and its properties. First of all, water is the most abundant chemical compound present in living organisms. It makes up anywhere from 60 to 95 percent of the total weight of different cells, tissues, and organisms. Most tissues contain about 75 percent water. Second, water constitutes the continuous phase of living organisms: it pervades all portions of every cell and every tissue. Third, the properties of water and its ionization products, H^+ and OH^- ions, profoundly influence the properties of many important components of cells, such as enzymes, proteins, nucleic acids, and lipids. We often take water for granted as a bland, inert liquid. Actually, however, it is a highly reactive substance with rather unusual properties.

Physical Properties of Water

Water has a higher melting point, boiling point, and heat of vaporization than most common liquids (Table 1-1). These properties indicate that there are strong forces of attraction between adjacent water molecules, which give it great internal cohesion. For example, the heat of vaporization, which is defined in Table 1-1, is a direct measure of the amount of energy required to overcome the attractive forces between adjacent molecules in a liquid so that they can escape from each other and enter the gaseous state.

Table 1-1 Heat of vaporization of some common liquids, given as the number of calories of heat energy required to convert 1.0 gram of a liquid at its boiling point (atmospheric pressure) into its gaseous state at the same temperature.

	Calories per gram
Water	540
Methanol	263
Ethanol	204
Propanol	164
Acetone	125
Benzene	94
Chloroform	59

Why does liquid water show such strong inter-molecular attraction? The answer lies in the structure of the water molecule. Each of the two hydrogen atoms of the water molecule shares an electron pair with the oxygen atom. The geometry of the shared electron pairs in the outer shell of the oxygen atom causes the molecule to be V-shaped (Figure 1-1). The strong electron-withdrawing tendency of the oxygen atom gives it a local negative charge at the apex of the V, and gives the two bare hydrogen nuclei local positive charges. Although the water molecule is electrically neutral and has no net charge, its positive and negative charges are widely separated, with the result that the molecule is an *electrical dipole*. It is this fact that is largely responsible for the attractive forces between water molecules. A strong electrostatic attraction occurs between the local negative charge on the oxygen atom of one water molecule and the local positive charge on the hydrogen atom of an adjacent water molecule. This type of electrostatic interaction is called a *hydrogen bond* (Figure 1-1). Because of the nearly tetrahedral arrangement of the electrons about the oxygen atom, each water molecule theoretically can hydrogen-bond with 4 neighboring water molecules. In liquid water at room temperature each water molecule is believed to hydrogen-bond with an average of 3.4 other water molecules. But in ice each water molecule hydrogen-bonds with the maximum of 4 other water molecules to yield a regular lattice structure (Figure 1-2). In contrast, the molecules present in other common liquids, such as ethanol or benzene, show little or no tendency to attract each other. Little energy is therefore required to separate molecules of benzene from each other.

Properties of Hydrogen Bonds

Hydrogen bonds are much weaker than covalent bonds. The hydrogen bonds in liquid water are estimated to have a bond energy of only about 4.5 kcal per mole, compared with 110 kcal/mole for the $H—O$ electron-pair bonds in water molecules. (Note: Bond energy is the energy required to break a bond.) Nevertheless, hydrogen bonds are sufficiently strong to give water its great internal cohesion. Although at any given time most of the molecules in liquid water are hydrogen-bonded, the half-life of each hydrogen bond is less than a millionth of a second. Consequently, liquid water is not viscous, but very fluid; at the same time it also has strong internal cohesion. The apt term "flickering clusters" has been

Figure 1-1
The dipolar nature of the water molecule. Because of the nearly tetrahedral arrangement of the valence electron pairs the two hydrogen atoms have localized partial positive charges (δ^+) and the oxygen atom partial negative charges (δ^-). Two water molecules joined by a hydrogen bond are shown at the bottom.

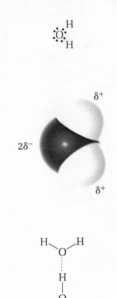

Figure 1-2
The structure of ice. (Below)
Tetrahedral hydrogen bonding around
a water molecule in ice. Molecules 1
and 2 and the central molecule are in
the plane of the paper; molecule 3 is
above it, and molecule 4 is behind it.
(Right) Regular lattice structure of ice.

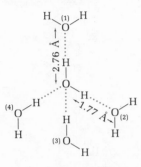

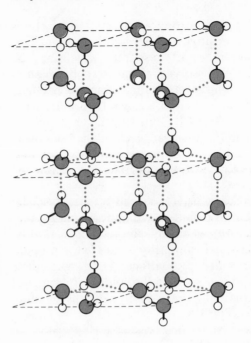

applied to the short-lived icelike groups of water molecules in liquid water.

Hydrogen bonds are not unique to water. They tend to form between an electronegative atom, such as oxygen, nitrogen, or fluorine, and an electropositive hydrogen atom covalently bonded to another electronegative atom. Hydrogen bonds may form between two molecules or between two parts of the same molecule. Some examples of biologically important hydrogen bonds are shown in Figure 1-3.

Hydrogen bonds tend to show strong directionality (Figure 1-3); they are thus capable of holding the two bonded molecules or groups in a very specific geometrical arrangement. We shall later see that this property of hydrogen bonds confers very precise three-dimensional structure to protein and nucleic acid molecules.

Solvent Properties of Water

Water is a much better solvent than most common liquids, largely because of its dipolar nature. Most crystalline salts readily dissolve in water but are nearly insoluble in nonpolar liquids such as chloroform or benzene. Since the crystal lattice of a salt such as sodium chloride is held together by very strong electrostatic attractions between alternating positive and negative ions, considerable energy is required to pull these ions away from each other. However, water dissolves crystalline sodium chloride because the strong electrostatic attraction between water dipoles and the Na^+ and Cl^- ions, to form the very stable hydrated Na^+ and Cl^- ions, greatly exceeds the tendency of Na^+ and Cl^- to attract each other. Water also dissolves many simple organic compounds having carboxyl or amino groups, which tend to ionize by interaction with water.

A second class of substances readily dissolved by water includes neutral organic compounds having polar functional groups, such as sugars, simple alcohols, aldehydes, and ketones. Their solubility is due to the propensity of water molecules to hydrogen-bond with their polar functional groups, such as the hydroxyl groups of sugars and alcohols, and the carbonyl groups of aldehydes and ketones (Fig. 1-3).

The third class of substances dispersed by water are _amphipathic_ compounds, those which contain both hydrophobic and hydrophilic groups. A simple example is the sodium salt of the long-chain fatty acid oleic acid. Because its long hydrocarbon chain is intrinsically insoluble in water, there is very little tendency for sodium oleate (a soap) to dissolve in water in the form of a truly molecular solution. However, it readily disperses in water to form aggregates called _micelles_, in which the negatively charged carboxyl groups of oleate are exposed to the water phase and the nonpolar, insoluble hydrocarbon chains are hidden within the structure (Figure 1-4). Such soap micelles remain evenly suspended in water because they are all negatively charged and thus tend to repel each other. Micelles may contain hundreds or thousands of molecules. The characteristic internal location of the nonpolar groups in micelles is the result of the tendency of the surrounding water molecules to hydrogen-bond with each other and to associate with the hydrophilic carboxyl groups, thus forcing the hydrocarbon chains into the interior of the micelle, where they have no contact with water. Water "likes" water more than it "likes" hydrocarbon chains, which cannot form hydrogen bonds. We use the term _hydrophobic bond_ or _hydrophobic interaction_ to refer to the association of the hydrophobic portions of amphipathic molecules in such micelles.

Many cell components are amphipathic and tend to form structures in which the nonpolar, hydrophobic

Figure 1-3
Some hydrogen bonds of biological importance.

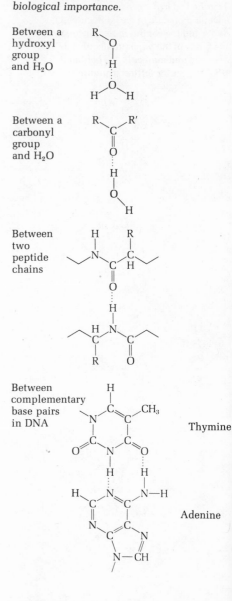

Between a hydroxyl group and H_2O

Between a carbonyl group and H_2O

Between two peptide chains

Between complementary base pairs in DNA

Thymine

Adenine

Figure 1-4
Formation of a soap micelle in water. The nonpolar tails of the sodium oleate are hidden from the water, whereas the negatively charged carboxyl groups are exposed.

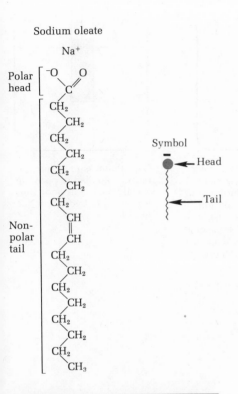

Sodium oleate

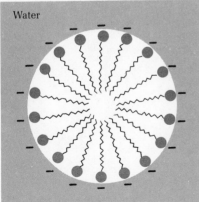

Sodium oleate micelle

parts are hidden from water, in particular, the phospholipids, the proteins, and the nucleic acids.

Effects of Solutes on Water; Colligative Properties

The properties of liquid water, particularly a group of four attributes called *colligative properties,* are profoundly modified by dissolved solutes. The term colligative means "bound together" and refers to the fact that these four properties have a common basis. They are (1) the freezing point, (2) the boiling point, (3) the vapor pressure, and (4) the osmotic pressure.

In a solution of 1.0 gram molecular weight (1.0 mole) of an ideal nonvolatile solute in 1,000 grams of water (i.e., a 1.0 molal solution) at a pressure of 760 mm of mercury, the presence of the solute depresses the freezing point of the water by 1.86°C, elevates its boiling point by 0.543°C, and yields an osmotic pressure of 22.4 atmospheres in an appropriate apparatus (Figure 1-5). An ideal solute is one which neither dissociates into two or more components or associates to reduce the total number of solute particles. The colligative properties depend only on the _number_ of solute molecules per unit volume of solvent and are independent of their chemical structure. This is because one mole of any nonionizing compound contains 6.03×10^{23} molecules (Avogadro's number). Thus 1.0 molal aqueous solutions of glycerol (mol wt 92) or glucose (mol wt 180) can be expected to have the same freezing point (−1.86°C), boiling point (100.54°C), and osmotic pressure (22.4 atmospheres), because both contain the same number of molecules per liter of water. But a 0.1 molal solution of glucose would have a freezing point depression only 0.1 as great; it would freeze at −0.186°C because it has only one-tenth the number of molecules per liter as a 1.0 molal solution. A 0.1 molal solution of NaCl, which is completely dissociated as Na+ and Cl− ions, would, on the other hand, be expected to show a freezing point of −0.372°C, since it contains twice as many solute particles per liter as a 0.1 molal solution of glucose. The colligative laws and constants hold accurately and quantitatively only in very dilute solutions.

These effects of solutes on the properties of water have considerable biological importance. For one thing, they permit fish to remain active in water at freezing temperature; the total solute concentration in the blood of

Figure 1-5
Osmosis and osmotic pressure.

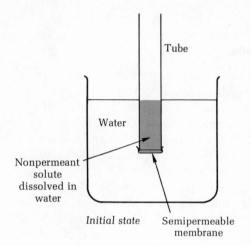

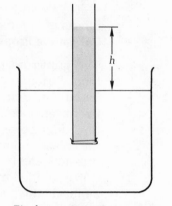

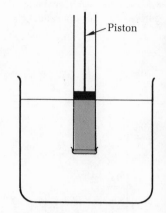

Tube

Water

Nonpermeant
solute
dissolved in
water

Initial state   Semipermeable
                          membrane

Final state. Water has moved
into solution of nonpermeant
compound. At equilibrium, the
height of column of solution
h just counterbalances the
osmotic pressure, the tendency
of water to flow toward a zone
where its concentration or
activity is less.

Osmotic pressure is force that
must be applied to piston to
exactly oppose osmotic flow.
It is equal to hydrostatic
head *h*.

Piston

the fish is sufficiently high to prevent the blood from freezing. Moreover, the concentration of those solutes in the blood which are incapable of passing across capillary membranes, particularly the proteins, gives the blood a higher osmotic pressure than the extracellular fluid. As a consequence, water from the latter tends to diffuse into the blood capillaries, thus keeping the vascular system full and preventing it from collapse.

Another way in which dissolved solutes influence the properties of water is by disturbing the hydrogen bonding among water molecules. The presence of an ionic solute such as NaCl causes a distinct change in the structure of liquid water since each Na^+ and Cl^- ion is surrounded by a shell of water dipoles. These hydrated ions have a geometry somewhat different from the clusters of hydrogen-bonded water molecules; they are more highly ordered and regular in structure. Dissolved salts thus tend to "break" the normal structure of water and change its solvent properties. We shall see later that the solubility of proteins is profoundly decreased by dissolved neutral salts, in such a way that we can use this effect to separate different proteins from each other.

Ionization of Water

Water has a slight tendency to undergo ionization. Because of the small mass of the hydrogen atom and the fact that its single electron is tightly held by the oxygen atom in the water molecule, it can leave the oxygen atom to which it is covalently bound and combine with the oxygen atom of an adjacent water molecule to yield a hydronium (H_3O^+) ion and a hydroxide (OH^-) ion

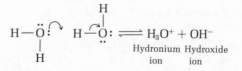

In a liter of pure water at 25°C, there are at any given time only 1.0×10^{-7} mole of H_3O^+ ions and an equal amount of OH^- ions, as shown by electrical-conductivity measurements. Henceforth, when we use the term "hydrogen ion" or the symbol H^+, it will be assumed that they refer to the hydronium ion. "Bare" hydrogen ions (i.e., protons) do not exist in water; they are always hydrated.

The Law of Mass Action and the Equilibrium Constant

The ionization of water may be described by the reaction

$$H_2O \rightleftharpoons H^+ + OH^-$$

which indicates that the reaction is reversible. In order to express the ionization of water in quantitative terms, we must review briefly some properties of reversible chemical reactions.

The position of equilibrium of any given chemical reaction is given by the *equilibrium constant,* which can be calculated if we know the concentrations of all reactants and products. For the generalized reaction

$$A + B \rightleftharpoons C + D$$

the equilibrium constant is easily derived by making use of the principle of mass action. By this principle the rate of the reaction proceeding from left to right (v_1) is proportional to the *product* of the active concentrations of the reactants A and B,

$$v_1 = k_1(A)(B)$$

where k_1 is a proportionality constant, and the parentheses ( ) indicate "active" concentration. The velocity v_2 of the reaction from right to left is, similarly,

$$v_2 = k_2(C)(D)$$

Since equilibrium is defined as the condition in which no further change in concentration is occurring and the rates of the forward and reverse reactions are equal, at equilibrium we will have the equality

$$k_1(A)(B) = k_2(C)(D)$$

Rearranging we will have

$$\frac{k_1}{k_2} = \frac{(C)(D)}{(A)(B)}$$

The ratio of the two constants k_1/k_2 can be replaced by a single new constant K_{eq}, the *equilibrium constant*

$$K_{eq} = \frac{(C)(D)}{(A)(B)}$$

This constant is fixed and characteristic for each chemical reaction at a specified temperature. It defines the composition of the final equilibrium mixture of any reaction no matter what the starting amounts of reactants and products.

The term "active" concentration used above requires comment. It refers to the *effective* concentration of each reacting species, which usually is less than its actual molar concentration, because the reacting molecules tend to be hindered somewhat in their capacity to react by surrounding molecules. The "active" concentration or *activity*, a_A, of a reactant A is given by the equation

$$a_A = [A] \times \gamma_A$$

in which [A] is the molar concentration and γ_A is the *activity coefficient* of A, a factor which corrects the molar concentration to give the "active" concentration. The activity coefficient of a compound varies with the composition of its surroundings, but approaches 1.0 at infinite dilution. In biochemistry it is customary to use molar concentrations rather than active concentrations in calculating the equilibrium constant. For the reaction

$$A + B \rightleftharpoons C + D$$

we would then write the equilibrium expression as

$$K'_{eq} = \frac{[C][D]}{[A][B]}$$

The brackets [ ] specify molar concentration and the prime sign in the symbol K_{eq} indicates that the equilibrium constant is based on molar concentrations.

The Ion Product of Water

The ionization of water

$$H_2O \rightleftharpoons H^+ + OH^-$$

is a reversible reaction for which we can write the expression

$$K'_{eq} = \frac{[H^+][OH^-]}{[H_2O]}$$

This expression can now be simplified since the concentration of H_2O is relatively very high (it is equal to the number of grams of H_2O in a liter divided by the gram molecular weight, or $1000/18 = 55.5$ M) and thus is essentially constant in relation to the very low concentration of H^+ and OH^- ions (1×10^{-7} M). We may accordingly substitute 55.5 in the equilibrium-constant expression to yield

$$K'_{eq} = \frac{[H^+][OH^-]}{55.5}$$

which on rearranging becomes

$$55.5 \times K'_{eq} = [H^+][OH^-]$$

If we now designate the term $55.5 \times K'_{eq}$ as K_w we will have

$$K_w = [H^+][OH^-] = 1 \times 10^{-14} \qquad (1)$$

K_w is called the *ion product* of water; it has the value 1.0×10^{-14} at 25°C. In pure water the concentrations of

H^+ and OH^- are exactly equal and have the value $1 \times 10^{-7} M$

$$K_w = [1 \times 10^{-7}][1 \times 10^{-7}] = 1 \times 10^{-14}$$

Note that whenever the concentration of H^+ is very high, as in a solution of an acid, the OH^- concentration must be low, and vice versa. From the ion product of water we can calculate the H^+ concentration if we know the OH^- concentration, or vice versa. For example, let us calculate the concentration of OH^- ions in a solution of 0.01 M HCl, a completely dissociated acid. We will have, from Equation (1),

$$0.01 \times [OH^-] = 1 \times 10^{-14}$$

Solving for $[OH^-]$ we will have

$$[OH^-] = \frac{1 \times 10^{-14}}{1 \times 10^{-2}}$$
$$= 1 \times 10^{-12} M \ OH^-$$

The pH Scale

K_w, the ion product of water, is the basis for the pH scale (Table 1-2), a means of designating the actual concentration of H^+ (and thus of OH^-) ions in any aqueous solution in the range of acidity between 1.0 M H^+ and 1.0 M OH^-. The term pH is defined by the expression

$$pH = \log_{10} \frac{1}{[H^+]} = -\log_{10} [H^+]$$

In a precisely neutral solution at 25°C, where the H^+ ion concentration is 1.0×10^{-7} M, the pH would be given by

$$pH = \log_{10} \frac{1}{1 \times 10^{-7}}$$
$$pH = \log_{10} (1 \times 10^7)$$
$$pH = 7.0$$

The value of 7.0 for the pH of a precisely neutral solution is thus not an arbitrarily chosen figure; it is derived from the absolute value of the ion product of water at 25°C.

It is especially important to note that the pH scale is logarithmic, not arithmetic. To say that two solutions

Table 1-2 The pH scale

$[H^+]$ (M)	pH	$[OH^-]$ (M)
1.0	0	10^{-14}
0.1	1	10^{-13}
0.01	2	10^{-12}
0.001	3	10^{-11}
0.0001	4	10^{-10}
0.00001	5	10^{-9}
10^{-6}	6	10^{-8}
10^{-7}	7	10^{-7}
10^{-8}	8	10^{-6}
10^{-9}	9	10^{-5}
10^{-10}	10	10^{-4}
10^{-11}	11	0.001
10^{-12}	12	0.01
10^{-13}	13	0.1
10^{-14}	14	1.0

Table 1-3 pH of some fluids

	pH
Seawater	7.0–7.5
Blood plasma	7.4
Interstitial fluid	7.4
Intracellular fluids	
Muscle	6.1
Liver	6.9
Gastric juice	1.2–3.0
Pancreatic juice	7.8–8.0
Saliva	6.35–6.85
Cow's milk	6.6
Urine	5–8
Tomato juice	4.3
Grapefruit juice	3.2
Soft drink (cola)	2.8
Lemon juice	2.3

differ in pH by 1 pH unit means only that one solution has ten times the hydrogen-ion concentration of the other, but does not tell us the absolute magnitude of the difference. Table 1-3 gives the pH of some fluids.

Measurement of pH

Measurement of pH is one of the most important and frequently used procedures in biochemistry since the pH determines many important features of the structure and activity of biological macromolecules, and thus of the behavior of cells and organisms. The primary standard for measurement of H^+ ion concentration (and thus of pH) is the *hydrogen electrode*. This is a specially treated platinum electrode that is immersed in the solution whose pH is to be measured. The solution is in equilibrium with gaseous hydrogen at a known pressure and temperature. The electromotive force at the electrode responds to the equilibrium

$$H_2 \rightleftharpoons 2H^+ + 2e^-$$

and can be used to calculate the H^+ ion concentration.

The hydrogen electrode is too cumbersome for general use. In the standard laboratory pH meter a glass electrode is used: it is directly sensitive to H^+ ion concentration in the absence of hydrogen gas. The response of the glass electrode must be calibrated against standard buffers of precisely known pH. Another way of measuring pH is by the use of acid-base indicators.

Acids and Bases

According to the *Brönsted-Lowry theory* of aqueous acid-base reactions, an acid is defined as a *proton donor* and a base as a *proton acceptor*. A *conjugate acid-base pair* is defined as a proton donor and its corresponding proton acceptor; for example, acetic acid (CH_3COOH) and the acetate anion (CH_3COO^-) form a conjugate acid-base pair. An acid may react with water, through transfer of a proton, to yield its conjugate base and the hydronium ion:

$$HA + H_2O \rightleftharpoons H_3O^+ + A^-$$

Each acid has a characteristic affinity for its proton in relation to the proton affinity of water. Those with a relatively high affinity are weak acids, those with a relatively low affinity are stronger acids. The tendency of an acid to give up a proton to water is given by the dissociation constant

$$K = \frac{[H_3O^+][A^-]}{[HA][H_2O]}$$

This expression is usually simplified by eliminating the water terms to

$$K' = \frac{[H^+][A^-]}{[HA]}$$

The acidic or protonic dissociation constants, often designated K_a', of some acids and bases are given in Table 1-4. Note that acids differ in their tendency to donate a proton. The stronger acids in Table 1-4, such as formic and lactic acid, have high dissociation constants, whereas the weaker acids, such as the ion $H_2PO_4^-$, have lower dissociation constants. Among the weakest acids in Table 1-4 is the NH_4^+ ion, which has only a very slight tendency to donate a proton, as shown by its very low dissociation constant. The conjugate base corresponding to the very weak acid NH_4^+ is the base NH_3, which tends to attract protons strongly.

Table 1-4 also gives values for the expression pK', which is defined by the equation

$$pK' = \log_{10} \frac{1}{K'} = -\log_{10} K'$$

The symbol p denotes "negative logarithm of" in both pH and pK'. The more strongly dissociated the acid, the lower its pK'. The stronger the base, the higher the pK' of its protonated form.

Table 1-4 Apparent dissociation constant and pK' of some acids (25°C)

Acid (proton donor)	K' (M)	pK'
HCOOH (formic acid)	1.78×10^{-4}	3.75
CH_3COOH (acetic acid)	1.74×10^{-5}	4.76
CH_3CH_2COOH (propionic acid)	1.35×10^{-5}	4.87
$CH_3CHOHCOOH$ (lactic acid)	1.38×10^{-4}	3.86
H_3PO_4 (phosphoric acid)	7.25×10^{-3}	2.14
$H_2PO_4^-$ (dihydrogen phosphate ion)	6.31×10^{-8}	7.20
HPO_4^{2-} (monohydrogen phosphate ion)	3.98×10^{-13}	12.4
H_2CO_3 (carbonic acid)	1.70×10^{-4}	3.77
HCO_3^- (bicarbonate ion)	6.31×10^{-11}	10.2
NH_4^+ (ammonium ion)	5.62×10^{-10}	9.25

Titration Curves of Acids

Figure 1-6 shows the titration curves of three acids. They are plots of the pH after each measured addition of a standard base of known concentration to the acid in question. The shapes of such titration curves are very similar from one acid to another. However, the important difference is that the curves are displaced vertically along the pH scale.

The titration curve reveals much information. The initial pH of the solution of an acid HA, before any base has been added, gives us the concentration of H^+ ions in the solution, from which we can calculate the dissociation constant of the acid being titrated. As each addition of base is made the hydroxide ions so added combine with the free protons to yield H_2O, thus causing the acid HA to dissociate further to maintain its own dissociation equilibrium. At the midpoint of the titration, at which exactly 0.5 equivalent of base has been added, exactly

Figure 1-6
Titration curves of three acids. The predominating ionic species are shown in boxes.

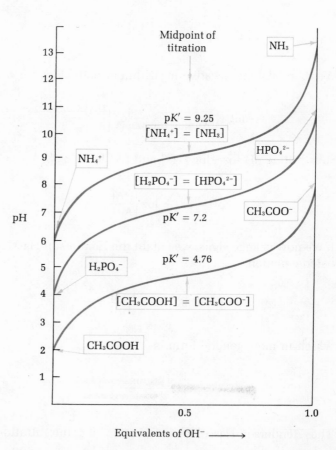

one-half of the original acid has undergone dissociation, so that [HA] now equals [A⁻]. At this midpoint the pH of the solution is exactly equal to the pK' of the acid titrated, as is evident on comparing the values in Table 1-4 and Figure 1-6. As we continue the titration by adding further NaOH, the remaining HA ultimately is completely converted to A⁻ by the removal of protons and their reaction with the added OH⁻.

The shape of the titration curve of any acid is expressed by the *Henderson-Hasselbalch equation*, which is important not only for the understanding of buffer action, which we shall consider below, but also for an understanding of acid–base balance in cells and tissues of the mammalian organism. It is simply derived from the expression for the dissociation constant of an acid,

$$K' = \frac{[H^+][A^-]}{[HA]}$$

First we solve for [H⁺]:

$$[H^+] = K' \frac{[HA]}{[A^-]}$$

We then take the negative logarithm of both sides:

$$-\log[H^+] = -\log K' - \log \frac{[HA]}{[A^-]}$$

Substituting pH for $-\log$ [H⁺] and pK' for $-\log K'$ we get

$$pH = pK' - \log \frac{[HA]}{[A^-]}$$

If we now change signs, we obtain the Henderson-Hasselbalch equation:

$$pH = pK' + \log \frac{[A^-]}{[HA]}$$

which in more general form is

$$pH = pK' + \log \frac{[\text{proton acceptor}]}{[\text{proton donor}]}$$

The Henderson-Hasselbalch equation fits the titration

curve of all weak acids. From it we can now see why the pK' of a weak acid is equal to the pH of the solution at the midpoint of the titration. At this point [HA] = [A⁻] and we have

$$pH = pK' + \log 1.0$$
$$pH = pK' + 0$$
$$pH = pK'$$

The Henderson-Hasselbalch equation also makes it possible to calculate the pK' of any acid from the molar ratio of proton-donor and proton-acceptor species at a given pH, to calculate the pH of a conjugate acid-base pair of a given pK' and a given molar ratio, and to calculate the molar ratio of proton donor and proton acceptor given the pH and pK'. Two sample problems follow.

1. Calculate the pK' of lactic acid, given the fact that when the concentration of free lactic acid is 0.010 M and the concentration of lactate ion is 0.087 M, the pH is 4.80.

$$pH = pK' + \log \frac{[\text{lactate}]}{[\text{lactic acid}]}$$

$$pK' = pH - \log \frac{[\text{lactate}]}{[\text{lactic acid}]}$$

$$= 4.80 - \log \frac{0.087}{0.010}$$

$$= 4.80 - \log 8.7$$
$$= 4.80 - 0.94$$
$$= 3.86$$

2. Calculate the pH of a mixture of 0.1 M acetic acid and 0.2 M sodium acetate. The pK' of acetic acid is 4.76.

$$pH = pK' + \log \frac{[\text{acetate}]}{[\text{acetic acid}]}$$

$$= 4.76 + \log \frac{0.2}{0.1}$$

$$= 4.76 + 0.301$$
$$= 5.06$$

Buffers

Figure 1-6 shows that the titration curve of each acid has a relatively flat zone extending about 1.0 pH on either side of its midpoint. In this zone, the pH of the system changes relatively little when increments of H^+ or OH^- are added. This is the zone in which a conjugate acid-base pair acts as a *buffer*, that is, a system which tends to resist change in pH when small amounts of H^+ or OH^- are added. At pH values outside this zone, there is much less capacity to resist changes in pH. The buffering power of any conjugate acid–base pair is maximum at the exact midpoint of its titration curve, that is, when the ratio [proton acceptor]/[proton donor] = 1.0, as is evident from Figure 1-6. Thus the acetic acid–acetate pair is most effective in buffering capacity at pH 4.76 whereas the lactic acid–lactate pair is most effective at pH 3.86.

Intracellular and extracellular fluids of living organisms contain conjugate acid-base pairs which act as buffers. The major intracellular buffer system is the conjugate acid-base pair $H_2PO_4^- - HPO_4^{2-}$ (pK' 7.2). The blood plasma and interstitial fluid of vertebrates also contains very active buffers. The extraordinary buffering power of blood plasma may be shown by the following comparison. If 1 ml of 10 N HCl is added to 1.0 liter of physiological saline (0.15 M NaCl) at pH 7.0, the pH of the saline will fall to pH 2.0, since NaCl solutions have no buffering power. However, if 1 ml of 10 N HCl is added to 1 liter of blood plasma, the pH will decline only slightly, from pH 7.4 to about pH 7.2.

The major buffer present in blood plasma is the bicarbonate buffer system ($H_2CO_3 - HCO_3^-$). While this conjugate acid-base pair functions as a buffer in the same way as other conjugate acid-base pairs, the proton-donor species carbonic acid is in reversible equilibrium with dissolved CO_2

$$H_2CO_3 \rightleftharpoons CO_2(diss) + H_2O$$

If such an aqueous system is in contact with a gas phase, then the dissolved CO_2 will, in turn, equilibrate between the gaseous and aqueous phases

$$CO_2(diss) \rightleftharpoons CO_2(gas)$$

Physiologically, the bicarbonate system can buffer effectively near pH 7.0 because the proton donor H_2CO_3 in the blood is in labile equilibrium with a relatively large reserve capacity of gaseous CO_2 in the lungs. Under any conditions in which the blood must absorb excess OH^-, the H_2CO_3 which is used up and converted to HCO_3^- is quickly replaced from the large pool of gaseous CO_2 in the lungs.

The pH of blood plasma in vertebrates is held at remarkably constant values. The blood plasma of man normally has a pH of 7.40. Should the pH-regulating mechanisms fail, as may happen in disease, and the pH of the blood fall below 7.0 or rise above 7.8, irreparable damage may occur. We may ask: What molecular mechanisms in cells are so extraordinarily sensitive that a change in H^+ concentration of as little as 3×10^{-8} M (approximately the difference between blood at pH 7.4 and blood at pH 7.0) can be lethal? Although many aspects of cell structure and function are influenced by pH, it is the catalytic activity of enzymes that is especially sensitive. The typical curves in Figure 1-7 show that enzymes have maximal activity at a characteristic pH, called the *optimum pH*, and that their activity often declines sharply on either side of the optimum. Thus biological control of the pH of cells and body fluids is of central importance in all aspects of intermediary metabolism and cellular function.

Figure 1-7
The effect of pH on the activity of some enzymes. Each enzyme has a characteristic pH–activity profile.

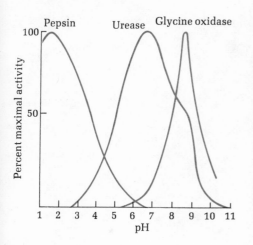

The Fitness of the Aqueous Environment for Living Organisms

Living organisms have effectively adapted to their aqueous environment and have even evolved means of exploiting the unusual properties of water. The high specific heat of water is useful to the cell since it allows water to act as a "heat buffer," allowing the temperature of the cell to remain relatively constant as the temperature of the environment fluctuates. Furthermore, the high heat of evaporation of water is exploited as an effective means for vertebrates to lose heat by evaporation of sweat. The high degree of internal cohesion of liquid water, due to hydrogen bonding, is exploited by higher plants to transport dissolved nutrients from the roots up to the leaves during the process of transpiration. Even the fact that ice has a lower density than liquid water and therefore floats has important biological consequences in the life of aquatic organisms. But most fundamental to all living organisms is the fact that many important biological properties of cell macromolecules, particularly the proteins and nucleic acids, derive from their interactions with water molecules of the surrounding medium.

Summary

Water is the most abundant compound in living organisms. Its relatively high freezing point, boiling point, and heat of vaporization are the result of strong intermolecular attractions in the form of hydrogen bonding between water molecules. Liquid water has considerable short-range order and consists of "flickering clusters" of very short half-life. The polarity and hydrogen-bonding properties of the water molecule make it a potent solvent for many ionic compounds and neutral molecules. Water also disperses amphipathic molecules, such as soaps, to form micelles, clusters of molecules in which the hydrophobic groups are hidden from exposure to water and the charged groups are located on the external surface.

Water ionizes very slightly to form hydronium (H_3O^+) and hydroxide (OH^-) ions. In dilute aqueous solutions, the concentrations of H^+ and OH^- ions are inversely related by the expression $K_w = [H^+][OH^-] = 1 \times 10^{-14}$ (25°C). The hydrogen-ion concentration of biological systems is expressed in terms of pH, defined as $pH = -\log [H^+]$. The pH of aqueous solutions is measured by means of the glass electrode.

Acids are defined as proton donors and bases as proton acceptors. A conjugate acid-base pair consists of a proton donor HA and its corresponding proton acceptor A^-. The tendency of an acid HA to donate protons is expressed by its dissociation constant ($K' = [H^+][A^-]/[HA]$), or by the function pK', defined as $-\log K'$. The pH of a solution of a weak acid is quantitatively related to its pK' and to the ratio of the concentrations of its proton-donor and proton-acceptor species by the Henderson-Hasselbalch equation. A conjugate acid-base pair can act as a buffer and resist changes in pH; its capacity to do so is greatest at the pH numerically equal to its pK'. The most important biological buffer pairs are H_2CO_3–HCO_3^- and $H_2PO_4^-$–HPO_4^{2-}. The catalytic activity of enzymes is strongly influenced by pH.

References

DAWES, E. A.: *Quantitative Problems in Biochemistry,* 2d ed., The Williams & Wilkins Company, Baltimore, 1968. Problems and solutions in acid-base biochemistry.

LEHNINGER, A. L.: *Biochemistry: The Molecular Basis of Cell Structure and Function,* Worth Publishers, New York, 1970. Chapter 2 gives a more detailed treatment of the properties of water.

EISENBERG, D., and W. KAUZMANN: *The Structure and Properties of Water,* Oxford University Press, Fair Lawn, N.J., 1969. Physical chemistry of water.

Problems

1. Calculate the pH of the following solutions:
   (a) 0.01 N NaOH
   (b) 0.001 N HCl
   (c) 1×10^{-4} N HCl
   (d) 0.03 N NaOH
   (e) 3.4×10^{-3} N NaOH

2. What is the actual hydrogen ion concentration in the following fluids:
   (a) Lemon juice (pH 2.3)
   (b) Cola drink (pH 2.8)
   (c) Saliva (pH 6.5)
   (d) Gastric juice (pH 1.8)
   (e) Milk (pH 6.6)

3. The pH of a 0.01 M solution of a given acid is 3.80. Calculate its apparent dissociation constant K'_{diss}.

4. What is the pK' of the acid of Problem 3?

5. It is desired to prepare a solution which can serve as a buffer at pH 4.0. Which of the following would be best for this purpose?
   (a) Propionic acid (pK' 4.87)
   (b) Lactic acid (pK' 3.86)
   (c) Phosphoric acid (pK'_I 2.14; pK'_{II} 7.20)

6. What ratio of the proton donor and proton acceptor species of the acid selected in problem 5 would yield a buffer having a pH of exactly 4.0?

7. How would you prepare such a buffer starting from a 0.1 M solution of the acid you selected and a 0.1 N solution of sodium hydroxide?

8. Calculate the pH of the solution resulting when 20.0 ml of 0.05 M lactic acid is mixed with 10.0 ml of 0.02 M NaOH.

CHAPTER **2** AMINO ACIDS AND PEPTIDES

The primary building blocks of all proteins, regardless of the species of origin, are a group of 20 different α-amino acids. Each of these has a distinctive side chain which lends it chemical individuality. Indeed, this group of 20 amino acids may be regarded as the alphabet of protein structure.

We shall now survey the chemical and physical properties of amino acids, since they determine many of the important properties of protein molecules. We shall also examine the structure of peptides, chains of amino acids joined by peptide bonds.

The Common Amino Acids of Proteins

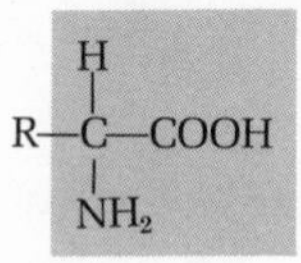

All of the 20 amino acids commonly found in proteins have as common denominators at least one carboxyl group and an amino group on the α-carbon atom (Figure 2-1). The α-amino group is free or unsubstituted in all the amino acids except one, proline. Each amino acid also has a characteristic side chain or R group. The different amino acids have been assigned three-letter symbols, which are used in indicating the composition and sequence of amino acids in polypeptide chains (Table 2-1).

The R groups vary in structure, size, and in their tendency to interact with water, a reflection of their polarity. The most meaningful way of classifying the various amino acids is on the basis of the polarity of their

Figure 2-1
The general structural formula for the α-amino acids found in proteins. The shaded portion is common to all α-amino acids.

35

R groups in water near pH 7.0. There are four main classes: (1) nonpolar or hydrophobic, (2) polar but uncharged, (3) positively charged, (4) negatively charged.

Amino Acids with Nonpolar R Groups

The nonpolar R groups in this class of amino acids are hydrocarbon in nature and thus tend to be hydrophobic (Figure 2-2). The group includes five amino acids with aliphatic R groups (alanine, leucine, isoleucine, valine, and proline), two with aromatic rings (phenylalanine and tryptophan), and one containing sulfur (methionine).

Of this group, proline requires special mention in that its α-amino group is substituted with a portion of its R group to yield a cyclic structure.

Amino Acids with Uncharged Polar R Groups

The uncharged polar R groups of these amino acids (Figure 2-2) are more soluble in water than those of the nonpolar amino acids because they contain functional groups that can hydrogen-bond with water. This group includes serine, threonine, tyrosine, asparagine, glutamine, cysteine, and glycine. The polarity of serine, threonine, and tyrosine is contributed by their hydroxyl groups; that of asparagine and glutamine by their amide groups; and that of cysteine by its sulfhydryl or thiol group. The R group of glycine, a single hydrogen atom, is too small to influence the high degree of polarity of the α-amino and α-carboxyl groups.

Asparagine and glutamine are derivatives of two other amino acids found in proteins. They are the amides of aspartic acid and glutamic acid, to which they are easily hydrolyzed by acid or base. Two amino acids of this class, cysteine and tyrosine, have R groups that tend to dissociate H^+ ions. The sulfhydryl group of cysteine and the phenolic hydroxyl group of tyrosine are slightly ionized at pH 7.0.

Cysteine requires special mention for another reason. It may occur in proteins in two forms, either as cysteine itself or as cystine, in which two cysteine molecules are linked together by a disulfide bridge formed by oxidation of their thiol groups (Figure 2-3). Cystine plays a special role in protein structure since its two cysteine half-molecules may be present in two different polypeptide chains, which are thus cross-linked by the disulfide linkage.

Table 2-1 Amino acid symbols

Amino acid	Three-letter symbol
Alanine	Ala
Arginine	Arg
Asparagine	Asn
Aspartic acid	Asp
Asn + Asp	Asx
Cysteine	Cys
Glutamine	Gln
Glutamic acid	Glu
Gln + Glu	Glx
Glycine	Gly
Histidine	His
Isoleucine	Ile
Leucine	Leu
Lysine	Lys
Methionine	Met
Phenylalanine	Phe
Proline	Pro
Serine	Ser
Threonine	Thr
Tryptophan	Trp
Tyrosine	Tyr
Valine	Val

Figure 2-2
The 20 amino acids common in proteins. They are shown with their amino
groups and carboxyl groups ionized, as they would occur at pH 7.0.

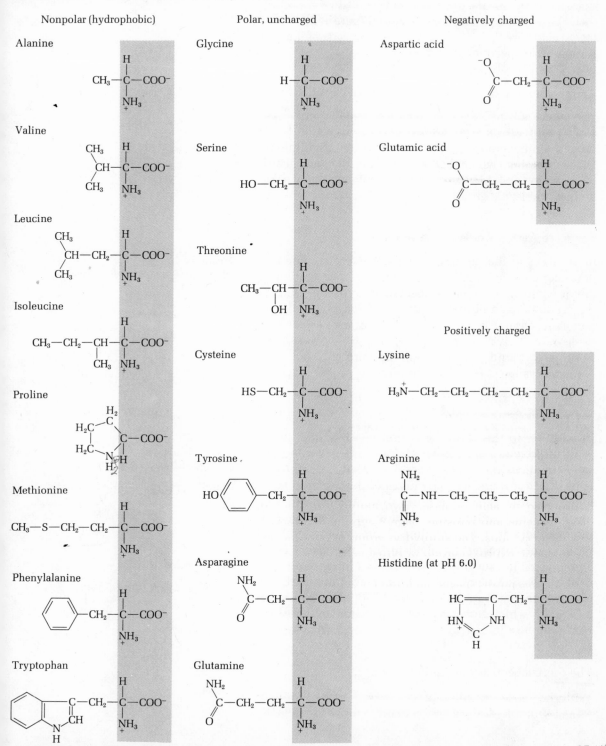

Amino Acids with Negatively Charged (Acidic) R Groups

The two amino acids whose R groups have a net negative charge at pH 7.0 are *aspartic acid* and *glutamic acid*, each with a second carboxyl group (Figure 2-2). These amino acids are the parent compounds of asparagine and glutamine, respectively (see above).

Amino Acids with Positively Charged (Basic) R Groups

The amino acids (Figure 2-2) in which the R groups have a net positive charge at pH 7.0 are *lysine*, which has a second amino group at the ε position on its aliphatic chain, *arginine*, which has a positively charged guanidine group, and *histidine*, which contains the weakly ionized imidazole group.

The Rare Amino Acids of Proteins

In addition to the 20 amino acids that are common in proteins, there are a few others that have been found as minor components of some specialized types of proteins (Figure 2-4). Each of them is derived from one of the 20 common amino acids. Among these are *4-hydroxyproline*, a derivative of proline, and *5-hydroxylysine*. Both are found in the fibrous protein collagen. *N-methyllysine* is found in the muscle protein myosin.

Nonprotein Amino Acids

In addition to the 20 common and several rare amino acids of proteins, many other amino acids occur biologically in either free or combined form, but never in proteins. Most of these are derivatives of the α-amino acids found in proteins, such as *ornithine* and *citrulline* (Figure 2-5), which are derivatives of arginine and serve as intermediates in the formation of urea, the major end product of nitrogen metabolism in mammals. Some amino acids in this group have their amino group in the β or γ position, such as *β-alanine*, an important precursor of the vitamin panthothenic acid, and *γ-aminobutyric acid*, which serves as a chemical transmitter of the nerve impulse in certain portions of the nervous system.

The Stereochemistry of Amino Acids

With the single exception of glycine, all amino acids obtained from hydrolysis of proteins under sufficiently

Figure 2-3
Cystine.

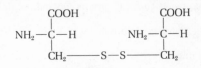

Figure 2-4
Three "rare" amino acids of proteins.

4-Hydroxyproline

5-Hydroxylysine

$$NH_2CH_2CHCH_2CH_2CHCOOH$$
$$\quad\quad\quad OH \quad\quad\quad NH_2$$

ε-N-Methyllysine

$$CH_3NHCH_2CH_2CH_2CH_2CHCOOH$$
$$\quad\quad\quad\quad\quad\quad\quad\quad NH_2$$

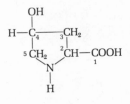

Figure 2-5
*Some naturally occurring amino
acids not found in proteins.*

Ornithine

$NH_2CH_2CH_2CH_2CHCOOH$
$\quad\quad\quad\quad\quad\quad\quad\quad NH_2$

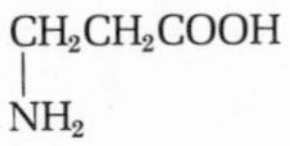

Citrulline

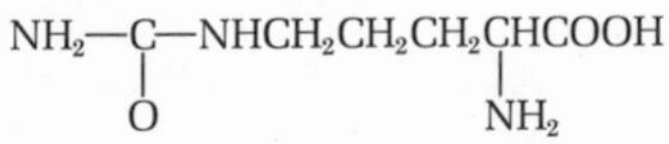

$NH_2-C-NHCH_2CH_2CH_2CHCOOH$
$\quad\quad\quad|\quad\quad\quad\quad\quad\quad\quad\quad\quad\quad|$
$\quad\quad\quad O\quad\quad\quad\quad\quad\quad\quad\quad\quad\quad NH_2$

β-Alanine

CH_2CH_2COOH
$|$
NH_2

γ-Aminobutyric acid

$CH_2CH_2CH_2COOH$
$|$
NH_2

Figure 2-6
*An asymmetric carbon atom. The four
substituent groups W, X, Y, and Z are
arranged to yield two isomers which are
nonsuperimposable mirror images. They
are shown in two conventions. In projection
formulas the horizontal bonds are assumed
to project forward from the plane of the
page and the vertical bonds behind it. In
perspective formulas wedges are used to
designate bonds projecting forward and
dotted lines those projecting behind.*

Projection formulas

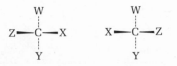

Perspective formulas

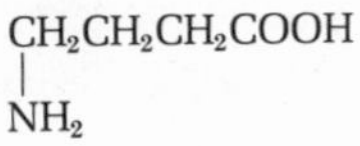

mild conditions show optical activity, that is, they can rotate the plane of plane-polarized light. Optical activity is given by all compounds capable of existing in two forms whose structures are nonsuperimposable mirror images of each other (Figure 2-6). This condition is met by compounds having an asymmetric carbon atom, that is, one with four different substituents. Because of the tetrahedral nature of valence bonds of the carbon atom, the four different substituent groups can occupy two different arrangements in space around the central carbon atom. Such a compound will have two different optical isomers, also called *stereoisomers* or *enantiomers*: one will rotate the plane of polarized light to the left (counterclockwise) and is called the *levorotatory* isomer [designated (−)], and the other will rotate the plane to the same extent, but to the right (clockwise), and is called the *dextrorotatory* isomer [designated (+)].

When a compound has two or more nonidentical asymmetric carbon atoms it will have 2^n possible stereoisomers, where n is the number of asymmetric carbon atoms. Glycine has no asymmetric carbon atom, and therefore cannot exist in stereoisomeric forms. All the rest of the amino acids commonly found in proteins have one asymmetric carbon, except threonine and isoleucine, which possess two and thus have four stereoisomers.

The optical activity of a stereoisomer is expressed quantitatively by its *specific rotation*, determined from measurements of the optical rotation of a given concentration in a tube of a given length in a polarimeter.

$$[\alpha]_D^{25°} = \frac{\text{observed rotation (degrees)}}{\text{length of tube (dm)} \times \text{concentration (grams/ml)}}$$

The temperature and the wavelength of the light employed (usually the D line of sodium, 5893 Å) must be specified.

However, there is a more fundamental and systematic way of classifying and naming stereoisomers than designating them as levorotatory or dextrorotatory. All optically active compounds are classified according to the absolute configuration of the four different substituents in the tetrahedron around the asymmetric carbon atom. For this purpose an arbitrarily chosen compound is used as a reference or standard, to which all other optically active compounds are compared. This reference compound is the three-carbon sugar *glyceraldehyde*, the smallest sugar to have a single asymmetric carbon atom (the chemistry of sugars will be considered in Chapter 5).

By convention, the two possible stereoisomers of glyceraldehyde are designated L and D (note the use of small capital letters). They possess the configurations shown in Figure 2-7, which have been established by x-ray analysis. Directly underneath the stereoisomers of glyceraldehyde are shown the two corresponding stereoisomers of the amino acid alanine. We see that the substituent amino group on the asymmetric carbon atom of L-alanine is sterically related to the substituent hydroxyl group on the asymmetric carbon atom of L-glyceraldehyde, the carboxyl group of L-alanine is related to the aldehyde group of L-glyeraldehyde, and the R group of L-alanine is related to the —CH_2OH group of L-glyceraldehyde. In a similar way, the absolute configuration of D-alanine is related to that of D-glyceraldehyde. Thus the stereoisomers of all the naturally occurring amino acids can be structurally related to the two stereoisomers of glyceraldehyde. All stereoisomers that are related to L-glyceraldehyde are designated L, and those that are related to D-glyceraldehyde are designated D, _regardless of the direction of rotation of plane-polarized light given by the isomers_. The symbols D and L thus refer to _absolute configuration_, not direction of rotation. Table 2-2 shows the specific rotation of some L- and D-amino acids. It is clear that some L-amino acids are levorotatory and some are dextrorotatory at pH 7.0.

Whenever the absolute configuration of a compound having an asymmetric carbon atom is known, it is the convention to designate it by D or L; specification of the direction of rotation is then unnecessary. But if the absolute configuration of an optically active compound has not been established, then by convention such compounds may be designated (+) or (−) to indicate direction of rotation.

All naturally occurring amino acids found in proteins, with the exception of glycine, are L-amino acids. D-Amino acids do occur biologically, but never in proteins. An example is D-glutamic acid, an important component in the structure of bacterial cell walls and of certain antibiotics.

In general, organic compounds having optical activity are not normally found as the products of nonbiological chemical reactions. For example, when a compound with an asymmetric carbon atom is synthesized by the organic chemist, it is optically inactive. This is because it is formed as a _racemate_, an equimolar mixture of the D and L stereoisomers, which is symbolized by the prefix DL-. On the other hand, the amino acids synthesized in the

Figure 2-7

Relationship of stereoisomers of alanine to stereoisomers of glyceraldehyde. They are shown in both projection and perspective formulas.

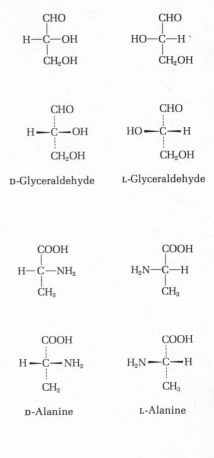

D-Glyceraldehyde   L-Glyceraldehyde

D-Alanine   L-Alanine

Table 2-2 Specific rotation of some L- and D-amino acids

	$[\alpha]_D^{25°}$
L-Alanine	+1.8
D-Alanine	−1.8
L-Glutamic acid	+12.0
D-Glutamic acid	−12.0
L-Lysine	+13.5
D-Lysine	−13.5
L-Leucine	−11.0
L-Phenylalanine	−34.5
L-Aspartic acid	+5.0

Figure 2-8
The ninhydrin test.

Ninhydrin

Blue pigment formed on
reaction with α-amino acids

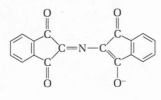

Figure 2-9
Formation of 2,4-dinitrophenyl derivatives
of amino acids

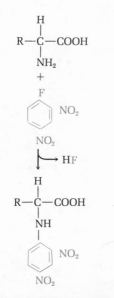

2,4-Dinitrophenylamino acid

cell are optically active because the enzymes forming them have stereochemical specificity.

Racemates can be resolved into their D and L isomers only by laborious fractionation procedures. Optically active amino acids can be converted into racemic mixtures by boiling with a strong base. However, exposure of optically active amino acids to boiling acid does not cause their racemization. For this reason, if we wish to preserve the optical activity of the amino acid components of a protein, the latter is hydrolyzed by boiling with strong acid.

The Chemical Reactions of Amino Acids

As is true for all organic compounds, the chemical reactions of amino acids are those characteristic of their functional groups, such as the amino and carboxyl groups, as well as other functional groups present in the amino acid side chains. We shall not examine all the organic reactions of amino acids. However, there are two reactions which are practical and widely used for the detection and measurement of amino acids. The first is the ninhydrin reaction (Figure 2-8), which is used to estimate amino acids quantitatively in very small amounts. Heating with excess ninhydrin yields a blue product with all amino acids having a free α-amino group, whereas proline, in which the α-amino group is substituted, yields a derivative having a characteristic yellow color.

A second important reaction of the α-amino group is with the reagent 1-fluoro-2,4-dinitrobenzene (abbreviated FDNB), introduced by Sanger for the quantitative labeling of amino groups in amino acids and peptides. In mildly alkaline solution, FDNB reacts with α-amino acids to yield yellow 2,4-dinitrophenyl derivatives (Figure 2-9). As we shall see, this reaction is extremely valuable in the identification of the amino-terminal amino acid of polypeptide chains (Chapter 3).

The Acid-Base Properties of Amino Acids

Knowledge of the acid-base properties of amino acids is extremely important in understanding many properties of proteins. Moreover, the entire art of separating, identifying, and quantitating the different amino acids, which are necessary steps in determining the amino acid composition and sequence of proteins, is based on their characteristic acid-base behavior.

Amino acids crystallize from neutral aqueous solutions in fully ionized form, as *dipolar ions*, or *zwitterions*, rather than in their un-ionized forms (Figure 2-10), a fact suggested by the rather high melting points of crystalline amino acids, which exceed 200°C. Although such dipolar ions are electrically neutral and do not move in an electrical field, they have opposite electrical charges at their two poles. When they crystallize, the crystal lattice is held together by strong electrostatic forces between positive and negatively charged groups of adjacent molecules, resembling the stable crystal lattice of NaCl. Very high temperatures must be applied to such an ionic lattice to separate the positive and negative charges from each other. In contrast, most simple un-ionized organic compounds of low molecular weight have relatively low melting points, consonant with their relatively "soft" and unstable nonionic crystal lattices.

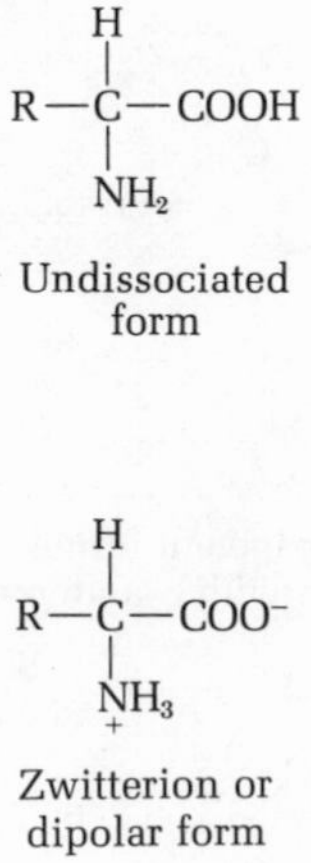

When a crystalline zwitterionic amino acid such as alanine is dissolved in water, it can act either as an acid (proton donor) or as a base (proton acceptor).

As an acid:

$$\overset{+}{N}H_3CH(CH_3)COO^- \rightleftharpoons H^+ + NH_2CH(CH_3)COO^-$$

As a base:

$$H^+ + \overset{+}{N}H_3CH(CH_3)COO^- \rightleftharpoons \overset{+}{N}H_3CH(CH_3)COOH$$

Substances having this property are *amphoteric* (Gr. *amphi*, both), and are often called *ampholytes*, a term abbreviated from "amphoteric electrolytes." The acid-base behavior of ampholytes, including the amino acids, is most simply treated in terms of the Brönsted-Lowry theory of acids and bases (Chapter 1). A simple mono-amino monocarboxylic α-amino acid such as alanine is considered to be a diprotic acid when it is fully protonated, that is, when both its carboxyl group and amino group have accepted protons. In this form it can donate two protons during its complete titration with a base. The course of such a two-stage titration with sodium hydroxide can be represented in the following equations, which indicate the nature of each ionic species involved:

$$^+NH_3CHRCOOH \longrightarrow {}^+NH_3CHRCOO^- + H^+$$

$$^+NH_3CHRCOO^- \longrightarrow NH_2CHRCOO^- + H^+$$

Figure 2-10
Undissociated and zwitterion forms of amino acids.

Undissociated
form

Zwitterion or
dipolar form

Figure 2-11 shows the titration curve of alanine. It has two distinct stages, corresponding to the titration of the two protons of the fully protonated species. Each leg of the curve resembles the typical titration curve of a mono-protic acid (Chapter 1). Thus each leg has a midpoint at which the pH of the system is equal to the pK' of the protonated group being titrated. The first leg of the curve, with a midpoint at pH 2.34, corresponds to the removal of a proton from the carboxyl group. The second leg, with its midpoint at pH 9.69, corresponds to the removal of a proton from the $—NH_3^+$ group. At pH 2.34 equimolar concentrations of proton donor ($^+NH_3CHRCOOH$) and proton acceptor ($^+NH_3CHRCOO^-$) species are present. At pH 9.69, equimolar concentrations of $^+NH_3CHRCOO^-$ and $NH_2CHRCOO^-$ are present. Each of the two legs of the biphasic curve can be expressed mathematically by the Henderson-Hasselbalch equation (Chapter 1). We can therefore calculate the ratios of the ionic species at any pH, given the values for pK_1' and pK_2'.

At pH 6.02, there is a point of inflection between the two separate legs of the titration curve of alanine. There is no <u>net</u> electrical charge on the molecule at this pH, and the molecule will not move in an electrical field.

Figure 2-11
The titration curve of alanine. The ionic species predominating at various pH values are shown in boxes.

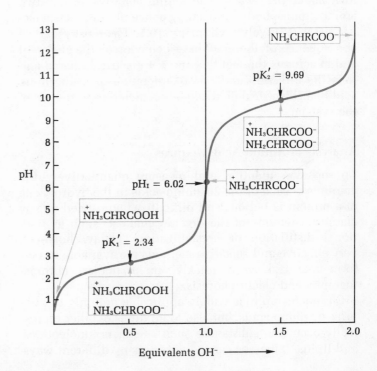

This pH value is called the *isoelectric pH* (symbolized pH_I). It is the arithmetic mean of the two pK' values

$$pH_I = \tfrac{1}{2}(pK'_1 + pK'_2)$$

Thus the isoelectric pH of alanine, 6.02, is the mean of 2.34 (pK'_1) and 9.69 (pK'_2). At any pH above the isoelectric pH the amino acid has a net negative charge and at any pH below the isoelectric pH it has a net positive charge.

All amino acids having a single α-amino group and a single carboxyl group, with no other ionizing groups, have titration curves closely resembling that of alanine. This group, which includes glycine, alanine, leucine, isoleucine, phenylalanine, and valine, among others, is characterized by very similar values for pK'_1, about 2.2, and for pK'_2, about 9.7.

Those amino acids with an ionizable R group have more complex titration curves, with three legs corresponding to the three dissociation steps. However, the third or extra leg usually merges to some extent with one of the others. For example, the amino acids aspartic acid and glutamic acid have two carboxyl groups and thus two corresponding proton dissociation stages below pH 7.0 (Figure 2-12). Their isoelectric pH is thus relatively low, about $pH_I = 4.0$. The amino acid lysine has two amino groups, both dissociating above pH 7.0; its isoelectric pH is relatively high, $pH_I = 10.5$. The most practical consequence of the acid–base behavior of the different amino acids is the fact that the net electrical charge and thus the direction and rate of migration of each amino acid in an electrical field can be predicted from the pH of the system.

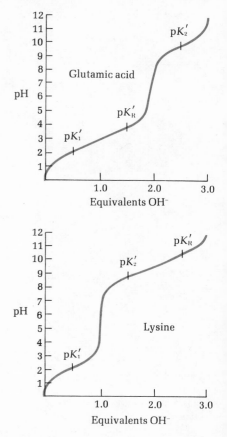

Figure 2-12
The titration curves of glutamic acid and lysine. The pK' of the R group is designated pK'_R.

Analysis of Amino Acid Mixtures

To separate, identify, and measure quantitatively the amounts of each of the 20 amino acids in the hydrolyzate of a protein is tedious and difficult when carried out by classical methods of the organic chemist, such as fractional distillation or crystallization. Today, however, very effective and sensitive methods are available to perform this task very quickly, in particular, chromatography and electrophoresis.

Chromatography is widely applicable not only to mixtures of amino acids, but also to mixtures of other biologically occurring substances such as sugars, nucleotides, and lipids. It may be carried out in many different ways, but its physical basis is the *partition principle,* which is best illustrated by an example. If a solution of an amino acid in water is shaken with an equal volume of some immiscible solvent, such as *n*-butanol, until equilibrium is reached, the amino acid will become partitioned between the liquid phases in a characteristic ratio, represented by the *partition coefficient*. Each amino acid has a distinctive partition coefficient for any given pair of immiscible solvents at a given temperature. By many successive partitions of a mixture of amino acids between two liquid phases it is possible to bring about separation of the mixture. This process is very time-consuming when done by hand, but by *partition chromatography* an enormous number of such partition steps can be carried out in a long column packed with granules of a hydrated inert substance such as starch. Each starch granule contains a layer of tightly bound water, which serves as a stationary aqueous phase past which an immiscible solvent flows as it passes down the column by gravity. Thus the starch column makes possible an enormous number of separate partition steps of microscopic dimensions. The different amino acids in a mixture will move down the column at different rates depending on their partition coefficients between the flowing phase and the absorbed aqueous phase on the particles. The liquid appearing at the bottom of the column, called the *eluate,* is caught in small fractions with an automatic fraction collector and is analyzed by means of the quantitative ninhydrin reaction. A plot of the amount of amino acid in each tube will show a series of peaks, each corresponding to a different amino acid.

The partition principle is also involved in *filter-paper chromatography* of amino acids. The cellulose of the filter-paper fibers is hydrated. As a solvent containing an amino acid mixture ascends in the vertically held paper by capillary action (or descends, in descending chromatography), many microscopic distributions of the amino acids occur between the flowing phase and the stationary water phase bound to the paper fibers. At the end of the process, the different amino acids have moved different distances from the origin. Paper chromatography of a mixture of amino acids may also be carried out in two directions successively on a square of filter paper, using two different solvent systems. A two-dimensional map of the different amino acids then results. In order to locate the position of the amino acids on one-dimensional or two-dimensional chromatograms the paper is dried, sprayed with ninhydrin solution, and heated to develop the color of the ninhydrin derivatives of the amino acids. Authentic specimens of known amino acids are used as "markers" to establish their location on paper chromatograms.

Ion-exchange chromatography is also based on the partition principle. The different amino acids are sorted out by the differences in their acid-base behavior. In this process, the column is filled with a synthetic resin containing fixed charged groups. There are two major classes of ion-exchange resins: those with fixed anionic groups, which are called *cation exchangers*, and those with fixed cationic groups, *anion exchangers*.

Amino acids are usually separated on columns of cation-exchange resins in which the fixed anionic groups are first "charged" with Na^+. An acid solution (pH 3.0) of the amino acid mixture is then applied to the column and allowed to percolate through. At pH 3.0 the amino acids are largely cations with net positive charges, but they differ in the extent to which they are so ionized. The cationic amino acids will therefore displace the bound Na^+ ions from the resin particles. At pH 3.0, the most basic amino acids (lysine, arginine, and histidine) will displace Na^+ first and will be bound to the resin most tightly, and the most acid (glutamic acid, aspartic acid) will be bound the least. The different amino acids will therefore move down the resin column at different rates. The eluate fractions collected are analyzed quantitatively by means of the ninhydrin reaction. The entire procedure has been automated, so that elution, collection of fractions, analysis of each fraction, and the recording

Figure 2-13
Automatically recorded chromatographic analysis of amino acids on an ion-exchange resin. The elution is carried out with different buffers of successively higher pH. The effluent is caught in small volumes, and the amino acid content of each tube is automatically analyzed. The area under each peak is proportional to the amount of each amino acid in the mixture. [Redrawn from D. H. Spackman, W. H. Stein, and S. Moore, *Analyt. Chem.*, **30**:1190 (1958). Reprinted by permission of the copyright owner.]

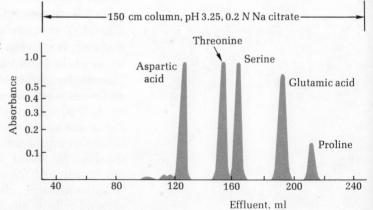

of data are performed automatically in an *amino acid analyzer* (Figure 2-13)

Another method of separating amino acids is *paper electrophoresis*. In this process, a drop of a solution of the amino acid mixture is placed on a filter-paper sheet, across which a high-voltage electrical field is applied. Because of their different pK' values, the amino acids migrate in different directions and at different rates, depending on the pH of the system and the voltage applied. For example, at pH 1.0 histidine, arginine, and lysine have a charge of +2 and move more rapidly to the negatively charged cathode than all the other amino acids, which have a charge of +1. At pH 6.0 the positively charged amino acids (lysine, arginine, histidine) will move to the cathode and the negatively charged amino acids (aspartic acid and glutamic acid) to the anode. All others will remain at the origin since they are isoelectric at this pH.

Peptides

Two amino acid molecules may be joined to yield a *dipeptide* through a *peptide bond*, formed by removal of a water molecule from the carboxyl group of one amino acid and the α-amino group of the other by the action of strong condensing agents

$$NH_2-\overset{\overset{\displaystyle R^1}{|}}{CH}-COOH + H_2N-\overset{\overset{\displaystyle R^2}{|}}{CH}-COOH \xrightarrow{-H_2O} NH_2-\overset{\overset{\displaystyle R^1}{|}}{CH}-\overset{\overset{\displaystyle}{\underset{\underset{\displaystyle O}{\parallel}}{C}}}{}-N-\overset{\overset{\displaystyle R^2}{|}}{CH}-COOH$$

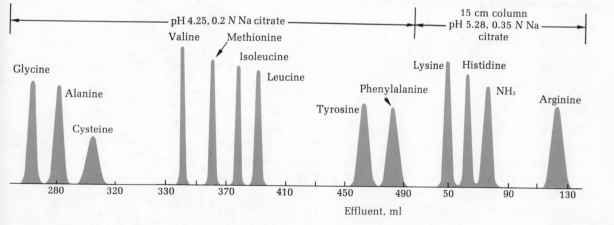

Three amino acids may be joined in a similar manner to form a *tripeptide*. Figure 2-14 shows the structure of a pentapeptide. Peptides are named from the sequence of their constituent amino acids, beginning from the amino-terminal end. When many amino acid residues are joined in a long chain, it is called a *polypeptide*.

Peptides of varying length are formed on partial hydrolysis of the very long polypeptide chains of proteins. In addition there are other biologically occurring peptides that are not derived from proteins, such as the tripeptide glutathione (Figure 2-15), in which one of the peptide bonds is unusual in that it involves an amino group other than that at the α-position. Oxytocin and vasopressin, hormones secreted by the posterior pituitary gland, are large cyclic peptides.

Peptides contain only one free α-amino group and one free α-carboxyl group at their ends. These groups ionize as they do in simple amino acids. All the other α-amino and α-carboxyl groups are joined in peptide bonds which cannot ionize. The R groups of the various amino acid residues in peptides may be regarded as side chains projecting from the backbone of the chain. Since the R groups of some amino acids can lose or gain protons, the acid–base behavior of a peptide can be predicted from its free α-amino and α-carboxyl groups at each end and the nature and number of its ionizing R groups. Peptides also have a characteristic isoelectric pH at which they do not move in an electrical field. Peptides of differing amino acid composition can be separated from each other by chromatography or electrophoresis on the basis of the difference in their acid-base behavior.

There are two important chemical reactions of peptides. In one of these reactions, the peptides may be hydrolyzed by boiling with strong acid or base to yield their constituent amino acids in free form

Figure 2-14
Structure of a pentapeptide. Peptides are named beginning with the NH$_2$-terminal residue. The peptide bonds are shown with color background.

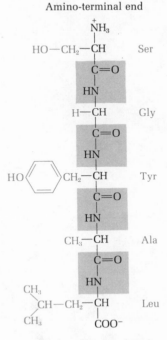

Serylglycyltyrosinylalanylleucine

$$NH_2\!-\!\underset{\underset{R^1}{|}}{CH}\!-\!\underset{\underset{O}{\|}}{C}\!-\!NH\!-\!\underset{\underset{R^2}{|}}{CH}\!-\!COOH + H_2O \longrightarrow NH_2\!-\!\underset{\underset{R^1}{|}}{CH}\!-\!COOH + NH_2\!-\!\underset{\underset{R^2}{|}}{CH}\!-\!COOH$$

The other important reaction of peptides, used in determining their amino acid sequence, is with 2,4-dinitrofluorobenzene. We have seen that this reagent reacts with the α-amino group of a free amino acid to yield a 2,4-dinitrophenyl derivative. It also reacts with the α-amino group of a peptide, whether short or long, to

Figure 2-15
Three peptides having biological activity.

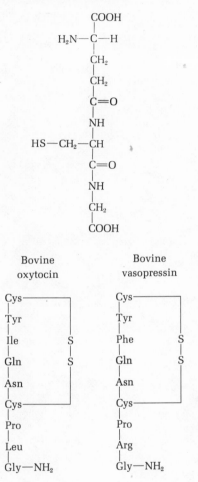

Glutathione
(γ-glutamylcysteinylglycine)

Bovine
oxytocin

Bovine
vasopressin

"label" the amino-terminal amino acid with the dinitrophenyl group to form a dinitrophenyl peptide, which is yellow in color. We shall see how this reaction is used for determination of amino acid sequence in Chapter 3.

Summary

The 20 amino acids commonly found as hydrolysis products of proteins contain an α-carboxyl group, an α-amino group, and a distinctive R group substituted on the α-carbon atom. The amino acids are classified on the basis of the polarity of their R groups. The nonpolar class includes alanine, leucine, isoleucine, valine, proline, phenylalanine, tryptophan, and methionine. The polar neutral class includes glycine, serine, threonine, cysteine, tyrosine, asparagine, and glutamine. The negatively charged (acidic) class contains aspartic acid and glutamic acid, and the positively charged (basic) class contains arginine, lysine, and histidine.

Monoaminomonocarboxylic amino acids are diprotic acids ($\overset{+}{N}H_3CHRCOOH$) at low pH. As the pH is raised to about 6, the isolectric point, the proton is lost from the carboxyl group to form the dipolar or zwitterion species $\overset{+}{N}H_3CHRCOO^-$, which is electrically neutral. Further increase in pH causes loss of the second proton, to yield the ionic species $NH_2CHRCOO^-$. Amino acids possessing ionizable R groups may exist in additional ionic species, depending on the pH. The α-carbon atom of the amino acids (except glycine) is asymmetric and thus can exist in at least two stereoisomeric forms; only the L-stereoisomers, which are related to L-glyceraldehyde, are found in proteins.

Amino acids form colored derivatives with ninhydrin. Complex mixtures of amino acids can be separated, identified, and estimated by means of chromatography on paper or on ion-exchange columns, which depend on the partitioning of the amino acid between two phases.

Amino acids may be joined covalently through the peptide bond to form peptides, which are also formed on incomplete hydrolysis of proteins. The acid-base behavior of a peptide is a function of its NH₂-terminal amino group, its COOH-terminal carboxyl group, and those R groups that ionize. Peptides may be hydrolyzed to yield free amino acids. The free amino group of a peptide can react with 2,4-dinitrofluorobenzene to yield a characteristic yellow derivative.

References

EDSALL, J. T. and WYMAN, J.: *Biophysical Chemistry*, Vol. I, Academic Press, New York, 1958. Physical chemistry and acid-base behavior of amino acids.

LEHNINGER, A. L.: *Biochemistry*, Worth Publishers, New York, 1970. Chapters 4 and 5 contain more detailed treatment of amino acids and peptides.

MEISTER, A.: *Biochemistry of the Amino Acids*, 2d ed., 2 vols., Academic Press, New York, 1965. Comprehensive treatment of structure, occurrence, and metabolism of amino acids.

Problems

1. Calculate the pH of a mixture of 10 ml of 0.1 M $^+NH_3CH_2COO^-$ and 10 ml of 0.05 M NaOH.

2. At what hydrogen ion concentration will an amino acid having pK'_I 2.48 and pK'_{II} 9.42 remain stationary in an electrical field?

3. Write the structure of the following peptides
   Ala · Leu · Tyr
   Lys · Gly · Ile · Glu
   Ser · Cys · Arg · Gln

4. Write the structure of the product of the reaction of 2,4-dinitrofluorobenzene with
   (a) Valine
   (b) Alanylvaline
   (c) Glutamylglycine

5. A mixture of alanine, valine, aspartic acid, threonine, and lysine at pH 6.0 was subjected to paper electrophoresis
   (a) Which amino acid(s) moved rapidly toward the anode?
   (b) Which moved rapidly toward the cathode?
   (c) Which remained at or near the origin?

6. What is the net charge (+, 0, or −) of alanine, glutamic acid, and lysine at (a) pH 1, (b) pH 6, (c) pH 12?

CHAPTER **3** PROTEINS

Figure 3-1
Crystals of horse cytochrome c, a protein functioning in electron transport.

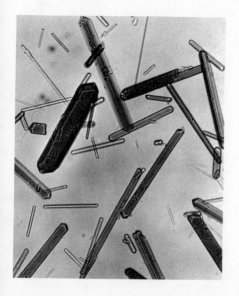

Proteins are the most abundant macromolecules in most cells and constitute 50 percent or more of their dry weight. They are fundamental to all biology since they are the instruments by which genetic information is expressed. They are also very versatile cell components: some are enzymes, some serve as structural components, and some have hormonal activity.

Proteins represent perhaps the most complex and varied class of macromolecules in the cell. Because of many recent advances in the study of proteins, we now have new insights into their structure and behavior.

Composition of Proteins

Many proteins have been isolated in pure crystalline form (Figure 3-1). All contain carbon, hydrogen, nitrogen, and oxygen, and nearly all contain sulfur. Some proteins contain additional elements, particularly phosphorus, iron, zinc, and copper. On acid hydrolysis all proteins yield α-amino acids as end products. Table 3-1 shows the amino acid composition of two typical proteins. The amino acids of any given protein do not occur in equal amounts, nor do all proteins contain all the 20 common α-amino acids.

Proteins are divided into two major classes on the basis of their composition: *simple* and *conjugated*. *Simple proteins* are those which on hydrolysis yield only amino acids and no other organic or inorganic hydrolysis products. They usually contain about 50 percent carbon,

51

7 percent hydrogen, 23 percent oxygen, 16 percent nitrogen, and from 0 to 3 percent sulfur. Conjugated proteins are those yielding not only amino acids but also other organic or inorganic components. The non-amino acid portion of a conjugated protein is called its prosthetic group. Conjugated proteins may be classified on the basis of the chemical nature of their prosthetic groups (Table 3-2).

Polypeptide Chains of Proteins

In protein molecules, the successive amino acid residues are covalently linked by peptide bonds to form long unbranched chains, called polypeptides. These chains may have anywhere from 100 to several hundred amino acid units or residues. However, proteins are not merely random polymers of varying length. All molecules of any given type of protein are identical in amino acid composition, sequence, and length of the polypeptide chain.

Some proteins contain only one polypeptide chain, but others, called oligomeric proteins, have two or more polypeptide chains. For example, the protein ribonuclease, which is an enzyme catalyzing hydrolysis of ribonucleic acid, has a single polypeptide chain, whereas hemoglobin, the oxygen-carrying pigment of the red blood cell, has four polypeptide chains.

The Size of Protein Molecules

The molecular weights of proteins can be determined by means of various physical-chemical methods. Some characteristic values are given in Table 3-3. Most range from about 12,000 to 1,000,000 or more. It is not possible to make generalizations about the size of proteins, even

Table 3-1   Amino acid composition of two proteins.

Amino acid	Human cyto-chrome c	Bovine chymotryp-sinogen A
Ala	6	22
Arg	2	4
Asn	5	15
Asp	3	8
Cys	2	10
Gln	2	10
Glu	8	5
Gly	13	23
His	3	2
Ile	8	10
Leu	6	19
Lys	18	14
Met	3	2
Phe	3	6
Pro	4	9
Ser	2	28
Thr	7	23
Trp	1	8
Tyr	5	4
Val	3	23
Total residues	104	245

Table 3-2   Conjugated proteins

Class	Prosthetic group	Examples
Nucleoproteins	Nucleic acids	Tobacco mosaic virus
Lipoproteins	Lipids	β-Lipoprotein of blood
Glycoproteins	Carbohydrates	γ-Globulin of blood
Phosphoproteins	Phosphate groups	Casein of milk
Hemoproteins	Heme (iron porphyrin)	Hemoglobin, cyto-chrome c
Metalloproteins	Iron	Ferritin
	Zn	Alcohol dehydrogenase

Table 3-3 Sizes of some proteins

	Molecular weight	No. of residues	No. of chains
Insulin (bovine)	5,733	51	2
Ribonuclease (bovine pancreas)	12,640	124	1
Lysozyme (egg white)	13,930	129	1
Myoglobin (horse heart)	16,890	153	1
Chymotrypsin (bovine pancreas)	22,600	241	3
Hemoglobin (human)	64,500	574	4
Serum albumin (human)	68,500	~550	1
Hexokinase (yeast)	96,000	~800	4
γ-Globulin (horse)	149,900	~1,250	4
Glutamate dehydrogenase (bovine liver)	336,000	~3000	6

among proteins having the same type of function. Different enzymes, for example, have molecular weights that vary over a wide range. We can calculate the approximate number of amino acid residues in a simple protein containing no prosthetic group by dividing its molecular weight by 120. [The average molecular weight of the 20 different amino acids in proteins is about 138, but since a molecule of water (mol wt 18.0) is removed to create each peptide bond, the average amino acid residue weight is about 120]. Table 3-3 also gives the approximate number of amino acid residues for some proteins of different sizes and function. Ribonuclease, cytochrome c, and myoglobin, which are among the best-known small proteins, contain between 100 and 155 amino acid residues.

Globular and Fibrous Proteins

Proteins can also be divided into two great classes on the basis of their physical characteristics: *globular* and *fibrous*.

Globular proteins are soluble in aqueous systems and diffuse readily. Their polypeptide chain or chains are tightly folded into compact spherical or globular shapes. Globular proteins usually have a mobile or dynamic function. Nearly all enzymes, for example, are globular proteins, as are blood proteins having a transport function, such as serum albumin and hemoglobin.

Fibrous proteins, on the other hand, are water-insoluble and physically tough. They serve as structural or protective elements in the organism. Typical fibrous proteins are *α-keratin*, the major component of hair, feathers, nails, and skin (leather is almost pure α-keratin), and *collagen*, the major component of tendons. In fibrous proteins the polypeptide chains are arranged in extended, parallel form along a single axis, to yield tough stringy fibers or sheetlike structures.

The Functional Diversity of Proteins

Proteins have many different biological functions. Table 3-4 gives some representative examples of different types of proteins, classified according to biological function. The largest and most important class of proteins are those having catalytic activity, the enzymes. Over 1500 different enzymes are known, each catalyzing a different kind of chemical reaction.

Some proteins serve as nutrient stores, such as *ovalbumin* and the proteins of plant seeds, which furnish the embryo with a supply of amino acids. *Ferritin*, which contains 30 percent iron, is the main storage form of iron in the spleen. Some proteins have a transport function. They bind and transport specific types of molecules via the blood. *Serum albumin* binds free fatty acids tightly and thus serves to transport these molecules between adipose tissue and other organs in vertebrates. *Serum β-lipoprotein*, a conjugated protein which consists of about 21 percent protein and 79 percent lipid, serves to transport lipids via the bloodstream. *Hemoglobin* of vertebrate erythrocytes transports oxygen from the lungs to the tissues; the oxygen is bound to the iron atoms of the four heme groups in the molecule.

Other proteins function as essential elements in contractile and motile systems. *Actin* and *myosin* are long, filamentous proteins which serve as the major elements of the contractile system of muscle.

Some proteins have a defensive or protective function. Antibodies in the blood of vertebrates combine with and inactivate foreign proteins which gain access to the bloodstream. Many bacteria secrete toxins, proteins that are toxic to certain other organisms; examples are diphtheria toxin and the toxin of the anaerobic bacterium *Clostridium botulinum*, which is responsible for some types of food poisoning. Venoms of some poisonous snakes contain toxic enzymes. Some plant proteins are extremely toxic to higher animals, for example, ricin of the castor bean.

Some proteins serve as hormones and have intense biological activity. Among these are *somatotrophin*, the

Table 3-4 Classification of proteins according to biological function

Class	Examples
Enzymes	Ribonuclease Trypsin
Storage proteins	Ovalbumin (egg) Casein (milk) Gliadin (wheat) Ferritin
Transport proteins	Hemoglobin Serum albumin Myoglobin β_1-Lipoprotein
Contractile proteins	Actin Myosin
Protective proteins of blood	Antibodies Fibrinogen Thrombin
Toxins	Botulinus toxin Diphtheria toxin Snake venoms Ricin
Hormones	Insulin Somatotrophin Adrenocortico-trophin
Structure proteins	Keratins Fibroin Collagen Elastin Mucoproteins

growth hormone of the anterior pituitary gland. *Insulin*, secreted by the pancreas, is a hormone regulating glucose metabolism. Its deficiency in humans causes diabetes mellitus. Other polypeptide hormones are *adrenocortico-trophic hormone* (ACTH), secreted by the anterior pituitary gland to stimulate the adrenal cortex, and *parathyroid hormone*, which regulates calcium and phosphate metabolism.

Another major class of proteins consists of those serving a structural role. In higher animals, the fibrous protein collagen is the major extracellular structural protein in connective tissue and in bone. Collagen fibrils also aid in forming a structural continuum binding a group of cells together to form a tissue. Other structural proteins include elastin of yellow elastic tissue and α-keratin. Cartilage is made up of a combination of collagen with a complex acidic polysaccharide. The mucoproteins endow mucous secretions and the synovial fluid in the joints of vertebrates with a slippery, lubricating quality.

It is extraordinary that all proteins, including those having intense biological or toxic effects, contain the same basic set of 20 amino acids, which by themselves have no intrinsic biological activity or toxicity. It is the specific sequence of the amino acids in the polypeptide chains of different proteins which ultimately determines their biological activity.

Amino Acid Sequence of Polypeptide Chains

The sequence of amino acids in the covalent backbone of a protein is known as its *primary structure*. The first step in establishing the amino acid sequence of a polypeptide chain is complete hydrolysis of all the peptide bonds to yield all the amino acids in free form. The polypeptide chain is heated with excess 6 M HCl at 100° to 120°C for 24 hours in a sealed glass tube. However, this acid treatment has side effects on three specific amino acids. It hydrolyzes the amide groups of glutamine and asparagine to yield glutamic acid and aspartic acid, respectively, plus ammonia. From the amount of ammonia in the hydrolyzate, the sum of the glutamine and asparagine can be determined. Acid hydrolysis also destroys tryptophan. In order to determine the tryptophan content of a polypeptide a separate sample is hydrolyzed by boiling with sodium hydroxide. Although this treatment destroys several amino acids, it leaves tryptophan intact.

The complete amino acid composition of such hydroly-
zates of polypeptide chains is then determined by use of
automated ion-exchange chromatography (Chapter 2).

The next step is to identify the amino-terminal and the
carboxyl-terminal residues of the polypeptide chain (Figure 3-2). The first successful method, developed by
F. Sanger, utilizes the reaction between the free α-amino
group of the amino-terminal residue of the chain with
2,4-dinitrofluorobenzene (Chapter 2) to form a yellow
2,4-dinitrophenyl (DNP) derivative. When such a DNP
derivative of a polypeptide is subjected to hydrolysis
with 6 N HCl, all the peptide bonds are hydrolyzed.
However, the bond between the 2,4-dinitrophenyl group

Figure 3-2
*Identification of the NH₂-terminal amino acid residue of peptides.
The Sanger procedure is shown below; the Edman method to the right.
Note that the peptide chain remains intact after removal of the NH₂-
terminal residue by the Edman method; it may then be subjected to
another round of this procedure. Long polypeptide chains can be
"sequenced" automatically by the Edman method.*

The Sanger procedure

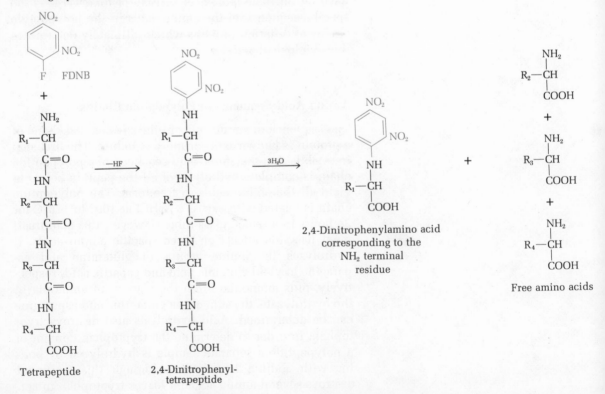

Tetrapeptide   2,4-Dinitrophenyl-
                      tetrapeptide

and the α-amino group of the amino-terminal amino acid is relatively stable to acid hydrolysis. Consequently, the hydrolyzate will contain all the residues of the polypeptide chain as free amino acids except the amino-terminal amino acid, which will be present as a 2,4-dinitrophenyl derivative (Figure 3-2). This can be easily separated from the unsubstituted amino acids and identified by chromatographic comparison with authentic DNP derivatives of the different amino acids.

More recently, greatly improved and more quantitative methods for determination of the amino-terminal amino acid residue have been developed, in particular the method of P. Edman, in which the terminal amino acid is converted into a phenylthiohydantoin derivative. The basis of this method is shown in Figure 3-2.

Several methods have been used to identify the carboxyl-terminal residue. In one, the terminal carboxyl group of the polypeptide is reduced to a hydroxyl group.

The Edman method

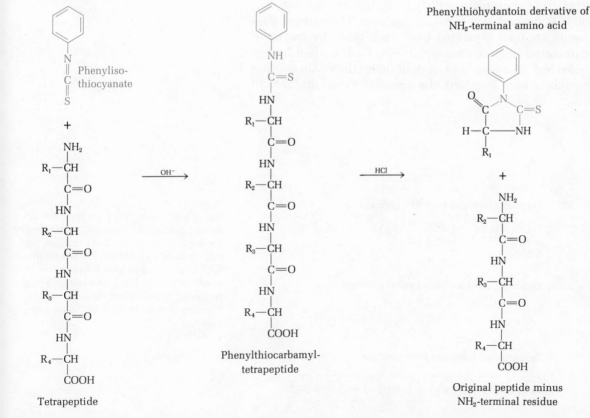

Phenylthiohydantoin derivative of NH₂-terminal amino acid

Phenylthiocarbamyl-tetrapeptide

Tetrapeptide

Original peptide minus NH₂-terminal residue

If the polypeptide chain is then completely hydrolyzed, the hydrolyzate will contain one molecule of an α-amino alcohol corresponding to the original carboxyl-terminal amino acid. This can be easily identified by chromatographic methods. All the other residues will be found as free amino acids. Another method involves use of the enzyme *carboxypeptidase*, which specifically hydrolyzes the carboxyl-terminal peptide bond of polypeptides. Measurement of the relative rates of release of amino acids by the action of this enzyme permits identification of the carboxyl-terminal residue.

The next step in determination of the amino acid sequence is to fragment the intact polypeptide into smaller pieces. The method of choice is enzymatic hydrolysis using the proteolytic enzyme *trypsin*. This enzyme catalyzes the hydrolysis of only those peptide bonds in which the carbonyl group is donated by either a lysine or an arginine residue, regardless of the length or amino acid sequence of the chain. The number of fragments produced by trypsin cleavage can thus be predicted from the total number of lysine or arginine residues in the chain. Moreover, all the fragments resulting from trypsin cleavage will have lysine or arginine residues at the carboxyl-terminal position. The various fragments are then separated from each other by means of chromatography or electrophoresis. Each is then hydrolyzed and its amino acid content determined. On another sample of each fragment, the amino-terminal amino acid

Fragments resulting from first cleavage

T-U	D-E-F	L-M-N	P-Q-R-S
A-B-C	O	G-H-I-J-K	
L-M-N			

Fragments resulting from second cleavage

E-F-G	U	H-I
N-O-P-Q	A-B-C-D	R-S-T
J-K-L-M		

Sequence reconstructed from overlaps

A–B–C–D–E–F–G–H–I–J–K–L–M–N–O–P–Q–R–S–T–U

Figure 3-3
Use of overlapping peptides from two different fragmentation procedures to establish the sequence of amino acids in a polypeptide chain of 21 residues. For simplicity each amino acid residue is designated by a single letter. The black brackets show the fragments from the first cleavage and the brackets in color those from the second cleavage.

Figure 3-4
The structure of bovine insulin, showing the amino acid sequence of the two chains and the cross-linkages.

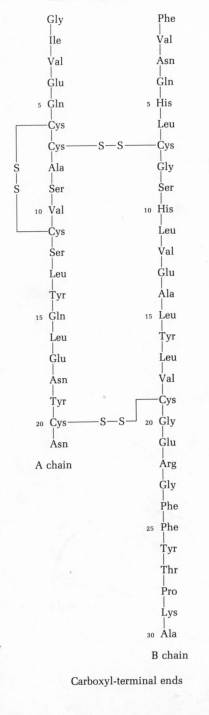

Amino-terminal ends

A chain

B chain

Carboxyl-terminal ends

is then identified by reaction with 2,4-dinitrofluorobenzene or the Edman reagent. With this information, the amino acid sequence of all the dipeptide and tripeptide fragments resulting from trypsin cleavage will be known. The larger peptides from the trypsin cleavage can then be sequenced by repeated Edman degradation steps. In this manner the amino acid sequence of each of the fragments obtained by trypsin cleavage of the chain can be established.

It remains to determine the proper ordering of the various trypsin fragments of the original polypeptide chain. To do this another sample of the polypeptide must be fragmented but at different points in the chain so that the peptide fragments resulting from the second procedure "overlap" those resulting from the first. If trypsin is used for the first cleavage, the second cleavage must be carried out with some other proteolytic enzyme, such as *chymotrypsin*, which hydrolyzes peptide bonds in which the carbonyl group is contributed by phenylalanine, tyrosine, or tryptophan. Chymotrypsin and other proteolytic enzymes, such as pepsin of gastric juice, are not as specific as trypsin. Polypeptide chains may also be fragmented with cyanogen bromide, which cleaves those peptide bonds whose carbonyl group is contributed by methionine residues. The fragments resulting from the second cleavage are then subjected to the same procedure of sequence determination as the first set. From the amino acid sequence of the second set of fragments, overlaps of specific sequences can be found so that the first set of fragments can be arranged into the proper sequence, as is shown in Figure 3-3.

The general approach described above was first devised by F. Sanger in his pioneering work on the amino acid sequence of insulin, which he completed in 1953 and for which he was awarded the Nobel Prize. Figure 3-4 shows the complete amino acid sequence of bovine insulin, which consists of two polypeptide chains: the A chain, which has 21 amino acid residues, and the B chain, which has 30. The two chains, which are cross-linked by disulfide bridges of cystine residues (Chapter 2), were first cleaved by oxidation before their sequence was determined. Soon other investigators reported the sequence of amino acids in adrenocorticotrophin, the hormone of the anterior pituitary gland that stimulates the adrenal cortex. This hormone has a single chain of 39 residues with a molecular weight of about 4600. Somewhat later, the first successful sequence analysis of an enzyme protein was achieved, namely, that of ribonuclease, which has 124 amino acid residues in a single chain (Figure 3-5). Ribonuclease contains four intrachain —S—S— cross-linkages.

The next important landmark was the identification of the amino acid sequences of the two types of polypeptides in crystalline hemoglobin. This was the first sequence analysis of an oligomeric or multichain protein. Hemoglobin contains four polypeptide chains, two identical α chains (141 residues) and two identical β chains (146 residues). This feat was carried out by groups in the United States and another in Germany. Among the longest polypeptide chains for which complete amino acid sequences have been deduced to date are those of bovine trypsinogen (229 residues), bovine chymotrypsinogen (245 residues), and the polypeptide chains of glyceraldehyde 3-phosphate dehydrogenase of lobster muscle (333 residues). Altogether the amino acid sequences of over 50 different proteins are known.

Figure 3-5
Amino acid sequences of human adrenocorticotrophin and bovine ribonuclease (below). The position of S—S cross-linkages in bovine ribonuclease is also shown.

Human adrenocorticotrophin

Ser-Tyr-Ser-Met-Glu-His-Phe-Arg-Trp-Gly- $_{10}$
Lys-Pro-Val-Gly-Lys-Lys-Arg-Arg-Pro-Val- $_{20}$
Lys-Val-Tyr-Pro-Asp-Ala-Gly-Glu-Asp-Gln- $_{30}$
Ser-Ala-Glu-Ala-Phe-Pro-Leu-Glu-Phe $_{39}$

Bovine ribonuclease

Lys-Glu-Thr-Ala-Ala-Ala-Lys-Phe-Glu-Arg- $_{10}$
Gln-His-Met-Asp-Ser-Ser-Thr-Ser-Ala-Ala- $_{20}$
Ser-Ser-Ser-Asn-Tyr-Cys-Asn-Gln-Met-Met- $_{30}$
Lys-Ser-Arg-Asn-Leu-Thr-Lys-Asp-Arg-Cys- $_{40}$
Lys-Pro-Val-Asn-Thr-Phe-Val-His-Glu-Ser- $_{50}$
Leu-Ala-Asp-Val-Gln-Ala-Val-Cys-Ser-Gln- $_{60}$
Lys-Asn-Val-Ala-Cys-Lys-Asn-Gly-Gln-Thr- $_{70}$
Asn-Cys-Tyr-Gln-Ser-Tyr-Ser-Thr-Met-Ser- $_{80}$
Ile-Thr-Asp-Cys-Arg-Glu-Thr-Gly-Ser-Ser- $_{90}$
Lys-Tyr-Pro-Asn-Cys-Ala-Tyr-Lys-Thr-Thr- $_{100}$
Gln-Ala-Asn-Lys-His-Ile-Ile-Val-Ala-Cys- $_{110}$
Glu-Gly-Asn-Pro-Tyr-Val-Pro-Val-His-Phe- $_{120}$
Asp-Ala-Ser-Val $_{124}$

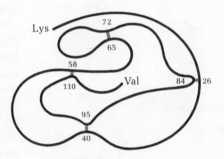

Separation and Purification of Proteins

Because different proteins differ in amino acid content and sequence, they contain different proportions of amino acids with acidic and basic R groups. Proteins will therefore differ in their acid–base titration curves and in their isoelectric points, the pH at which they do not move in an electric field (Table 3-5). These differences are exploited in separating proteins from each other. For example, since proteins are least soluble at their isoelectric pH, one protein can be separated from others in a mixture by precipitating it at its isoelectric pH with neutral salts like ammonium sulfate or neutral solutes, such as ethanol or acetone, which cause proteins to become less soluble. Proteins may also be separated from each other by electrophoresis at a given pH, which may cause some to move toward the positive pole or anode, some to the cathode, and some to remain stationary (Figure 3-6).

Table 3-5 The isoelectric points of some proteins

	Isoelectric pH
Pepsin	< 1.0
Egg albumin	4.6
Serum albumin	4.9
Urease	5.0
β-Lactoglobulin	5.2
γ_1-Globulin	6.6
Hemoglobin	6.8
Myoglobin	7.0
Chymotrypsinogen	9.5
Cytochrome c	10.65
Lysozyme	11.0

Figure 3-6

Separation of blood plasma proteins by electrophoresis. Each protein will move toward the positive pole (anode) at a characteristic rate, depending on the magnitude of its negative charge. The migration of each protein can be detected by observing the movement of its boundary by an optical method. Each peak in the optical pattern (below) corresponds to a specific protein (A = serum albumin). [Redrawn from R. Alberty, J. Chem. Educ., 25:619 (1948).]

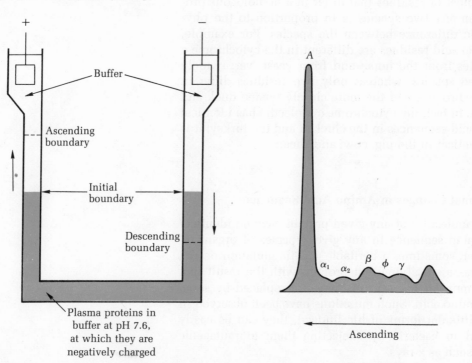

Another important means of separating proteins is chromatography on columns of hydrated materials. The separation may be on the basis of electrical charge at a given pH, using columns of diethylaminoethylcellulose (abbreviated DEAE cellulose), or on the basis of particle radius or size, as may be carried out on columns of the hydrated polysaccharide derivative called Sephadex.

Species Specificity of Proteins

An important conclusion from study of amino acid sequence is that homologous proteins from different species are similar, but not identical. _Homologous proteins_ are those with identical functions, such as the hemoglobins in the erythrocytes of different species of vertebrates, which function to carry oxygen in the red blood cell. Another is _cytochrome c_, an iron-containing protein which carries electrons during biological oxidations in all animal, plant, and many microbial cells. The amino acid sequence of cytochrome _c_'s from many different species have been studied. This protein has a molecular weight of about 12,500 and has about 100 amino acid residues. Thirty-five amino acid residues are identical in the cytochrome _c_ of all species (Figure 3-7). The number of residues that differ in a homologous protein from any two species is in proportion to the phylogenetic difference between the species. For example, 48 amino acid residues are different in the cytochrome _c_ molecules from the horse and from yeast, very widely separated species, whereas only two residues differ in the cytochrome _c_'s of the more closely related duck and chicken. In fact, the cytochrome _c_ molecule has identical amino acid sequences in the chicken and the turkey; it is also identical in the pig, cow, and sheep.

Mutational Changes in Amino Acid Sequence

All the molecules of any given protein have an identical amino acid sequence in any given species of organism. However, sometimes a heritable genetic mutation occurs in the gene specifying that protein, with the result that one or more amino acid residues are replaced by some other amino acid. Such mutations have been observed in many different forms of life; indeed, they can be easily induced in bacteria by subjecting them to mutagenic agents such as x-rays.

Figure 3-7 (right)
The 35 invariant amino acid residues in cytochrome c. The following species were used to compile these data: horse, man, hog, chicken, yeast, cow, sheep, tuna, Macacus mulatta monkey, Samia cynthia moth, dog, kangaroo, rattlesnake, snapping turtle, turkey, duck, pigeon, king penguin, screwworm fly, Neurospora crassa mold, Candida krusei yeast, donkey, and chimpanzee.
The cytochrome c's of invertebrates, plants and fungi have four to eight additional amino acids at the NH$_2$-terminal end.

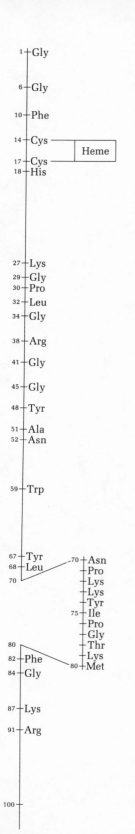

Many mutations in the hemoglobin molecule have been detected in human patients. The hemoglobin isolated from individuals with sickle-cell anemia is different from normal hemoglobin. In sickle-cell anemia, the erythrocytes assume a crescentlike shape instead of the flat, disklike conformation of normal erythrocytes. L. Pauling and H. Itano found that the electrophoretic mobility of hemoglobin isolated from sickle cells was slightly different from that of normal hemoglobin. Later V. Ingram discovered that sickle-cell hemoglobin differs from normal hemoglobin in only a single amino acid residue. He found that the α chains of the normal and sickle cell hemoglobin are identical, but that the glutamic acid residue at position 6 in the β chain of normal hemoglobin is replaced by a valine residue in sickle-cell hemoglobin. Since glutamic acid has an acidic R group and valine an uncharged R group, sickle-cell hemoglobin has a slightly different electrical charge at neutral pH. Sickle-cell anemia is thus a molecular disease of genetic origin. The amino acid replacement is the result of a mutation in the DNA molecule that codes the synthesis of the hemoglobin β chain. An abnormal hemoglobin of one type or another is found in 6 out of every 10,000 humans.

The Three-Dimensional Configuration of Protein Molecules

One of the most difficult problems in the study of proteins is to determine how their polypeptide chain(s) are coiled or folded in their native state. By means of x-ray diffraction analysis very precise information has been obtained on a number of fibrous and globular proteins. The first successful analysis of protein structure by x-ray methods was made by L. Pauling. He deduced that the polypeptide chains of the fibrous protein *α-keratin* are regularly coiled to form a structure called the *α helix* (Figure 3-8). There are 3.6 amino acid residues in each turn of the helix. The R groups extend outward from the backbone of the coiled polypeptide chain. The α helix is a very stable structure since its successive loops are linked to each other by hydrogen bonds. The α-keratins of hair and wool consist of "ropes" of three or seven such α-helical coils twisted around each other (Figure 3-8). Their adjacent polypeptide chains are held together by many —S—S— cross-linkages.

In *fibroin*, the fibrous protein of silk, the polypeptide chains are in an extended zigzag configuration, the

Figure 3-8
The α helix.

Ball-and-stick model of α-helix,
showing intrachain hydrogen bonds.

α-Helical coils in
hair and wool keratins.

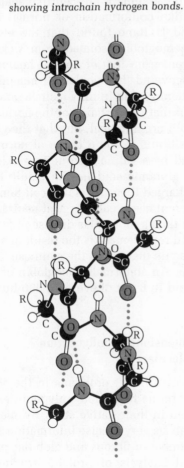

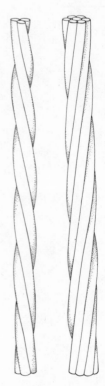

β configuration. Such chains are arranged alongside each other to form a structure called a *pleated sheet* (Figure 3-9), in which the adjacent polypeptide chains run in opposite directions. The adjacent chains of the pleated sheet are held together by hydrogen bonding.

In the fibrous protein *collagen*, which is rich in proline and glycine residues, x-ray diffraction analysis indicates that three polypeptide chains are twisted around each other in a *triple helix* (Figure 3-10).

Figure 3-9
The β configuration of the polypeptide chain and the pleated sheet.

Edge view

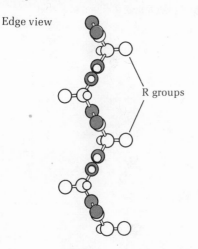

R groups

Hydrogen bonds

Top view

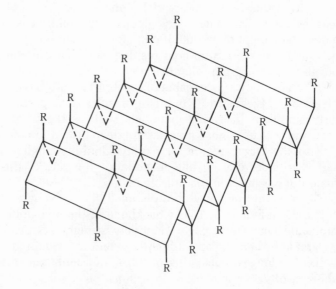

Schematic representation of three parallel chains in β structure, showing the pleated-sheet arrangement. All the R groups project above or below the plane of the sheet.

The specific configurations of the polypeptide chains in these three types of fibrous proteins are stable only because they have specific amino acid sequences. For example, an α helix tends to form spontaneously only in the case of polypeptide chains in which consecutive R groups are relatively small and uncharged, as is true in the chains of α-keratin. If such a sequence contains one or more amino acids with a bulky or charged R group, the α helix will become strained and deviate from the dimensions shown in Figure 3-8. The β configuration is formed by chains with many glycine and alanine residues. The triple helix of collagen requires regularly spaced proline, hydroxyproline, and glycine residues. We have earlier seen that the term *primary structure* refers to the sequence of amino acids in the covalent backbone of the polypeptide chain. We may now define the term *secondary structure*: it refers to the specific geometrical arrangement of the polypeptide chain along one axis. The α helix, the β conformation and the collagen helix are examples of various types of secondary structure. The term *tertiary structure* is used to refer to the three-dimensional structure of globular proteins, in which the polypeptide chain is tightly folded and packed into a compact spherical form.

The tertiary structure of globular proteins is much more difficult to deduce than the secondary structure of fibrous proteins, in which the polypeptide chains are arranged along one axis. The first important breakthrough came from the x-ray studies of J. Kendrew and his colleagues on sperm whale myoglobin. Myoglobin has a molecular weight of 16,900. It contains a single polypeptide chain of 153 amino acid residues and has an iron-porphyrin heme group identical with that of hemoglobin. Like hemoglobin, it is capable of combining with oxygen reversibly. Myoglobin is found in muscles. It is particularly abundant in diving mammals such as the whale, seal, and walrus, whose muscles are deep brown in color due to the large amount of oxymyoglobin. Myoglobin serves to store oxygen and to enhance the rate of diffusion of oxygen into the muscle cell.

The structure of myoglobin deduced by x-ray analysis is shown in Figure 3-11. The backbone of the myoglobin molecule consists of eight relatively straight segments separated by bends. Each straight segment is a length of α helix, the longest consisting of 23 amino acids and the shortest of 7. X-ray studies have since been carried out on a number of other single-chain globular proteins, such as cytochrome *c* and the enzyme *lysozyme*. From these studies several general conclusions have been drawn regarding the three-dimensional conformation or tertiary structure of the polypeptide chain in globular proteins. (1) The conformations of different globular proteins are not identical; each has its own characteristic conformation. (2) The polypeptide chains of globular proteins are very compactly folded. In their interior there is room for few if any water molecules. (3) Nearly all the polar or

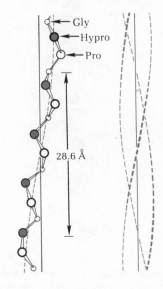

Figure 3-10
The supercoiled arrangement of polypeptide chains in collagen. Hypro designates hydroxyproline.

Gly
Hypro
Pro

28.6 Å

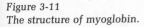
Figure 3-11
The structure of myoglobin.

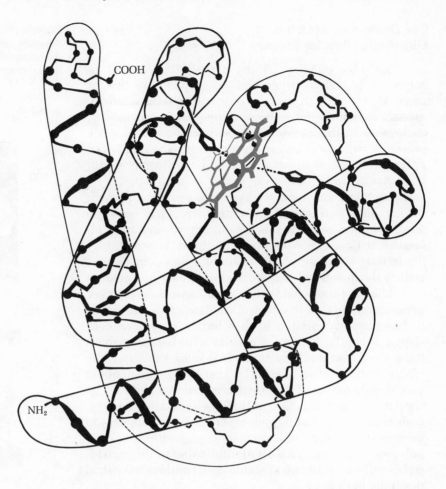

COOH

NH₂

hydrophilic amino acid R groups are located on the outer surface of globular proteins, exposed to water. (4) Nearly all the nonpolar or hydrophobic R groups are in the interior of the molecule, hidden from exposure to water (as is true in soap or phospholipid micelles). (5) Proline residues are often present at "bends" in the polypeptide chain.

Although it is not yet possible to make precise correlations between amino acid sequence and the tertiary structure of globular proteins, there is little doubt that their three-dimensional structure is ultimately determined by amino acid sequence. It is significant that homologous proteins, such as hemoglobins of various species, which possess many common amino acid sequences, also have similar three-dimensional structures.

The Quaternary Structure of Oligomeric Globular Proteins

Most globular proteins having a molecular weight in excess of 50,000 are oligomeric and consist of two or more separate polypeptide chains. The characteristic manner in which the individual polypeptide chains fit each other in the native conformation of an oligomeric protein is called its *quaternary structure*. The number of polypeptide chains in some well-studied oligomeric proteins was given in Table 3-3.

Hemoglobin was the first oligomeric protein for which the complete tertiary and quaternary structures were determined by x-ray analysis by M. Perutz and his colleagues at Cambridge. Figure 3-12 shows schematically the tertiary structure of the separate α and β chains, as well as the quaternary structure of the intact hemoglobin molecule. The four chains, two α and two β, fit together in an approximately tetrahedral configuration. The α and β chains are very similar in their tertiary structure, consisting of similar lengths of α helix with bends of about the same angles and directions. It is remarkable that the tertiary configuration of α and β chains of hemoglobin is very similar to that of the single chain of myoglobin. The similar biological function of these two proteins, namely, their capacity to bind oxygen reversibly, thus appears to be conferred by the similar tertiary configurations of their polypeptide chains. This similarity is due to the considerable number of identical amino acid residues at critical positions in their chains.

Denaturation of Proteins

When globular proteins in their native state are subjected briefly to heat, to extremes of pH, or to certain solutes such as urea or guanidine, they undergo *denaturation*, a process in which the biological activity of the protein is lost. For example, if the protein happens to be an enzyme, its catalytic activity is lost on such denaturing treatments. Furthermore, a globular protein usually becomes insoluble on denaturation; an example is the white of a freshly opened egg, which coagulates on heating.

When proteins are denatured the backbones of their polypeptide chains remain intact; no peptide bonds are broken. What happens during denaturation is unfolding of the polypeptide chain, which loses its characteristic three-dimensional configuration and becomes randomly

Figure 3-12
Structure of the α chain, the β chain, and the hemoglobin molecule, as deduced from x-ray diffraction analysis. [From A. F. Cullis, H. Muirhead, A. C. T. North, M. F. Perutz, and M. G. Rossmann, Proc. Roy. Soc., London A, 265:161 (1962).]

α Chain

β Chain

Hemoglobin molecule from "The Hemoglobin Molecule," by M. F. Perutz. Copyright © November 1964 by Scientific American, Inc. All rights reserved.

Figure 3-13
Denaturation of a protein and its reversal.

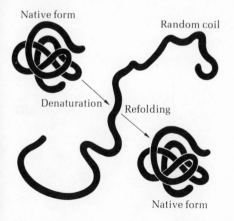

Native form

Random coil

Denaturation Refolding

Native form

coiled. It is therefore clear that the biological activity of proteins is not directly due to their amino acid sequence, but rather to the native three-dimensional configuration of the polypeptide chain. That the native configuration of a protein is determined by its amino acid sequence has been proven in the case of the enzyme ribonuclease, which loses its catalytic activity when it is denatured by acid or heat. When the unfolded polypeptide chain of ribonuclease is allowed to stand at pH 7.0 and at room temperature it spontaneously regains its native three-dimensional configuration as well as its catalytic activity (Figure 3-13). It is now believed that the native, biologically active three-dimensional configuration of a given protein is its most stable configuration under biological conditions of temperature and pH and that this configuration is the automatic consequence of its specific amino acid sequence.

Summary

Proteins, the most abundant organic biomolecules in cells, are macromolecules containing one or more polypeptide chains, each having 100 or more α-amino acid residues. Simple proteins contain only polypeptide chains; conjugated proteins contain other organic or inorganic components in addition. Globular proteins are water soluble and have dynamic functions, such as enzymes, hormones and antibodies; fibrous proteins are insoluble and serve a structural role, such as collagen of connective tissue.

Complete hydrolysis of a protein yields all its amino acids in free form. To determine the primary structure (i.e., the sequence of amino acids) in a polypeptide chain, its NH_2-terminal and COOH-terminal residues are identified and its amino acid content determined. One sample of the chain is fragmented by using trypsin, which hydrolyzes those peptide bonds in which lysine or arginine contributes the carbonyl carbon. The resulting peptides are separated, their amino acid content determined, and their sequence established. Another sample of the polypeptide chain is cleaved by a second method to yield a different set of fragments, which are also separated and analyzed. From the "overlaps" between the two sets of fragments the sequence of the fragments can be deduced. Each type of protein has a characteristic amino acid sequence. Functionally homologous proteins from different species, such as hemoglobins and cytochromes, possess the same amino acid residues at certain invariant positions in the chain, but the other residues vary.

The secondary structure of fibrous proteins (the manner in which the polypeptide chain is arranged along one dimension) has been determined by x-ray analysis. The three important types of secondary structure are the α helix (in α-keratin), the β conformation (in silk) and the collagen helix. The tertiary structure (the manner in which the polypeptide chain of globular proteins is folded into a compact globular shape) has also been established for some proteins by x-ray analysis. The quaternary structure is defined as the manner in which the individual polypeptide chains of an oligomeric protein such as hemoglobin are packed together. Ultimately it is the amino acid sequence that determines the secondary, tertiary, and quaternary structure of proteins. The unfolding of protein molecules by heat or extremes of pH, a process that causes loss of biological activity, is called denaturation.

References

DICKERSON, R. E. and I. GEIS: *The Structure and Action of Proteins*, Harper and Row, New York, 1969. A protein crystallographer and a scientific illustrator collaborated to produce this unique book, which emphasizes the conformation of protein molecules.

LEHNINGER, A. L.: *Biochemistry*, Worth Publishers, New York, 1970. Chapters 3–7 give considerably more information on proteins, as well as a large number of references.

SMITH, E. L.: "The Evolution of Proteins," *Harvey Lectures*, **62**: 231–256 (1967).

Problems

1. List the fragments formed by the action of trypsin on the following peptides:
   (a) Lys-Asp-Gly-Ala-Ala-Glu-Ser-Gly
   (b) Ala-Ala-His-Arg-Glu-Lys-Phe-Ile-Gly-Glu-Gly-Glu
   (c) Tyr-Cys-Lys-Ala-Arg-Arg-Gly
   (d) Phe-Ala-Glu-Ser-Ala-Gly-Lys
   Each of the fragments yielded by trypsin is then treated with 2,4-dinitrofluorobenzene, followed by hydrolysis of the peptide linkages. List the resulting 2,4-dinitrophenyl amino acids formed from each peptide.

2. List the peptides formed when the following polypeptide is treated with (a) trypsin (b) chymotrypsin:
   Val-Ala-Lys-Glu-Glu-Phe-Val-Met-Tyr-Cys-Glu-Trp-Met-Gly-Gly-Phe-Arg-Phe-Trp-Val-Lys-Ala-Gly-Ser-Phe-Gly

3. Predict the direction of migration [i.e., stationary (0), toward cathode (C), or toward anode (A)] of the following peptides during paper electrophoresis at pH 1.0, pH 6.5, and pH 11.1:
   (a) Lys-Gly-Ala-Gly, (b) Lys-Gly-Ala-Glu, (c) Gly-Ala-Glu, (d) Glu-Gly-Ala-Glu, (e) Gln-Gly-Ala-Lys.

4. In what direction will the following proteins migrate in an electrical field at the pH indicated? (a) Egg albumin at pH 5.0; (b) β-lactoglobulin at pH 5.0, at pH 7.0; (c) chymotrypsinogen at pH 5.0, at pH 9.5, at pH 11. (Use data in Table 3-5.)

5. How would you proceed to determine how many polypeptide chains are present in a molecule of an oligomeric protein of molecular weight 200,000?

CHAPTER 4 ENZYMES

Figure 4-1
Crystals of bovine chymotrypsin.

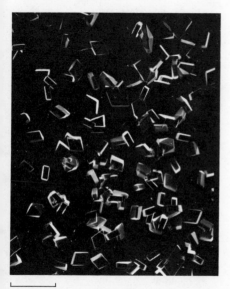

0.1 mm

The enzymes make up the largest and most highly specialized class of proteins. They catalyze the thousands of chemical reactions that collectively constitute the intermediary metabolism of cells.

Much of the history of biochemistry is the history of enzyme research. The activity of enzymes was first recognized in early studies of digestion in the stomach in the period 1780 to 1825. Later Louis Pasteur deduced that fermentation of sugar to alcohol by yeast is catalyzed by "ferments" or enzymes. In 1860 he postulated that they are inextricably linked with the structure and life of the yeast cell. It was therefore a major landmark in the history of enzyme research when in 1897 E. Büchner succeeded in extracting from yeast cells the enzymes that catalyze alcoholic fermentation. This achievement clearly demonstrated that these important enzymes which catalyze a major energy-yielding metabolic pathway can function independent of cell structure.

The first enzyme isolated in pure crystalline form was urease, isolated from extracts of the jack bean by J. B. Sumner in 1926. Sumner found that the crystals consisted of protein, and he suggested contrary to prevailing opinion, that all enzymes are proteins. His views were not immediately accepted, however, and it was not until the 1930's, during which J. Northrop isolated the digestive enzymes pepsin, trypsin, and chymotrypsin in crystalline form, that the protein nature of enzymes was firmly established. Today over 1500 enzymes are known. Many have been isolated in pure crystalline form (Figure 4-1).

General Properties of Enzymes

Enzymes catalyze chemical reactions that would otherwise occur only at extremely low rates. They do not change the equilibrium point of the reactions they catalyze. They are true catalysts in that they are not used up or permanently changed during catalysis. Since all known enzymes are proteins, any agency that can damage native protein structure also causes loss of catalytic activity. Thus, heating of enzymes, treatment with strong acids or bases, or exposure to denaturing agents, destroys their catalytic activity.

Enzymes have molecular weights ranging from about 12,000 to over 1 million. Some enzymes consist only of one or more polypeptide chains, but others contain in addition some other chemical component that is required for activity, called a *cofactor*. The cofactor may be either a metal such as Mg, Mn, Zn, or Fe (Table 4-1), or it may be a complex organic molecule, usually called a *coenzyme* (Chapter 8). Some enzymes require both a metal ion and a coenzyme. The cofactor is often tightly bound to the protein part of the enzyme. When this is the case, the cofactor is usually called a *prosthetic group*. Cofactors are generally stable to heat, whereas the protein part of the enzyme is labile to heat.

Table 4-1 Some enzymes containing or requiring metal ions as cofactors

Zn^{2+}
  Alcohol dehydrogenase
  Carbonic anhydrase
Mg^{2+}
  Phosphohydrolases
  Phosphotransferases
Mn^{2+}
  Arginase
Fe^{2+} or Fe^{3+}
  Cytochromes
  Catalase
Cu^{2+}
  Cytochrome oxidase
K^+
  Pyruvate phosphokinase
    (also requires Mg^{2+})

The Classification of Enzymes

In the past enzymes have usually been named according to the substance they act upon, called the *substrate*, or the nature of the reaction catalyzed. Thus *urease* catalyzes the hydrolysis of urea, and *arginase* catalyzes the hydrolysis of arginine. But in many cases enzymes have been given names that are not informative as to the reaction catalyzed, such as pepsin and trypsin, which catalyze the hydrolysis of proteins. Because of the great number of enzymes now known, a more systematic nomenclature and classification have been adopted. Enzymes are grouped into six major classes according to the type of reaction they catalyze (Table 4-2). Each enzyme also has an identifying classification number.

Table 4-2 International classification of enzymes

1. Oxido-reductases
   (Electron-transfer reactions)
2. Transferases
   (Transfer of functional groups)
3. Hydrolases
   (Hydrolysis reactions)
4. Lyases
   (Addition to double bonds)
5. Isomerases
   (Isomerization reactions)
6. Ligases
   (Formation of bonds with ATP cleavage)

Catalysis

A chemical reaction such as A → P takes place because a certain fraction of the A molecules at any given instant possess more energy than the rest of the population, sufficient to attain an "activated" state in which a chemical bond may be made or broken to form the product P (Figure 4-2). The term *activation energy* refers to the amount of energy in calories required to bring all the molecules in one mole of a substance at a given temperature to the reactive state.

The term *transition state* refers to the energy-rich state of the interacting molecules at the top of the activation barrier (Figure 4-2). The rate of a chemical reaction is proportional to the concentration of the transition-state species. There are two general ways in which the rate of a chemical reaction can be accelerated. One way is to increase the temperature, which increases thermal motion and energy and thus increases the number of molecules in the transition state. Usually the reaction rate is approximately doubled by a 10°C rise in temperature. The second way is to add a catalyst. Catalysts accelerate chemical reactions by lowering the activation energy. They combine with the reactants to produce a transition state having less free energy than the transition state of the uncatalyzed reaction (Figure 4-2). When the reaction products are formed, the free catalyst is regenerated again.

Figure 4-2
Energy diagram for a chemical reaction, uncatalyzed and catalyzed.

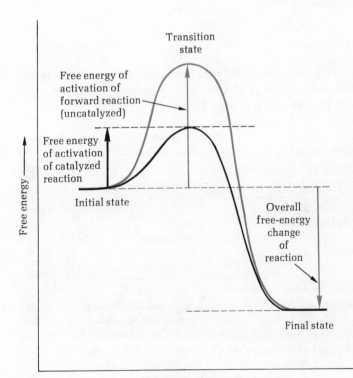

The Effect of Substrate Concentration on Enzyme Catalysts

If we examine the effect of varying the substrate concentration on the rate of an enzyme-catalyzed reaction, with the enzyme concentration held constant (Figure 4-3), we would find at relatively low concentrations of substrate that the rate of the reaction increases in proportion to the substrate concentration. But as the substrate concentration is increased further, the reaction rate increases by smaller and smaller amounts. Finally, when the substrate concentration is raised still higher, a point will be reached beyond which there is no further increase in reaction rate. No matter how high the substrate concentration is raised beyond this point, the reaction rate will remain essentially constant. At this plateau of the reaction rate, called the maximum rate (symbolized V_{max}), the enzyme is "saturated" with its substrate and can work no faster.

This saturation effect is exhibited by all enzymes. It has led to a general theory of enzyme action first proposed by L. Michaelis and M. Menten. They postulated that the enzyme E first combines reversibly with its substrate S to form an enzyme–substrate complex ES in a fast reaction

$$E + S \rightleftharpoons ES$$

The ES complex then breaks down in a second reversible reaction, which is slower, to regenerate the free enzyme and the reaction product P

$$ES \rightleftharpoons E + P$$

Since the second reaction is the rate-limiting one, the overall rate of the enzyme-catalyzed reaction must be proportional to the concentration of the enzyme–substrate complex ES. At any given instant in an enzyme-catalyzed reaction, the enzyme exists in two forms, in the free or uncombined form E and in the form of the ES complex. The rate of the catalyzed reaction will obviously be at a maximum when all of the enzyme is present as the ES complex and the concentration of free enzyme E is vanishingly small. This condition will exist at a very high concentration of the substrate. By the law of mass action, the equilibrium of the first reaction

$$E + S \rightleftharpoons ES$$

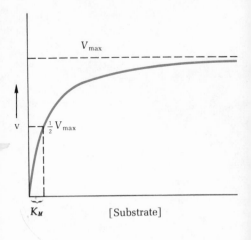

Figure 4-3
Effect of substrate concentration on initial reaction rate. K_M, the Michaelis constant, is the substrate concentration giving one-half the maximum velocity.

will be pushed to the right if we increase the concentration of S. If we increase it to high enough levels, essentially all the free enzyme E will have been converted into the ES form. In the second reaction of the catalytic cycle, which is the rate-limiting reaction for most enzymes, the ES complex breaks down to yield the product P and regenerate the free enzyme. But at a high concentration of S, the free enzyme so formed will immediately combine again with another molecule of S. Under these conditions a steady state is achieved in which the enzyme is always saturated with its substrate and the reaction rate is maximum.

The *Michaelis constant* (K_M) specifies the quantitative relationship between substrate concentration and V_{max} for different enzymes. It is defined as that substrate concentration at which a given enzyme yields one-half its maximum velocity. Table 4-3 gives the values of the Michaelis constant for a number of enzymes. In those enzymatic reactions having more than one substrate or product, each has its own characteristic K_M. In many cases the Michaelis constant is an inverse measure of the affinity of the enzyme for its substrate; the lower the K_M the higher the affinity. This relationship holds only for those enzymes in which the rate of the reaction

$$E + S \rightleftharpoons ES$$

is much greater than the rate of the breakdown of the ES complex to form E + P.

The characteristic shape of the substrate–saturation curve for an enzyme can be expressed mathematically by the *Michaelis-Menten equation*

$$v_0 = \frac{V_{max}[S]}{K_M + [S]} \tag{1}$$

where v_0 is the initial velocity of the reaction at substrate concentration $[S]$, V_{max} is the maximum velocity, and K_M is the Michaelis constant. This equation is basic to all aspects of the kinetics of enzyme action. If we know K_M and V_{max}, which can be easily derived from simple experiments as shown below, we can calculate the reaction rate at any given concentration of the substrate.

The Lineweaver-Burk Plot

The Michaelis–Menten equation can be transformed algebraically into other forms that are more useful in

Table 4-3 K_M for some enzymes

Enzyme and substrate	K_M, mM
Catalase (H_2O_2)	25
Hexokinase (glucose)	0.15
Chymotrypsin (glycyltyrosinylglycine)	108
Carbonic anhydrase (HCO_3^-)	9

$K = [S]$
when $V = \frac{1}{2} V_{max}$

plotting experimental data. One of the most widely used transformations is derived simply by taking the reciprocal of both sides of the Michaelis-Menten equation and rearranging. The result is the *Lineweaver-Burk equation*

$$\frac{1}{v} = \frac{K_M}{V_{max}} \times \frac{1}{[S]} + \frac{1}{V_{max}}$$

It represents a straight line obtained by plotting $1/v$ vs. $1/[S]$, with a slope of K_M/V_{max} and an intercept of $1/V_{max}$ on the $1/v$ axis (Figure 4-4). Such a "double-reciprocal" plot has the advantage that V_{max} does not have to be measured, but can be extrapolated. The intercept on the $1/[S]$ axis is $-1/K_M$. The Lineweaver-Burk plot is very useful for studying enzyme inhibition, as we shall see below.

Effect of pH on Enzymatic Activity

Enzymes have a characteristic *optimum pH* at which their activity is maximal (Table 4-4); above or below this pH the activity declines (Figure 4-5). The optimum pH of an enzyme is not necessarily identical with the pH of its normal intracellular surroundings, which may be on the ascending or descending slope of its pH–activity profile. Thus intracellular pH may help regulate the activity of enzymes.

Figure 4-4
Lineweaver-Burk plot.

Table 4-4   Optimum pH of some enzymes

	Optimum pH
Pepsin	1.5
Trypsin	7.7
Catalase	7.6
Arginase	9.7
Fumarase	7.8
Ribonuclease	7.8

Figure 4-5
pH–*Activity profiles of two enzymes.*

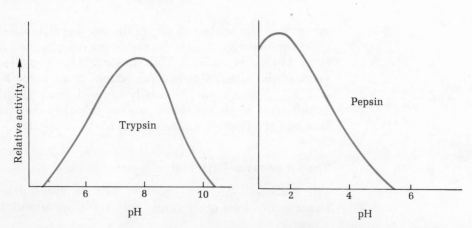

Quantitative Assay of Enzymatic Activity

The amount of an enzyme in a given solution or tissue extract can be assayed quantitatively in terms of the catalytic effect it produces. For this purpose it is necessary to know (1) the overall equation of the reaction catalyzed, (2) a simple analytical procedure for determining the disappearance of the substrate or the appearance of the reaction products, (3) whether the enzyme requires cofactors such as metal ions or coenzymes, (4) its dependence on substrate concentration, that is, the K_M for the substrate, (5) its optimum pH, and (6) a temperature zone in which it is stable and has high activity. Ordinarily enzymes are assayed at their optimum pH and temperature and with a saturating concentration of substrate. Under these conditions the initial reaction rate is usually proportional to enzyme concentration, at least over a given range of enzyme concentration (Figure 4-6).

By international agreement, 1.0 unit of enzyme activity is defined as that amount causing transformation of 1.0 micromole (10^{-6} mole) of substrate per minute at 25°C, under optimal conditions of measurement. The *specific activity* is the number of enzyme units per milligram of protein. It is a measure of enzyme purity; it increases during purification of an enzyme and becomes maximal and constant when the enzyme is in the pure state. The *turnover number* of an enzyme is the number of substrate molecules transformed per unit time by a single enzyme molecule (or by a single catalytic site) when the enzyme is the rate-limiting factor (Table 4-5). The enzyme carbonic anhydrase has the highest turnover number of any known enzyme, 36,000,000 per minute per molecule.

Substrate Specificity of Enzymes

Some enzymes have nearly absolute specificity for a given substrate and will not attack even very closely related molecules. For example, the enzyme aspartase catalyzes the reversible addition of ammonia to the double bond of fumaric acid, but no other unsaturated acids (Figure 4-7). Aspartase also has rigid stereo-specificity and geometrical specificity; thus, it will not deaminate D-aspartate nor will it add ammonia to maleate, the *cis* geometrical isomer of fumarate. At the other extreme are enzymes which have relatively broad specificity and act on many compounds having a common structural feature. For example, kidney phosphatase catalyzes hydrolysis of many different esters of phosphoric acid, but at varying rates. From study of the substrate

Figure 4-6
Effect of enzyme concentration on initial reaction rate when substrate concentration is saturating.

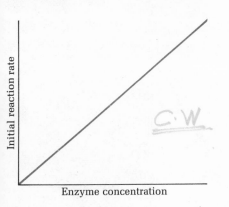

Table 4-5 Turnover numbers of some enzymes (per minute at 20–38°C)

Carbonic anhydrase	36,000,000
β-Amylase	1,100,000
β-Galactosidase	12,500
Phosphoglucomutase	1,240

specificity of enzymes the idea arose that there is a complementary or lock-and-key relationship between the substrate molecule and a specific area on the surface of the enzyme molecule called the *active site* or *catalytic site*, to which the substrate molecule is bound as it undergoes the catalytic reaction.

Two distinct structural features determine the specificity of an enzyme for its substrate. First, the substrate must possess the specific chemical bond or linkage that can be attacked by the enzyme. Second, the substrate usually must have some other functional group, a binding group, which binds to the enzyme and positions the substrate molecule so that the susceptible bond is properly located in relation to the catalytic site of the enzyme. As an example, Figure 4-8 shows the substrate specificity of chymotrypsin, which normally hydrolyzes those peptide bonds in proteins and simple peptides in which the carbonyl group is contributed by amino acids having an aromatic ring, that is, by tyrosine, tryptophan, and phenylalanine. However, as can be seen in Figure 4-8, chymotrypsin can also cleave amide and ester linkages. Moreover, it can also accept large alkyl groups instead of aromatic rings. Such experiments permit one to "map" the active site of an enzyme (Figure 4-8).

Enzyme Inhibition

Most enzymes can be poisoned or inhibited by certain specific chemical reagents. From the study of enzyme inhibition valuable information has been obtained on the substrate specificity of enzymes, the nature of the functional groups at the active site, and the mechanism of the catalytic event. Enzyme inhibitors also are useful in elucidating metabolic pathways in cells. Moreover, many drugs useful in medicine appear to function because they can inhibit certain enzymes in tissues.

There are different types of enzyme inhibition. Irreversible inhibition of an enzyme occurs when a functional group required for activity is destroyed or modified. An example of an irreversible inhibitor is the compound *diisopropyl fluorophosphate*, abbreviated DFP, which inhibits the enzyme choline esterase. Choline esterase catalyzes the reaction which takes place at the junction between certain cells in the nervous system. Animals poisoned by this substance, an ingredient of "nerve gas", become paralyzed because of the failure of nerve impulses to be transmitted properly. DFP com-

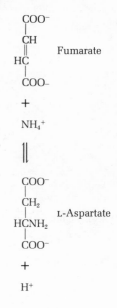

Figure 4-7
The aspartase reaction.

Figure 4-8
Substrate specificity of chymotrypsin.
From studies of modified substrates,
such as those shown at the far right,
it has been concluded that chymo-
trypsin is an acyl group transferase
rather than strictly a peptidase. The
minimum structural requirements of
chymotrypsin and a "map" of its active
site are shown below.

Some compounds hydrolyzed by
chymotrypsin. The hydrophobic
positioning group and the susceptible
bond are shown in color.

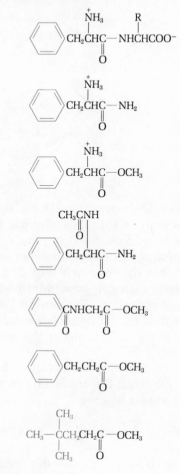

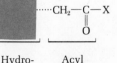

| Hydro-phobic positioning group | Acyl group containing susceptible bond |

"Map" of active site of chymotrypsin.

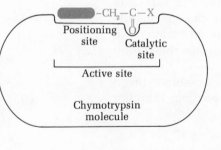

bines with the hydroxyl group of a vital serine residue in the active site of the choline esterase molecule (Figure 4-9), thus chemically modifying the enzyme so that it cannot bring about catalysis.

Competitive inhibition occurs when the inhibitor competes with the substrate for binding to the active site of the enzyme. This type of inhibition is reversible. The degree of inhibition depends on the relative concentrations of the inhibitor and the substrate. Usually a competitive inhibitor closely resembles the substrate in its structure but is unable to undergo the catalytic reaction. A classic example is the inhibition of the enzyme _succinate dehydrogenase_, which catalyzes the removal of

Figure 4-9
Inhibition of choline esterase by
diisopropyl fluorophosphate.

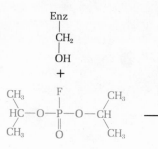

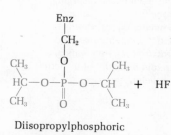

Diisopropylphosphoric
ester of enzyme
(inactive)

two hydrogen atoms from succinate to yield fumarate
(Figure 4-10). This enzyme is inhibited by malonate,
which resembles succinate in having two carboxyl
groups, but cannot undergo removal of hydrogen atoms.
When the enzyme is inhibited by malonate, the degree of
inhibition can be diminished by raising the succinate
concentration. The percent inhibition of the enzyme depends on the ratio of the concentrations of malonate to
succinate. The higher this ratio, the greater the inhibition.

In competitive inhibition the enzyme combines reversibly with the inhibitor I to yield an enzyme–inhibitor
complex EI

$$E + I \rightleftharpoons EI$$

in competition with the normal reaction with the substrate

$$E + S \rightleftharpoons ES$$

Since the enzyme cannot break down the inhibitor molecule, the presence of the latter prevents a given fraction
of the enzyme from combining with substrate to form
reaction products.

In noncompetitive inhibition increasing the substrate
concentration does not reverse the inhibition. In this
type of inhibition the inhibitor does not combine with
the substrate binding site but rather with some other
group on the enzyme molecule that is essential for its
function. The most common type of noncompetitive
inhibitor is represented by chemical reagents that can

Figure 4-10
Competitive inhibition of succinate
dehydrogenase by malonate. Note the
similarity in structure between succinate
and malonate, which at pH 7.0 are fully
dissociated anions.

Succinate dehydrogenase reaction

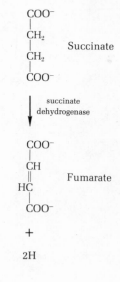

combine reversibly with an essential metal ion required to stabilize the enzyme in its active conformation. For example, some enzymes require bound Mg^{2+} for activity. When a complexing agent for Mg^{2+} is added such enzymes are inhibited. Another example of a noncompetitive inhibitor is cyanide, which combines reversibly with the iron atom of some iron-containing enzymes to yield an inactive form. In noncompetitive inhibition, the inhibitor forms inactive complexes with *both* the free enzyme and the enzyme-substrate complex.

Competitive and noncompetitive inhibition are distinguished from each other experimentally by determining the response of the reaction rate at a given concentration of enzyme to the relative concentrations of substrate and inhibitor. To do this the substrate concentration is held constant and the inhibitor concentration varied, or vice versa. Evaluation of the data obtained is most easily carried out by the use of Lineweaver-Burk plots (Figure 4-11). In noncompetitive inhibition V_{max} decreases, but

Figure 4-11
Lineweaver-Burk plots of competitive and noncompetitive inhibition. Competitive inhibitors yield a family of lines with a common intercept on the 1/v axis. Noncompetitive inhibitors yield plots with a common intercept on the 1/S axis. See text.

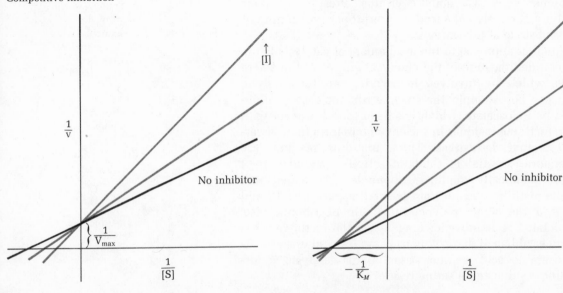

Competitive inhibition

Noncompetitive inhibition

K_M, the intercept on the 1/[S] axis, remains constant. In competitive inhibition K_M increases, but V_{max} stays constant.

Mechanism of Enzyme Action

The precise mechanism of catalysis is not yet known for any enzyme. However, one approach which has yielded valuable information is the study of nonenzymatic catalysis of model reactions similar to those which occur biologically. From such model catalysts, it has been found that acids and bases (that is, proton donors and proton acceptors) are very versatile catalysts, which can enhance the rates of many types of organic reactions, such as the hydrolysis of esters and of phosphorylated compounds, the addition of water to carbonyl groups, and the elimination of water from alcohols to yield unsaturated compounds. Some enzymes are known to contain proton-donating groups, such as $-NH_3^+$, carboxyl ($-COOH$), and sulfhydryl ($-SH$) groups, as well as proton-accepting groups, such as $-NH_2$ and carboxylate ($-COO^-$) groups.

Nucleophilic groups are also effective catalysts; they are functional groups rich in electrons which can donate an electron pair to the nucleus of some other atom. Typical nucleophilic groups are hydroxyl groups, sulfhydryl groups, and imidazole groups, which are also known to be present in proteins (Figure 4-12). It appears probable that such functional groups in the active sites of different enzymes serve as catalytic groups. Presumably their activity is greatly enhanced by the proper positioning of the substrate at the active site.

Another approach to the mechanism of catalysis is the direct identification of the chemical groups of the active sites which are involved in carrying out the catalytic process. For example, the enzyme ribonuclease is inhibited by iodoacetate, which has been found to react with two histidine residues in the enzyme to form their N-carboxymethyl derivatives. These histidine residues are therefore essential in catalytic activity. Degradation of such chemically modified ribonuclease revealed that these histidine residues are located at positions 12 and 119 in the single polypeptide chain of ribonuclease, which is 124 amino acids long. By similar methods it has been established that chymotrypsin also contains at least two amino acid residues essential for catalysts, a histidine residue and a serine residue.

Figure 4-12
Important nucleophilic groups of proteins.

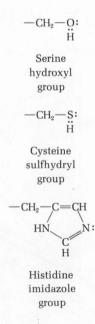

Serine
hydroxyl
group

Cysteine
sulfhydryl
group

Histidine
imidazole
group

A third approach is study of the structure of enzyme–substrate compounds. For example, it has been found that when chymotrypsin catalyzes the hydrolysis of certain esters of acetic acid, the acetyl group of the substrate becomes covalently bound in ester linkage to the hydroxyl group of a serine residue at position 195 in the enzyme molecule. This fact, together with the demonstration that a histidine residue is also required for activity, has led to a hypothetical mechanism of chymotrypsin catalysis (Figure 4-13).

Function of the Protein in Enzyme Catalysis

An important question arises: Why are enzyme molecules so large in relation to the size of their substrate or their active sites, which may occupy only a few percent of their surface area? Recent research has revealed some clues. For one thing, it has been found that the amino acid residues required for catalytic activity of some enzymes are not necessarily adjacent to each other in the polypeptide chain, but yet are near neighbors. For example, x-ray analysis has shown that the two essential histidine residues at positions 12 and 119 of the ribonuclease chain are located near each other at the active site by the looping and folding of the polypeptide chain.

Figure 4-13
A proposed mechanism for the action of chymotrypsin.

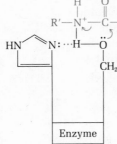
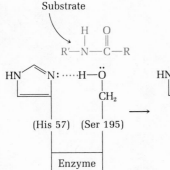
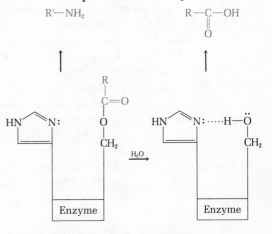

Ser 195 hydroxyl and imidazole of His 57 are hydrogen-bonded.

Imidazole of His 57 now hydrogen-bonds with amino group of substrate, orienting acyl carbon for attack of hydroxyl oxygen of Ser 195.

Transfer of acyl group to hydroxyl of Ser 195

H_2O molecule accepts acyl group. Both products now discharged.

Thus the native conformation of the entire molecule must be intact if the catalytic site is to have the proper configuration in space. Moreover, it has been found that most enzymes undergo a slight change in their three-dimensional conformation as they bind their specific substrates, a change that has been called the "induced fit" of the enzyme to the substrate. The enzyme reverts to its original conformation after the catalytic event has taken place and the product of the reaction has dissociated from the enzyme molecule. This conformational change during the catalytic cycle presumably makes possible the specific binding of the substrate at the catalytic site, the imposition of a strain on the substrate molecule, making it more susceptible to reaction, and also enables the enzyme to "kick off" the products of catalysis.

Isozymes

Many enzymes occur in more than one molecular form in the same species, tissue, or even within the same cell. Such multiple forms are called isoenzymes or isozymes. The best known example is lactate dehydrogenase, which occurs in animal tissues in five forms separable by electrophoresis. The five isozymes are made up of combinations of two different kinds of polypeptide chains, each of molecular weight 33,500; the M chains (M for muscle) and H chains (H for heart). These chains differ in amino acid composition and sequence. All five forms of the enzyme contain four polypeptide chains. That form of the enzyme which predominates in skeletal muscle has four identical M chains (designated M_4) and that in heart has four identical H chains (H_4). The lactate dehydrogenases from other tissues are hybrids consisting of mixtures of the M and H chains, i.e., M_3H, M_2H_2, and MH_3. The various isozymes of lactate dehydrogenase differ significantly in their maximum rates (V_{max}) and in their Michaelis constants. The M_4 isozyme predominates in muscles that are used only intermittently, as in the wing muscles of the chicken, whereas H_4 isozyme predominates in the heart and other muscles which are continuously active.

Many other enzymes are now known to occur in multiple forms. Another example is malate dehydrogenase, which occurs in two different forms, one in the mitochondria and the other in the cytosol fraction of liver cells. Isozymes are important in the regulation of metabolism and in the development of tissues.

Regulatory Enzymes (Allosteric Enzymes)

In the cell most enzymes function in sequential chains, called *multienzyme systems,* in which the product of one enzyme becomes the substrate of the next. Each multienzyme system carries out a specific metabolic task, such as the breakdown of glucose to lactate or the synthesis of an amino acid from a simpler precursor. In most multienzyme systems the first enzyme of the sequence functions to regulate the rate of the entire system and is called a *regulatory* or *allosteric enzyme.* Usually this enzyme is inhibited by the end product of the sequence, so that whenever the end product accumulates above a certain critical concentration, it inhibits the first or regulatory enzyme in the sequence, thus "turning off" that segment of metabolism. An example is the multienzyme sequence catalyzing the conversion of L-threonine to L-isoleucine, which occurs in five enzyme-catalyzed steps (Figure 4-14). The first enzyme of the sequence, L-threonine deaminase, is strongly inhibited by L-isoleucine, the end product of the sequence. Isoleucine is quite specific as an inhibitor; no other intermediate in the sequence is inhibitory. This type of inhibition is known as *end-product inhibition* or *feedback inhibition.*

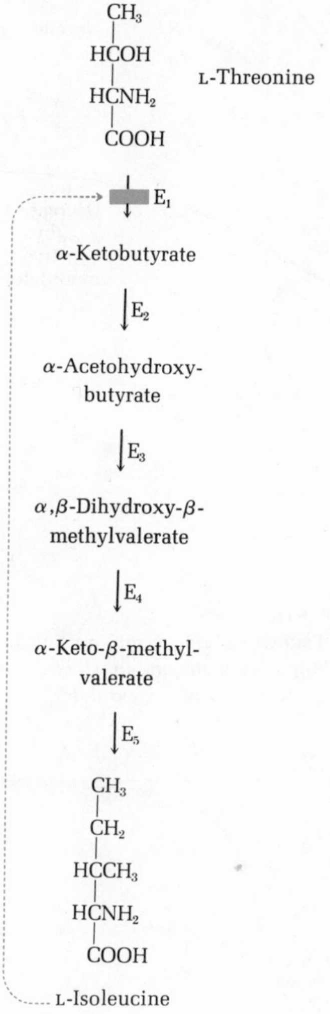

The properties of regulatory enzymes, such as threonine deaminase, are significantly different from those of nonregulatory enzymes. They are usually large in molecular weight and contain several subunits. They have been found to contain binding sites not only for their normal substrate but also for the regulating metabolite, which is called the *effector* or *modulator.* The binding site for the modulator is called the *allosteric site* (Gr., other space or site); it is highly specific for that metabolite. When the allosteric site is empty the enzyme functions at its normal catalytic rate. When it is occupied by the regulatory metabolite, the enzyme undergoes a change in its conformation to a less active or more active form, depending on whether the modulator is inhibitory or stimulatory.

The catalytic rate of a regulatory enzyme may be modified by its modulator in two different ways. In some regulatory enzymes, the binding of the modulator causes a change in V_{max}, but no change in the affinity of the enzyme, that is, in K_M. In others there is a change in K_M but no change in V_{max} (Figure 4-15).

Sometimes the substrate of a regulatory enzyme can itself act as an allosteric modulator. Such enzymes are

Figure 4-14
Feedback inhibition of threonine deaminase (E_1) by L-isoleucine. E_1, E_2, E_3, E_4, and E_5 are enzymes catalyzing the successive steps in the synthesis of isoleucine from threonine.

Figure 4-15

Two classes of regulatory enzymes. In one (left) the modulator changes K_M without altering V_{max}. In the other (right) the modulator changes V_{max} without altering K_M.

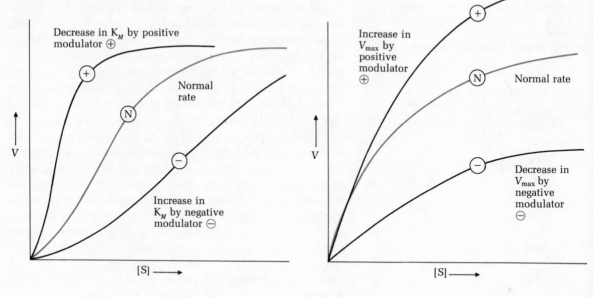

called *homotropic*, whereas regulatory enzymes in which the modulator is some metabolite other than the substrate, are called *heterotropic*. Figure 4-16 shows the effect of substrate concentration on the activity of a homotropic regulatory enzyme in which the substrate also acts as a positive or stimulatory modulator. The sigmoid curve shows that whenever the substrate concentration builds up beyond a certain level in the cell, this type of enzyme responds by a very sharp increase in rate. Binding of one substrate molecule enhances the binding of another at the allosteric site, causing the enzyme molecule to undergo a change in conformation to a more active form.

Figure 4-16

Plot of substrate concentration vs. velocity for a homotropic enzyme in which the substrate is also a positive modulator.

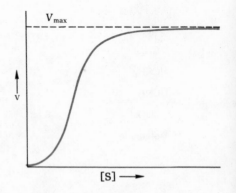

Summary

Enzymes are classified on the basis of the reactions they catalyze. Some enzymes are simple proteins; others contain prosthetic groups of metal ions, coenzymes, or both. At low substrate concentrations, the rate of an enzymatic reaction is proportional to the concentration of substrate, but as the substrate concentration is increased, a point is reached where the reaction rate becomes maximal (V_{max}) and independent of substrate concentration. In this zone, the enzyme is saturated. Each enzyme has a characteristic substrate concentration (K_M, the Michaelis-Menten constant) at which the reaction velocity is one-half V_{max}. The quantitative relationship between K_M, substrate concentration, and V_{max} is given by the Michaelis-Menten equation. Enzymes are assayed at their optimum pH by measuring the initial reaction rate under conditions in which the enzymes is saturated with substrate. One unit of activity is that amount causing transformation of 1.0 micromole of substrate per min at 25 °C.

Competitive inhibitors of enzymes are those whose action can be reversed by increasing the substrate concentration. They usually have a structural resemblance to the substrate and compete with it for binding at the active site. Noncompetitive inhibition cannot be reversed by the substrate; it results from the reversible interaction of the inhibitor with some other essential group of the molecule. Irreversible inhibitors produce a permanent modification of some essential functional group in the enzyme molecule. The active site of an enzyme can be "mapped" by examining its substrate specificity. Enzymes have a specific three-dimensional conformation, which undergoes change during the catalytic cycle, subjecting the substrate molecule to stress and making it more likely to react. Allosteric or regulatory enzymes are inhibited or stimulated by the binding of some other molecule, the effector or modulator, usually a metabolic product whose concentration in the cell is critical. Some enzymes exist in multiple forms called isozymes.

References

BERNHARD, S.: *The Structure and Function of Enzymes*, W. A. Benjamin, Menlo Park, Cal., 1968. Up-to-date paperback.

LEHNINGER, A. L.: *Biochemistry*, Worth Publishers, New York, 1970. Chapters 8 and 9.

DIXON, M., and E. C. WEBB: *Enzymes*, Longmans, London, 2nd edition, 1964. Classical text on general enzyme properties.

Enzyme Nomenclature, American Elsevier Publishing Company, New York, 1965. International nomenclature and classification.

Problems

1. The following experimental data were collected during a study of the catalytic activity of an enzyme. From these data determine by graphical analysis the values of K_M and V_{max} for this enzyme.

   Substrate concentration (mM) 1.5   2.0   3.0   4.0   8.0   16
   Product formed per min (mg) 0.21 0.24 0.28 0.33 0.40   0.45

2. The V_{max} of a given enzyme (mol wt 120,000) was found experimentally to be 28 μmoles per minute for a 10 μg sample of the enzyme. Calculate its turnover number.

3. At what substrate concentration will an enzyme having a maximum velocity of 30 μmoles substrate transformed per min per mg and a K_M of .005 M show one-quarter of its maximum rate?

4. From the following data on an enzymatic reaction, determine whether the inhibitor is acting competitively or noncompetitively.

Substrate concentration (mM)	2.0	3.0	4.0	10.0	15.0
Product formed/hr (μmoles) (no inhibitor)		13.9	17.9	21.3	31.3	37.0
Product formed/hr (μmoles) (inhibitor present)		8.8	12.1	14.9	25.7	31.3

CHAPTER **5** **CARBOHYDRATES**

Carbohydrates play four important roles in living organisms; (1) as an energy source, through their combustion, (2) as a source of carbon in the synthesis of other cell components, (3) as a major storage form of chemical energy, and (4) as structural elements of cells and tissues. Carbohydrates are polyhydroxy aldehydes or ketones or substances that yield such compounds on hydrolysis; most carbohydrates have the empirical formula $(CH_2O)_n$. The term carbohydrate originated because many substances of this group have empirical formulas corresponding to "hydrates" of carbon, in which the ratio of hydrogen to oxygen is 2:1. An example is glucose, $C_6H_{12}O_6$. Although this "hydrate" relationship holds for most compounds we designate as carbohydrates today, some do not show this ratio and some also contain nitrogen, phosphorus, and sulfur.

The most abundant carbohydrates in the biosphere are the plant polysaccharides cellulose and starch, which are polymers of glucose. Cellulose is the predominant extracellular structural component of the fibrous and woody tissues of plants and starch is the chief form of fuel storage in plants.

Classes of Carbohydrates

There are three major classes of carbohydrates: *monosaccharides, oligosaccharides,* and *polysaccharides. Monosaccharides,* or *simple sugars,* consist of a single polyhydroxy aldehyde or ketone unit. The most abundant monosaccharide is the six-carbon sugar glucose, which is the most important fuel molecule for most organisms and also serves as the building block of some of the most abundant polysaccharides, such as starch and cellulose.

Oligosaccharides consist of chains of two to ten monosaccharide units joined in glycosidic linkage. Among them are disaccharides and trisaccharides, which possess two and three monosaccharide units respectively. Polysaccharides consist of very long chains of monosaccharide units; they may be linear or branched. Most polysaccharides contain recurring monosaccharide units of only a single kind. For example, starch consists of recurring glucose units.

Families of Monosaccharides

Monosaccharides have the empirical formula $(CH_2O)_n$, where $n = 3$ or some larger number. In most monosaccharides each carbon atom except one contains a hydroxyl group; at the remaining carbon atom, there is a carbonyl oxygen. If the carbonyl group is at the end of the chain, the monosaccharide is an aldehyde and is called an aldose; if it is at any other position, the monosaccharide is a ketone and is called a ketose. The simplest monosaccharides are the three-carbon trioses glyceraldehyde (an aldotriose) and dihydroxyacetone (a ketotriose), which are shown in Figure 5-1.

Monosaccharides having four, five, six, and seven carbon atoms in their chains are called, respectively, tetroses, pentoses, hexoses, and heptoses. Each of these exists in two series, that is, aldotetroses and ketotetroses, aldopentoses and ketopentoses, aldohexoses and ketohexoses, and so on. In nature the hexoses are the most abundant monosaccharides. However, aldopentoses are important components of nucleic acids, and derivatives of trioses and heptoses are intermediates in carbohydrate metabolism. Most simple monosaccharides are white crystalline solids that are freely soluble in water but insoluble in nonpolar solvents. Most have a sweet taste.

Stereoisomerism of Monosaccharides

All the monosaccharides except dihydroxyacetone contain one or more asymmetric carbon atoms. The simplest aldose, glyceraldehyde, contains only one asymmetric carbon atom and thus is capable of existing in the form of two different stereoisomers which are mirror images of each other. However, the aldohexoses have four asymmetric carbon atoms and can exist in the form of $2^n = 2^4 = 16$ different stereoisomers. Figure 5-2 shows

Figure 5-1
The trioses.

Glyceraldehyde

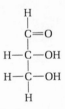

Dihydroxyacetone

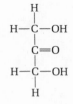

Figure 5-2
Projection formulas (pp. 39, 40) of the D-aldoses. Those printed in color are the most important and abundant.

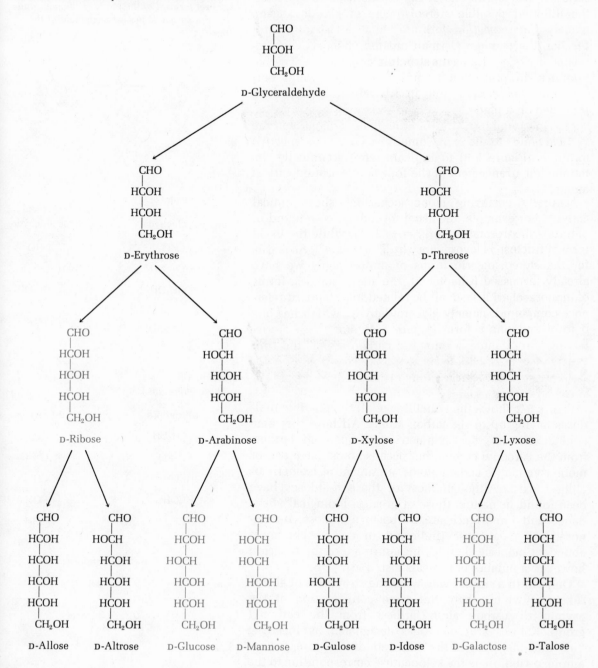

the structure of the different stereosiomers of aldotrioses aldotetroses, aldopentoses, and aldohexoses in the D series. It also shows one of the means used to represent the different possible stereoisomers of simple sugars, namely the _projection formula_, first devised by Emil Fischer, the eminent German organic chemist who pioneered in systematizing the structure and naming of carbohydrates. In projection formulas the four groups surrounding each carbon atom in a tetrahedral manner are projected on a plane, so that the horizontal bonds are assumed to extend forward from the plane of paper and the vertical bonds to the rear. Sometimes we use _perspective formulas_ (Figure 5-3) to indicate more graphically the tetrahedral arrangement of the four bonds around carbon atoms.

Naturally occurring monosaccharides show optical activity. For example, the usual form of glucose found in nature is dextrorotatory ($[\alpha]_D^{20} = +52.7°$), while the usual form of fructose is levorotatory ($[\alpha]_D^{20} = -92.4°$). As is true for the stereoisomeric forms of amino acids we have already discussed (Chapter 2), the stereoisomeric forms of monosaccharides can all be related to a standard reference compound, namely glyceraldehyde, which has one D form and one L form (Figure 5-4). For those sugars having two or more asymmetric carbon atoms, the convention has been adopted that the prefixes D and L refer to the asymmetric carbon atom _farthest removed from the carbonyl carbon atom._

Figure 5-2 shows the structures of all the possible D-aldoses having up to six carbon atoms. All have the same configuration at the asymmetric carbon atom farthest from the carbonyl carbon, but because most have two or more asymmetric carbon atoms, a number of isomeric D-aldoses exist. Although most of these D-aldoses have been found in nature, those of greatest biological abundance and importance are D-ribose, D-glucose, D-mannose, and D-galactose (indicated in color). The most abundant monosaccharides in nature are of the D series; however, a number of L sugars also exist.

One can in a similar way write the structures of all the D-ketoses which share the same configuration at the asymmetric carbon atom farthest from the carbonyl group. Ketoses are sometimes designated by inserting "ul" into the name of the corresponding aldose; for example, D-ribulose is the ketopentose corresponding to the aldopentose D-ribose. Biologically the most important ketoses are D-ribulose, D-fructose, and D-sedoheptulose, a seven-carbon sugar (Figure 5-5).

Figure 5-3
Perspective formula of an asymmetric carbon atom. The wedge-shaped bonds project forward of the plane of the page and the dotted bonds project to the rear of the page.

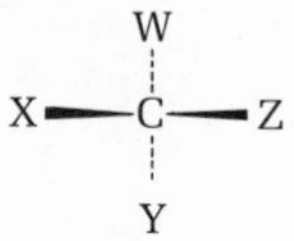

Figure 5-4
Stereoisomers of glyceraldehyde.

D-Glyceraldehyde   L-Glyceraldehyde

```
   CHO                          CHO
    |                            |
H—C—OH                      HO—C—H
    |                            |
  CH₂OH                        CH₂OH
```

Figure 5-5
Three important ketoses.

D-Ribulose   D-Fructose

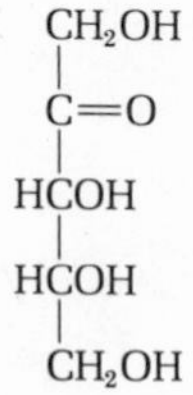

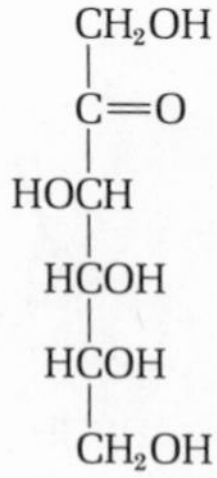

```
  CH₂OH                         CH₂OH
    |                             |
   C=O                           C=O
    |                             |
  HCOH                          HOCH
    |                             |
  HCOH                          HCOH
    |                             |
  CH₂OH                         HCOH
                                  |
                                CH₂OH
```

D-Sedoheptulose

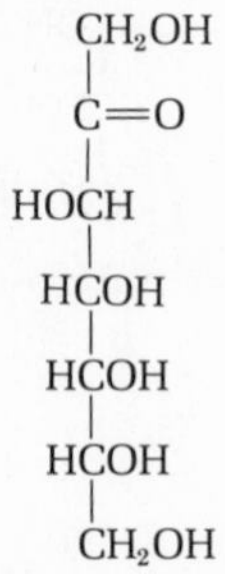

```
  CH₂OH
    |
   C=O
    |
  HOCH
    |
  HCOH
    |
  HCOH
    |
  HCOH
    |
  CH₂OH
```

Figure 5-6
D- and L-Glucose.

D-Glucose   L-Glucose

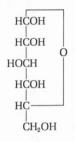

Figure 5-7
Pyranose ring forms of D-glucose.

α-D-Glucopyranose

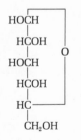

β-D-Glucopyranose

Pyran

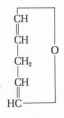

Aldoses and ketoses of the D series are mirror images or *enantiomers* of their L counterparts, as is shown for the case of D- and L-glucose (Figure 5-6). Two sugars differing only in the configuration around one specific carbon atom are called *epimers* of each other. Thus, D-glucose and D-mannose are epimers with respect to carbon atom 2, and D-glucose and D-galactose are epimers with respect to carbon atom 4 (see Figure 5-2).

Ring Structures of Glucose

We have written the structures of various aldoses and ketoses as open-chain forms (Figures 5-2 and 5-5). Such structures are correct for the trioses and tetroses. However, most monosaccharides having five or more carbon atoms in their backbones exist for the most part as cyclic compounds in which the carbonyl groups are "masked" and do not show their usual chemical characteristics. For example, we have seen that glucose has an aldehyde group, yet it has been found that glucose is relatively stable to reagents that normally react readily with aldehyde groups. For example, glucose is quite inert when exposed to air or oxygen, whereas most aldehydes tend to oxidize readily under these conditions.

A second property of D-glucose that suggested it can exist in a ring structure is the fact that it has two crystalline forms. If D-glucose is crystallized from water, a form called α-D-glucose results, for which the specific rotation is $[\alpha]_D^{20} = +112.2°$. If it is crystallized from pyridine, β-D-glucose results; its $[\alpha]_D^{20} = +18.7°$. However, the two forms do not differ in chemical composition. When α-D-glucose is dissolved in water its specific rotation gradually changes with time and reaches a stable value of 52.7°; when β-D-glucose is similarly treated, its rotation ultimately attains the same value. This change, called *mutarotation,* is due to the formation of an equilibrium mixture consisting of about one-third α-D-glucose and two-thirds β-D-glucose at 25°C. From various chemical considerations, it has been deduced that the α and β isomers of D-glucose are not open-chain structures, but rather six-membered ring structures formed by an addition reaction of the alcoholic hydroxyl group at carbon atom 5 to the aldehydic carbon atom. As can be seen from Figure 5-7, D-glucose can form two different ring structures. Such six-membered ring forms of sugars are called *pyranoses* because they are derivatives of the heterocyclic compound *pyran*. The systematic names for the ring forms of α-D-glucose are α-D-glucopyranose and β-D-glucopyranose (Figure 5-7).

The formation of pyranoses is the result of a reaction between an aldehyde group and an alcohol to form a derivative called a *hemiacetal* (Figure 5-8) which contains an asymmetric carbon atom and thus may exist in two stereoisomeric forms. D-Glucopyranose is an intramolecular hemiacetal in which the free hydroxyl group at carbon atom 5 has reacted with the aldehydic carbon atom 1, rendering the latter asymmetric (below). D-Glucopyranose therefore can exist as two different stereoisomers, designated α- and β-. The net result is that D-glucose behaves as though it has one more asymmetric center than is given by its open-chain formula. Isomeric forms of monosaccharides that differ from each other only in their configuration about the hemiacetal carbon atom, such as α-D-glucose and β-D-glucose, are *anomers,* and the carbonyl carbon atom is called the *anomeric carbon*. Only aldoses having five or more carbon atoms can form stable rings and exist in anomeric forms. Aldohexoses may also exist in forms which have five-membered rings; they are derivatives of *furan* and are called *furanoses*. However, the six-membered aldopyranose ring is much more stable than the aldofuranose ring and thus predominates in aldohexose solutions.

Ketohexoses also occur in α- and β-anomeric forms. In these compounds the alcoholic hydroxyl group on carbon atom 5 reacts with the carbonyl group at carbon atom 2, forming a five-membered furanose ring. The common ring form of D-fructose is β-D-fructofuranose (Figure 5-9).

Haworth projection formulas are commonly used to indicate the configuration in space of the ring forms of monosaccharides; the edge of the ring nearest the reader is usually represented by bold lines (Figure 5-10). However, the six-membered pyranose ring is not planar; in most sugars it exists in a chairlike configuration.

Important Chemical Properties of Simple Monosaccharides

Simple monosaccharides have some characteristic chemical properties which are useful in their isolation, identification, or analysis.

Action of Acids

Monosaccharides are stable to hot dilute mineral acids. This fact makes possible the quantitative recovery of

Figure 5-8
Formation of a hemiacetal.

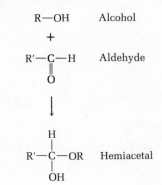

Figure 5-9
The furanose ring.

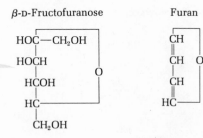

Figure 5-10
Haworth projection formulas.

α-D-Glucopyranose

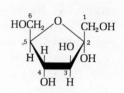

α-D-Fructofuranose

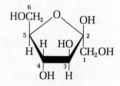

β-D-Glucopyranose

β-D-Fructofuranose

Figure 5-11
Furfural formation.

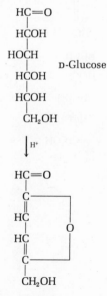

D-Glucose

5-Hydroxymethylfurfural

+

3 H₂O

most monosaccharides in intact form after the hydrolysis of polysaccharides. Hot concentrated acids, however, cause dehydration of sugars to yield *furfurals*, aldehyde derivatives of furan; for example, D-glucose yields 5-*hydroxymethylfurfural* (Figure 5-11). Furfurals condense with phenols to give characteristic colored products often used for colorimetric analysis of sugars.

Action of Bases

Dilute aqueous bases at room temperature promote the epimerization of sugars. They also cause rearrangements of the substituent groups about the anomeric carbon atom and its adjacent carbon atom, without affecting substituents at other carbon atoms. For example, treatment of D-glucose with dilute NaOH yields an equilibrium mixture of D-glucose, D-fructose, and D-mannose (Figure 5-12).

Acetylation

The free hydroxyl groups of monosaccharides and polysaccharides can be acylated to yield O-acyl derivatives, which are useful in structure determination. For example, treatment of α-D-glucose with excess acetic anhydride yields penta-O-acetyl-α-D-glucose (Figure 5-13).

Oxidation

Monosaccharides are readily oxidized by alkaline solutions of Cu^{2+}, Ag^+, or ferricyanide. Such reactions are the basis of widely used qualitative and quantitative tests for simple sugars with potentially free carbonyl groups. In Benedict's test the sugar is oxidized with alkaline cupric tartrate solution; the blue cupric ion is reduced to Cu^+, which precipitates from solution as the red insoluble Cu_2O. Sugars giving a positive reaction with Benedict's test are called reducing sugars; they include glucose, fructose, and mannose. Benedict's test is often used to detect the presence of glucose in the urine, a sign of the disease diabetes mellitus.

Reduction

Reduction of the carbonyl group of monosaccharides yields corresponding sugar alcohols. For example, reduction of D-glucose yields sorbitol. (Figure 5-13).

Reaction with Phenylhydrazine

Monosaccharides in acid solution at 100°C react with excess phenylhydrazine to form phenylosazones, derivatives which are easily crystallized and sometimes used for identification of sugars. The structure of D-glucose phenylosazone is given in Figure 5-13. D-Glucose, D-fructose, and D-mannose yield the same phenylosazone.

Figure 5-12
Isomerization of D-glucose by alkali.

```
HC=O
 |
HCOH
 |
HOCH            D-Glucose
 |
HCOH
 |
HCOH
 |
CH₂OH

  ‖

HOCH₂
 |
C=O
 |
HOCH            D-Fructose
 |
HCOH
 |
HCOH
 |
CH₂OH

  ‖

HC=O
 |
HOCH
 |
HOCH            D-Mannose
 |
HCOH
 |
HCOH
 |
CH₂OH
```

Figure 5-13
Some derivatives of D-glucose.

Penta-O-acetyl-α-D-glucose

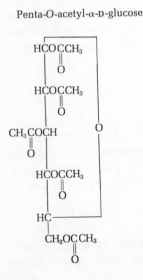

Sorbitol

```
CH₂OH
 |
HCOH
 |
HOCH
 |
HCOH
 |
HCOH
 |
CH₂OH
```

D-Glucose phenylosazone

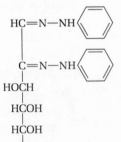

```
           HC=N—NH⟨⟩
            |
           C=N—NH⟨⟩
            |
           HOCH
            |
           HCOH
            |
           HCOH
            |
           CH₂OH
```

Formation of Glycosides

Aldohexoses readily react with alcohols in the presence of a mineral acid to form isomeric α- and β-glycosides, analogous to the α and β forms of glucose described above. D-Glucose yields, with methanol, methyl α-D-glucoside ($[\alpha]_D^{20} = +158.9°$) and methyl β-D-glucoside ($[\alpha]_D^{20} = -34.2°$), which are shown in Figure 5-14.

Biologically Important Derivatives of Monosaccharides

Some sugars occur in nature as derivatives in which one or more of the normal functional groups of the sugar are modified.

Phosphoric Acid Esters

Phosphoric acid esters of various monosaccharides are found in all cells; they serve as important intermediates in carbohydrate metabolism. Three representative sugar phosphates are shown in Figure 5-15.

Deoxy Sugars

Deoxy sugars lack one or more oxygen atoms. The most abundant deoxy sugar found in nature is 2-deoxy-D-ribose, the sugar component of deoxyribonucleic acid. Other deoxy sugars are important components of some bacterial cell walls (Figure 5-15).

Amino Sugars

In amino sugars an amino group replaces a hydroxyl group. Two amino sugars of wide distribution are D-glucosamine and D-galactosamine, in which the hydroxyl group at carbon atom 2 is replaced by an amino group (Figure 5-15). D-Glucosamine occurs in many polysaccharides of vertebrate tissues and is also a major component of chitin, a structural polysaccharide found in the exoskeletons of insects and crustaceans. D-Galactosamine is a major component of the polysaccharide of cartilage, chondroitin sulfate.

Sugar Acids

When the aldehyde carbon of D-glucose is oxidized to a carboxyl group the resulting product is D-gluconic acid, an intermediate in the metabolism of glucose in some organisms. When the carbon bearing the primary hydroxyl group is oxidized to a carboxyl group, the product is D-glucuronic acid, an important building block of

Figure 5-14
The methyl glucosides.

Methyl α-D-glucoside

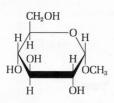

Methyl β-D-glucoside

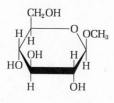

Figure 5-15
Important derivatives of sugars.

Sugar phosphates

α-D-Glucose 1-phosphoric acid   α-D-Glucose 6-phosphoric acid   α-D-Fructose 6-phosphoric acid

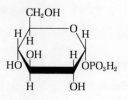

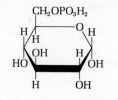

Deoxy sugar   Amino sugars

2-Deoxy-D-ribose   D-Glucosamine   D-Galactosamine

Sugar acids

D-Gluconic acid   D-Glucuronic acid   L-Ascorbic acid

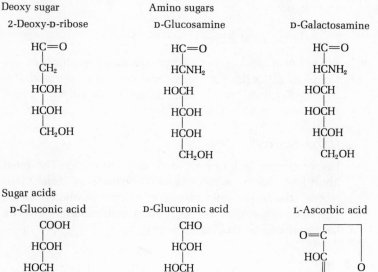

many polysaccharides. Another important sugar acid is *ascorbic acid,* or vitamin C (Figure 5-15). It is a very unstable compound and readily undergoes oxidation to dehydroascorbic acid. Lack of ascorbic acid in the diet of humans results in the deficiency disease scurvy. Ascorbic acid is present in large amounts in citrus fruit.

Disaccharides

Of the various oligosaccharides (2–10 monosaccharide units) found in nature, the most abundant are those with two monosaccharide units, namely, the disaccharides.

Figure 5-16
Disaccharides.

Maltose (β-form)
O-α-D-Glucopyranosyl-(1→4)-β-D-glucopyranose)

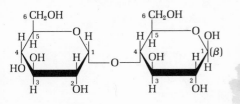

Cellobiose
O-β-D-Glucopyranosyl-(1→4)-β-D-glucopyranose)

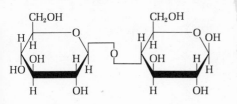

Lactose (α-form)
(O-β-D-Galactopyranosyl-(l→4)-
α-D-glucopyranose)

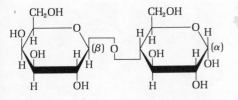

Sucrose
(α-D-Glucopyranosyl-(1→2)-β-D-fructo-
furanoside)

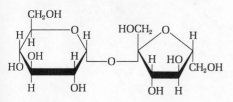

Higher oligosaccharides occur largely as degradation products of polysaccharides.

The most common disaccharides are maltose, cellobiose, lactose, and sucrose. Disaccharides can be hydrolyzed to yield their free monosaccharide building blocks by boiling with dilute acid.

Maltose, which is formed as an intermediate product of the action of amylases on starch, contains two D-glucose residues joined by a glycosidic linkage between carbon atom 1 of the first glucose residue and carbon atom 4 of the second glucose (Figure 5-16). The configuration of the anomeric carbon atom in the glycosidic linkage between the glucose residues is α and the linkage is thus symbolized α(1 → 4). Both the glucose residues are in pyranose form. Maltose may therefore be called 4-O-α-D-glucopyranosyl-D-glucopyranose. Maltose is a reducing sugar since it has one potentially free carbonyl group. The second glucose residue of maltose is capable of existing in α and β forms; the β form shown in Figure 5-16 is that formed by the action of β-amylase on starch (see below).

Another common disaccharide that contains two D-glucose units is *cellobiose,* the repeating disaccharide unit of cellulose. It has the glycosidic linkage β(1 → 4) (Figure 5-16). Both maltose and cellobiose yield D-glucose as end products after hydrolysis by acids.

The disaccharide *lactose* is found in milk but otherwise does not occur in nature. It yields D-galactose and D-glucose on hydrolysis. Since it possesses a potentially free carbonyl group on the glucose residue, lactose is a reducing disaccharide (Figure 5-16).

Sucrose, or cane sugar, is a disaccharide of glucose and fructose. It is extremely abundant in the plant world. In contrast to maltose and lactose, sucrose contains no free anomeric carbon atom; the anomeric carbon atoms of the two sugars are linked to each other (Figure 5-16). Sucrose does not react with phenylhydrazine to form osazones, nor is it a reducing sugar. The enzyme invertase, which is secreted into the intestine, catalyzes the hydrolysis of sucrose to yield D-glucose and D-fructose.

Polysaccharides

Most of the carbohydrates found in nature occur as polysaccharides of high molecular weight. On complete hydrolysis with acid or specific enzymes, these polysaccharides yield monosaccharides or monosaccharide derivatives.

Polysaccharides, which are also called *glycans*, differ in the nature of their recurring monosaccharide units, in the length of their chains, and in the degree of branching. They are divided into *homopolysaccharides*, which consist of only a single type of monomeric unit, and *heteropolysaccharides*, which contain two or more different monomeric units. An example of a homopolysaccharide is *starch*, which contains only D-glucose units. An example of a heteropolysaccharide is *hyaluronic acid*, which contains alternating residues of D-glucuronic acid and N-acetyl-D-glucosamine. Polysaccharides containing only glucose units, such as starch and glycogen, are sometimes called *glucans* and those consisting only of mannose units are *mannans*. The important polysaccharides are best described in terms of their biological function.

Storage Polysaccharides

Storage polysaccharides, of which starch is the most abundant in plants and glycogen in animals, are usually deposited in the form of large granules in the cytoplasm of cells. These granules have a diameter of 100–400 Å and consist of a number of polysaccharide molecules in close association (Figure 5-17).

Starch contains two types of polysaccharides, *α-amylose* and *amylopectin*. The former consists of long unbranched chains of D-glucose units connected by $\alpha(1 \rightarrow 4)$ linkages. Such chains vary in molecular weight from a few thousand to 500,000. In contrast, amylopectin is highly branched (Figure 5-18). The branches are about 12 glucose residues long and occur on the average at every twelfth glucose residue along the backbone. The glycosidic linkages joining successive glucose residues are $\alpha(1 \rightarrow 4)$, except for the branch points, which are $\alpha(1 \rightarrow 6)$ linkages.

Amylose may be hydrolyzed by the enzyme α-amylase, which is present in pancreatic juice and saliva and participates in the digestion of starch in the gastrointestinal tract. It hydrolyzes $\alpha(1 \rightarrow 4)$ linkages throughout the amylose chain in such a way as to yield ultimately a mixture of glucose and maltose. The polysaccharides of intermediate chain length that are formed during amylase action are called *dextrins*. Amylopectin is also attacked by α-amylase, but since this enzyme cannot hydrolyze the $\alpha(1 \rightarrow 6)$ linkages at the branch points, another enzyme, $\alpha(1 \rightarrow 6)$ *glucosidase*, which is also called *debranching enzyme*, also hydrolyzes the $\alpha(1 \rightarrow 6)$ link-

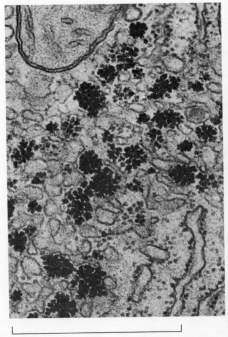

Figure 5-17
Electron micrograph of glycogen granules in a liver cell of the hamster.

1.0 μ

Figure 5-18
Structure of amylopectin and glycogen.

An $\alpha(1 \rightarrow 6)$ branch point in amylopectin

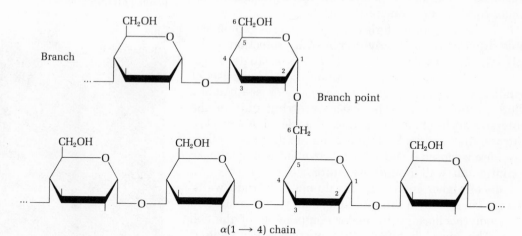

$\alpha(1 \rightarrow 4)$ chain

Schematic representation of the branched
structure of amylopectin and glycogen.
Each circle represents a glucose residue.

ages at the branch points. The combined action of α-amylase and $\alpha(1 \rightarrow 6)$ glucosidase can therefore completely degrade amylopectin.

The storage polysaccharide _glycogen_ is found in animal tissues; it is especially abundant in liver and muscle. Like amylopectin, glycogen is a branched polysaccharide of D-glucose in $\alpha(1 \rightarrow 4)$ linkage with branches linked by $\alpha(1 \rightarrow 6)$ bonds. Glycogen is also readily hydrolyzed by α-amylase and $\alpha(1 \rightarrow 6)$ glucosidase. Liver glycogen is an immediate precursor of blood glucose.

Many plants contain storage polysaccharides made up of D-fructose units. _Inulin,_ found in the artichoke, consists of D-fructose residues in $\beta(2 \rightarrow 1)$ linkage.

Structural Polysaccharides

Structural polysaccharides are manufactured within cells, but extruded to the exterior to form a wall or coat surrounding the cell. In plants the polysaccharide _cellulose_ is the major compound of the thick rigid cell walls. It makes up more than 50 percent of the total organic matter in the biosphere. Wood is about 50 percent cellulose, and cotton is nearly pure cellulose. On complete hydrolysis with strong acids, cellulose yields only D-glucose, but partial hydrolysis yields the reducing disaccharide _cellobiose_ described above (Figure 5-16).

The minimum molecular weight of cellulose has been estimated to vary from 50,000 to 500,000, equivalent to 300 to 3000 glucose residues. Cellulose molecules are organized in bundles of parallel chains, or fibrils, which confer rigidity and strength (Figure 5-19).

The cell walls of bacteria are rigid, porous, nonextensible structures which provide physical protection to the cell. The framework of the cell wall is a single, large, sacklike molecule, called a *peptidoglycan*. It consists of parallel polysaccharide chains covalently cross-linked by short peptide chains. The basic recurring unit in the polysaccharide chains is a disaccharide of N-acetyl-D-glucosamine and N-acetylmuramic acid. This netlike structure is closed on all sides by covalent bonds to make a continuous wall around the entire cell.

Cells of higher animal tissues do not have rigid walls, but are covered by a thin, flexible, sticky, cell coat. *Acid mucopolysaccharides* are major components of the cell coat. The most abundant acid mucopolysaccharide is *hyaluronic acid*, which has repeating units of a disaccharide composed of D-glucuronic acid and N-acetyl-D-glucosamine.

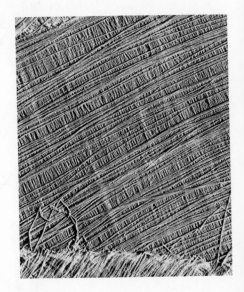

Figure 5-19
Electron micrograph of the cell wall of an alga (Chaetomorpha). *The wall consists of successive layers of cellulose fibers in parallel arrangement.*

Summary

Carbohydrates are polyhydroxylic aldehydes or ketones having the empirical formula $(CH_2O)_n$. They are classified as monosaccharides or sugars (a single aldehyde or ketone unit), oligosaccharides (several units), and polysaccharides, large linear or branched molecules consisting of many sugar units. Monosaccharides have at least one asymmetric carbon atom and thus exist in stereoisomeric forms. Most common naturally occurring sugars, such as ribose, glucose, fructose, and mannose, are of the D-series. Sugars having five or more carbon atoms may exist in the form of closed ring hemiacetals, either furanoses (5-membered ring) or pyranoses (6-membered ring). Furanoses and pyranoses may exist in anomeric α and β forms, which are interconverted by the process of mutarotation. Important monosaccharide derivatives include glycosides, phosphate esters, deoxy sugars, amino sugars, sugar acids, and sugar alcohols.

Disaccharides consist of two monosaccharides joined in glycosidic linkage. Maltose contains two glucose residues in $\alpha(1 \rightarrow 4)$ linkage, cellobiose contains two glucose residues, lactose contains galactose and glucose, and sucrose, a nonreducing sugar, contains glucose and fructose.

Polysaccharides (glycans) are classified functionally as either storage or structural polysaccharides. The most important storage polysaccharides are starch and glycogen; these are high molecular weight branched polymers of glucose having

$\alpha(1 \rightarrow 4)$ linkages in the main chains and $\alpha(1 \rightarrow 6)$ linkages at the branch points. The most important structural polysaccharide of plant cell walls is cellulose, having D-glucose units in $\beta(1 \rightarrow 4)$ linkage. The walls of bacterial cells contain peptidoglycans, cross-linked polysaccharides of alternating N-acetylmuramic acid and N-acetylhexosamine. Animal cells possess cell coats containing acid mucopolysaccharides, such as hyaluronic acid.

References

DAVIDSON, E. A.: *Carbohydrate Chemistry,* Holt, Rinehart, and Winston, Inc., New York, 1967. An excellent survey.

FLORKIN, M., and E. H. STOTZ (eds.): *Carbohydrates,* vol. 5 of *Comprehensive Biochemistry,* American Elsevier Publishing Co, New York, 1963. Reference treatise.

LEHNINGER, A. L.: *Biochemistry,* Worth Publishers, New York, 1970. Chapter 11 gives further details of carbohydrate chemistry.

Problems

1. Aldopentoses exist in how many stereoisomeric forms?

2. A freshly prepared 0.10 M solution (50 ml) of α-D-glucose is mixed with 50 ml of a freshly prepared 0.10 M solution of β-D-glucose at 20°. Calculate (a) the initial specific rotation $[\alpha]_D^{20}$ of the mixture, and (b) the specific rotation after several hours have elapsed.

3. Write the structures of the following as Haworth projection formulas:
   (a) α-D-galactopyranose
   (b) α-L-mannose (pyranose form)
   (c) D-sedoheptulose 7-phosphate (pyranose form)
   (d) β-D-ribose 5-phosphate (furanose form)

4. Write the structures and names of the products of the following treatments:
   (a) Excess acetic anhydride on α-D-galactose
   (b) Exhaustive methylation of α-L-mannose
   (c) Reduction of D-galactose
   (d) Exhaustive methylation of sucrose followed by hydrolysis.

CHAPTER **6** LIPIDS

Lipids are water-insoluble organic substances which are extractable from cells and tissues by nonpolar solvents such as chloroform, ether, and benzene. They serve two major functions: (1) as structural components of membranes and (2) as storage forms of metabolic fuel. There are several different classes of lipids, but all of them contain large nonpolar hydrocarbon-like structures which give them an oily or waxy, water-insoluble nature.

Fatty Acids and Their Properties

Fatty acids are long-chain aliphatic acids which are the building blocks of several classes of lipids and endow them with their fatty or oily nature. They do not normally occur in the free state in cells or tissues but are derived by hydrolysis of lipids. Many different fatty acids have been isolated from different species. All possess a long hydrocarbon chain and a terminal carboxyl group (Figure 6-1). The chain may be saturated or it may have one or more double bonds; a few fatty acids contain triple bonds. Fatty acids differ from each other primarily in chain length and in the number and position of their unsaturated bonds. Table 6-1 gives the structures of some important naturally occurring saturated and unsaturated fatty acids.

Nearly all fatty acids in nature have an even number of carbon atoms and have chains that are between 14 and 22 carbon atoms long; those having 16 or 18 carbons are the most abundant. In general, unsaturated fatty acids

Figure 6-1
Two fatty acids.

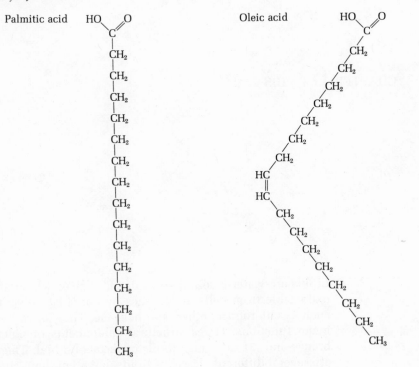

predominate over the saturated type. In most of the unsaturated fatty acids there is a double bond between carbon atoms 9 and 10; additional double bonds usually occur between that point and the methyl-terminal end of the chain. In fatty acids containing two or more double bonds, the double bonds are never conjugated ($-CH=CH-CH=CH-$), but are separated by one methylene group ($-CH=CH-CH_2-CH=CH-$). The double bonds of nearly all the naturally occurring unsaturated fatty acids are in the *cis* geometrical configuration, which produces a rigid bend in the aliphatic chain (Figure 6-1). The molecules of fatty acids with multiple double bonds, such as *arachidonic acid* (four double bonds) thus are kinked. Saturated fatty acid molecules are more flexible than unsaturated ones.

The saturated fatty acids from C_{12} to C_{24} are solids having a waxy consistency. The unsaturated fatty acids, on the other hand, are oily liquids at room temperature. All are insoluble in water, but become dispersed into micelles (Chapter 1) in dilute aqueous NaOH or KOH, which convert them into *soaps*, the name given to the Na^+ or K^+ salts of fatty acids. Soaps have the property of

Table 6-1 Some naturally occurring fatty acids

Carbon atoms	Structure	Systematic name	Common name	Melting point (°C)
Saturated fatty acids				
12	$CH_3(CH_2)_{10}COOH$	n-Dodecanoic	Lauric acid	44.2
14	$CH_3(CH_2)_{12}COOH$	n-Tetradecanoic	Myristic	53.9
16	$CH_3(CH_2)_{14}COOH$	n-Hexadecanoic	Palmitic	63.1
18	$CH_3(CH_2)_{16}COOH$	n-Octadecanoic	Stearic	69.6
20	$CH_3(CH_2)_{18}COOH$	n-Eicosanoic	Arachidic	76.5
24	$CH_3(CH_2)_{22}COOH$	n-Tetracosanoic	Lignoceric	86.0
Unsaturated fatty acids				
16	$CH_3(CH_2)_5CH{=}CH(CH_2)_7COOH$		Palmitoleic	− 0.5
18	$CH_3(CH_2)_7CH{=}CH(CH_2)_7COOH$		Oleic	13.4
18	$CH_3(CH_2)_4CH{=}CHCH_2CH{=}CH(CH_2)_7COOH$		Linoleic	− 5
18	$CH_3CH_2CH{=}CHCH_2CH{=}CHCH_2CH{=}CH(CH_2)_7COOH$		Linolenic	−11
20	$CH_3(CH_2)_4CH{=}CHCH_2CH{=}CHCH_2CH{=}CHCH_2CH{=}CH(CH_2)_3COOH$		Arachidonic	−49.5

emulsifying greasy, water-insoluble substances, through their ability to form a coating around grease droplets in which their hydrophobic tails extend into the droplets and the hydrophilic heads face toward the water.

The hydrocarbon tails of the saturated fatty acids are chemically rather inert. However, the tails of unsaturated fatty acids are quite reactive. Halogens add to their double bonds rather readily. For example, iodine (I_2) readily reacts with oleic acid to yield the saturated 9,10-diiodo derivative

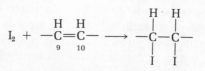

This reaction is used as the basis for measurement of the number of double bonds in a sample of fatty acids.

When exposed to air natural fatty acids having two or more double bonds tend to undergo a complex process called _autoxidation_, in which molecular oxygen attacks a double bond to yield a series of products which ultimately polymerize to form a hard resinous material. Linseed oil, used as a base for paints, is rich in highly unsaturated fatty acids, and undergoes this polymerization process as it "dries." Autoxidation of unsaturated fats in the tissues is also believed to occur in some diseases.

Neutral Lipids (Triacylglycerols)

The simplest and most abundant lipids containing fatty acids as building blocks are the _neutral lipids,_ which are also called _fats, triglycerides,_ or _triacylglycerols;_ the latter name is preferred in systematic chemical nomenclature. They are esters of the alcohol glycerol, with three fatty acid molecules (Figure 6-2). Unlike the phospholipids to be discussed later, they have no net electrical charge and are thus designated _neutral_ lipids. Triacylglycerols are the major components of depot or storage fats in plant and animal cells, especially in the adipose or fat cells of vertebrates.

Triacylglycerols occur in many different types, depending on the identity and position of the three fatty acid components esterified to glycerol. Those containing a single kind of fatty acid in all three positions are called _simple triacylglycerols;_ they are named after the fatty acids they contain. Examples are tristearoylglycerol, tripalmitoylglycerol, and trioleylglycerol; the corresponding trivial and more commonly used names are _tristearin, tripalmitin,_ and _triolein,_ respectively. Triacylglycerols containing two or more different fatty acids are called _mixed triacylglycerols._ Most natural fats are extremely complex mixtures of simple and mixed triacylglycerols.

Triacylglycerols containing saturated fatty acids, such as tristearin, are waxy white solids, whereas triacylglycerols containing three unsaturated fatty acids, such as triolein, a major compound of olive oil, are liquids at room temperature. Lard, which is prepared from the depot fat of the pig, is a mixture of triacylglycerols.

Triacylglycerols undergo hydrolysis when boiled with acids or bases or when acted upon by lipases such as are present in pancreatic juice. Hydrolysis of triacylglycerols with alkali, which is called _saponification,_ yields a mixture of fatty acid soaps and glycerol (Figure 6-3).

Phospholipids

Phospholipids, which are waxy solids, are found almost exclusively in cellular membranes and in the lipoproteins of blood plasma. Only traces are found in fat depots. Phospholipids thus serve primarily as structural elements and are never stored in large amounts. As their name implies, this group of lipids contains phosphorus in the form of phosphoric acid. The major phospholipids found in cells contain two fatty acid molecules which are

Figure 6-2
Triacylglycerols.

Glycerol

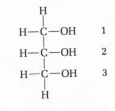

General structure of triacylglycerols

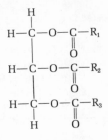

Tripalmitin

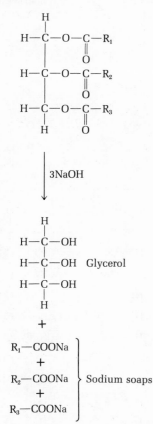

Figure 6-3
Saponification (hydrolysis) of a
triacylglycerol.

esterified to the first and second hydroxyl groups of glycerol. The third hydroxyl group, at carbon atom 3, is esterified with phosphoric acid. Phospholipids contain a second alcohol which is also esterified to the phosphoric acid to form a phosphodiester; the second alcohol group is thus located on the polar head of the phospholipid molecule. The general structural formula of the phospholipids is shown in Figure 6-4; RO— denotes the second alcohol group.

The most abundant fatty acids found in phospholipids have 16 or 18 carbon atoms. Usually one of the fatty acids is saturated and the other unsaturated; the latter is always esterified to the middle or 2-hydroxyl of the glycerol component.

Different types of phospholipids are named according to the second alcohol at their polar heads (Figure 6-4). The most abundant phospholipids are the closely related *phosphatidylethanolamine* (also called cephalin) and *phosphatidylcholine* (also called lecithin), which contain ethanolamine and choline, respectively, at their heads. Each of these can occur in different forms depending on the fatty acids they contain.

Other phospholipids include *phosphatidylserine*, containing as head group the amino acid serine; *phosphatidylinositol*, containing the cyclic alcohol inositol; and *cardiolipin*, in which two simple phospholipid molecules are bridged by a glycerol molecule.

Phospholipids differ from the triacylglycerols in having a highly polar hydrophilic head. All of them have a negative charge on the phosphate group at pH 7.0. In addition, the R groups may also contribute one or more electrical charges, as is shown in Figure 6-4. Because phospholipids have both nonpolar hydrocarbon tails and polar hydrophilic heads they are called *polar* lipids. As we shall see, this characteristic enables the phospholipids to form micelles in water. The neutral fats or triglycerides, on the other hand, cannot form micelles since they lack polar heads.

Phospholipids readily undergo hydrolysis, catalyzed by acids, bases, or enzymes. Dilute base removes the two fatty acid groups of phosphatidylcholine, leaving the rest of the molecule intact. Strong base causes cleavage of both the fatty acids as well as the choline, leaving glycerol 3-phosphate, which can then be cleaved to yield glycerol and phosphoric acid by boiling with HCl.

Phospholipases of different types catalyze hydrolysis of specific linkages in the phospholipid molecule (Figure 6-5).

Figure 6-4
Structure of some phospholipids.

	General structure	Phosphatidylethanolamine	Phosphatidylcholine	Phosphatidylserine
Polar heads				
Nonpolar tails				

Polar heads

General structure:
R
$|$
O
$|$
$O=P-O^-$
$|$
O

Phosphatidylethanolamine:
$\overset{+}{N}H_3$
$|$
CH_2
$|$
CH_2
$|$
O
$|$
$O=P-O^-$
$|$
O

Phosphatidylcholine:
$\overset{+}{N}(CH_3)_3$
$|$
CH_2
$|$
CH_2
$|$
O
$|$
$O=P-O^-$
$|$
O

Phosphatidylserine:
COO^-
$|$
$HC-\overset{+}{N}H_3$
$|$
CH_2
$|$
O
$|$
$O=P-O^-$
$|$
O

Nonpolar tails (glycerol backbone and two fatty acid chains):

General structure:
CH_2
$H-C$ ... $C-H$
O ... O
$C=O$ ... $C=O$
CH_2 ... CH_2
CH_2 ... CH_2
CH_2 ... CH_2
CH_2 ... CH_2
CH_2 ... CH_2
CH_2 ... CH_2
CH_2 ... CH_2
CH_2 ... CH
 ... $\parallel$
CH_2 ... CH
CH_2 ... CH_2
CH_2 ... CH_2
CH_2 ... CH_2
CH_2 ... CH_2
CH_2 ... CH_2
CH_3 ... CH_3

Phosphatidylethanolamine:
H ... CH_2
$H-C$ ... $C-H$
O ... O
$C=O$ ... $C=O$
CH_2 ... CH_2
CH_2 ... CH_2
CH_2 ... CH_2
CH_2 ... CH_2
CH_2 ... CH_2
CH_2 ... CH_2
CH_2 ... CH
 ... $\parallel$
CH_2 ... CH
CH_2 ... CH_2
CH_2 ... CH_2
CH_2 ... CH_2
CH_2 ... CH_2
CH_3 ... CH_3

Phosphatidylcholine:
H ... CH_2
$H-C$ ... $C-H$
O ... O
$C=O$ ... $C=O$
CH_2 ... CH_2
CH_2 ... CH_2
CH_2 ... CH_2
CH_2 ... CH_2
CH_2 ... CH_2
CH_2 ... CH_2
CH_2 ... CH
 ... $\parallel$
CH_2 ... CH
CH_2 ... CH_2
CH_2 ... CH_2
CH_2 ... CH_2
CH_2 ... CH_2
CH_3 ... CH_3

Phosphatidylserine:
H ... CH_2
$H-C$ ... $C-H$
O ... O
$C=O$ ... $C=O$
CH_2 ... CH_2
CH_2 ... CH_2
CH_2 ... CH_2
CH_2 ... CH_2
CH_2 ... CH
 ... $\parallel$
CH_2 ... CH
CH_2 ... CH_2
CH_2 ... CH_2
CH_2 ... CH_2
CH_2 ... CH_2
CH_3 ... CH_3

Figure 6-4 (continued)

Phosphatidylinositol

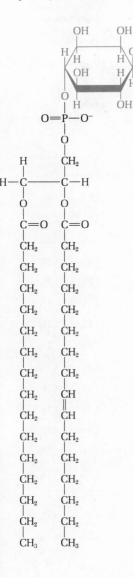

Figure 6-5
Sites of action of phospholipases on phosphatidylcholine. Phospholipase B is a mixture of phospholipases A_1 and A_2.

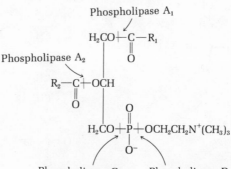

Sphingolipids

Sphingolipids also contain phosphoric acid but are usually classified separately because they contain the long-chain unsaturated amino alcohol *sphingosine* (Figure 6-6) or its saturated analog, dihydrosphingosine. Sphingolipids contain no glycerol. *Sphingomyelin*, the simplest and most common sphingolipid, contains one molecule of a fatty acid, one of sphingosine, one of phosphoric acid, and one of the alcohol choline (Figure 6-6).

Glycolipids

Glycolipids characteristically contain a sugar group, but no phosphoric acid. The simplest glycolipids are the *glycosyldiacylglycerols*, found in plants and microorganisms (Figure 6-7). Another group, the *cerebrosides*, may be classified either as glycolipids or as sphingolipids since they contain both a sugar and sphingosine (Figure 6-7). They are especially abundant in the membranes of brain and nerve cells, particularly in the myelin sheath. The *gangliosides* are carbohydrate-rich complex lipids with negatively charged heads of extremely large size and complexity. They are usually found on the outer surface of cell membranes, especially of nerve cells.

Figure 6-6
Sphingosine and the sphingolipid sphingomyelin.
The polar head of sphingomyelin is in color.

Figure 6-7
Structure of two glycolipids. Both contain
D-galactose (in color) as polar head.

Sphingosine   Sphingomyelin

Monogalactosyl
diacylglycerol

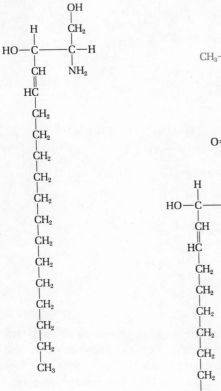

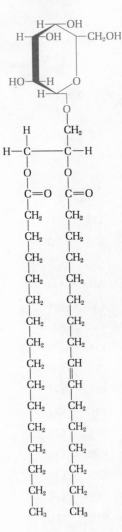

Figure 6-7 (continued)

A cerebroside.

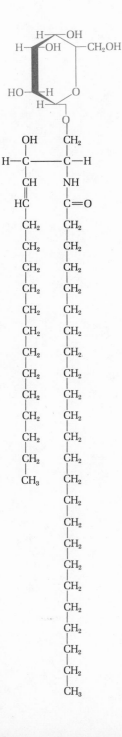

Waxes

Closely related to the triacylglycerols in structure and properties are the waxes, which are esters of higher fatty acids with long-chain monohydroxylic alcohols. Waxes are found as protective coatings on skin, fur, and feathers, on leaves and fruits of higher plants, and on the cuticle of the exoskeleton of many insects. Beeswax consists of palmitic acid esters of long-chain fatty alcohols. Leaf waxes contain esters of fatty acids and alcohols having from 26 to 34 carbon atoms.

Nonsaponifiable Lipids

The lipids discussed to this point are often called *saponifiable*, since they may be hydrolyzed by heating with alkali to yield soaps of their fatty acid components. Cells also contain another class of lipids, quantitatively minor, termed nonsaponifiable lipids, since they do not undergo hydrolysis to yield fatty acids. There are two major types of nonsaponifiable lipids, *steroids* and *terpenes*.

Steroids

Steroids are derivatives of the *perhydrocyclopentanophenanthrene* nucleus. Among the important naturally occurring steroids are the bile acids, the male and female sex hormones, the adrenocortical hormones, and various other steroids that have intense biological activity, such as toad poisons. While most of these occur in only trace amounts in cells, one type of steroid, the sterols, are extremely abundant. *Cholesterol* is the major sterol in animal tissues and occurs both in free and combined form. *Lanosterol*, which is found in the fatty coating of wool, also serves as an important intermediate in the biosynthesis of cholesterol. Some typical steroids are shown in Figure 6-8.

Terpenes

Many naturally occurring hydrocarbons or substituted hydrocarbons are constructed of covalently-joined multiples of the five-carbon hydrocarbon *isoprene*, or 2-methyl-1,3-butadiene (Figure 6-9). Terpenes may be either linear or cyclic molecules; some contain both linear and cyclic structures. The successive isoprene units of terpenes are usually linked in a head-to-tail arrangement, particularly in the linear segments. The double bonds in the linear segments of most terpenes are

Figure 6-8
Some steroids.

Perhydrocyclopentanophenanthrene
nucleus

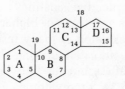

Cholesterol

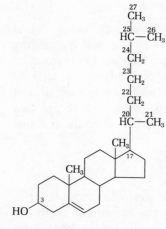

Lanosterol

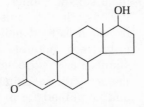

Cholic acid (a bile acid)

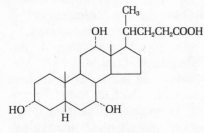

Testosterone (a male sex hormone)

in the stable *trans* configuration. The highly colored *β-carotene*, a red-brown oily substance which is especially abundant in carrots, consists of eight isoprene residues joined as shown in Figure 6-9. It is a precursor of vitamin A in animals. In fact, several vitamins are members of the terpene class (Chapter 7).

Lipoproteins

Polar lipids associate with certain specific proteins to form *lipoproteins*. The best known are the lipoproteins of

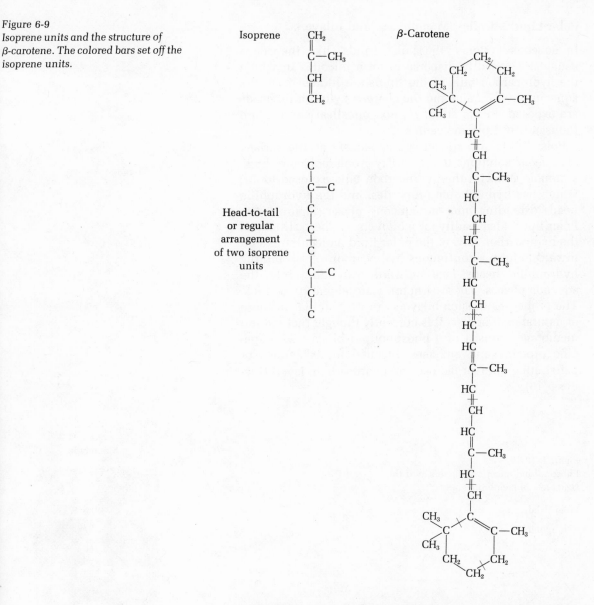

Figure 6-9
Isoprene units and the structure of
β-carotene. The colored bars set off the
isoprene units.

mammalian blood plasma, which may contain from 30 to
75% lipid. In these conjugated proteins, no covalent linkages exist between the lipid molecules and the protein
components. Lipoproteins usually contain both polar
and neutral lipids, as well as cholesterol and its esters.
They serve as vehicles for the transport of lipids from the
small intestine to the liver and from the liver to the fat
depots and other tissues. The lipoproteins of blood
plasma are classified on the basis of their density, which
in turn is a reflection of their lipid content. The greater
the lipid content, the lower the density.

Polar Lipid Micelles, Monolayers, and Bilayers

In aqueous systems, the polar lipids, like the soaps (Chapter 1), readily disperse to form micelles in which the hydrocarbon tails of the lipids are hidden from the aqueous environment and the charged hydrophilic heads are exposed on the surface. Such micelles may contain thousands of lipid molecules.

Polar lipids also spread spontaneously on the surface of aqueous solutions to form a layer one molecule thick, a _monolayer,_ with the hydrocarbon tails exposed to air, which has hydrophobic properties, and the hydrophilic heads extending into the aqueous phase (Figure 6-10). Polar lipids also readily form _bilayers._ In these structures the hydrocarbon tails of the polar lipid molecules extend inward to form a continuous hydrocarbon phase and the hydrophilic heads face outward, extending into the aqueous phases. These structures are about 70 Å thick. The properties of such bilayers are very similar to those of natural membranes. It is currently thought that natural membranes consist of a phospholipid bilayer with specific proteins and enzymes attached to its surface or penetrating into or across its hydrocarbon layer (Figure 6-10).

Figure 6-10
Phospholipid micelles, bilayers, and the structures of membranes.

Polar head
Nonpolar tails

Micelles

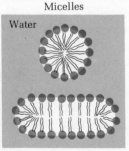

Bilayer

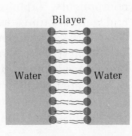

Membrane

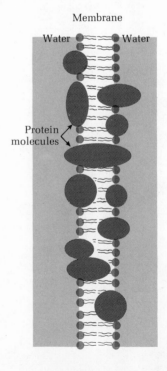

Summary

Lipids are water-soluble components of cells that can be extracted by nonpolar solvents. Some lipids serve as structural components of membranes and others as storage forms of fuel. Fatty acids, which are components of lipids, usually have an even number of carbon atoms; most are from 12 to 22 carbon atoms long. Fatty acids may be saturated or unsaturated; the double bonds of unsaturated fatty acids have the *cis* configuration. In most unsaturated fatty acids, one double bond is at the 9, 10 position. Sodium or potassium salts of fatty acids are called soaps.

Triacylglycerols (neutral fats) contain three fatty acid molecules esterified to the three hydroxyl groups of glycerol. Simple triacylglycerols contain only one type of fatty acid; mixed triacylglycerols contain at least two different types. Triacylglycerols serve primarily to store fuel in the form of fat droplets in cells. Waxes are esters of long-chain fatty acids with high-molecular-weight monohydroxylic alcohols. There are several types of polar lipids. They include the phosphoglycerides, which contain two fatty acid molecules esterified to the two free hydroxyl groups of glycerol 3-phosphate, and a second alcohol esterified to the phosphoric acid. Phosphoglycerides differ in the structure of the second alcohol. The most common phosphoglycerides are phosphatidylethanolamine and phosphatidylcholine. Sphingomyelin contains no glycerol but possesses two long hydrocarbon chains, one contributed by a fatty acid and the other by sphingosine, a long-chain aliphatic amino alcohol; it also contains phosphoric acid and choline. Glycolipids contain fatty acids and one or more carbohydrate molecules. All polar lipids possess polar or charged "heads" and nonpolar hydrocarbon "tails"; they spontaneously form micelles, monolayers, and bilayers. Phospholipid bilayers function as the structural core of cell membranes.

The class of nonsaponifiable lipids cannot be hydrolyzed to form soaps; they include steroids and terpenes. Steroids are derivatives of perhydrocyclopentanophenanthrene. Many steroids, such as the adrenal cortical and sex hormones, possess intense biological activity. Cholesterol is an important component of plasma membranes. Terpenes are constructed of two or more units of the hydrocarbon isoprene (2-methyl-1,3-butadiene); β-carotene is an important terpene.

References

ANSELL, G. B., and J. N. HAWTHORNE: *Phospholipids: Chemistry, Metabolism, and Function,* Elsevier, New York, 1964. Comprehensive treatise.

LEHNINGER, A. L.: *Biochemistry,* Worth Publishers, New York, 1970. Chapters 10 and 23 provide more detailed coverage.

MASORO, E. J.: *Physiological Chemistry of Lipids in Mammals*, Saunders, Philadelphia, 1968. A short textbook.

Problems

1. A sample of triacylglycerols from the depot fat of the rat was completely hydrolyzed. The fatty acids found were palmitic (P), oleic (O), and stearic (S). Symbolize the structure of the various molecular species of triacylglycerols that could conceivably be present.

2. Name the products of mild hydrolysis with dilute sodium hydroxide of
   (a) 1-stearoyl-2,3-dipalmitoylglycerol
   (b) 1-palmitoyl-2-oleyl phosphatidylcholine
   What should be the product of concentrated NaOH on (b)?

3. Electrophoresis at pH 7.0 was carried out on a mixture of lipids containing (a) phosphatidylcholine, (b) phosphatidylethanolamine, and (c) phosphatidylserine. How would you expect these compounds to move [toward the anode (A), toward the cathode (C), or remain at origin (O)]?

4. Predict the products of the action of phospholipase B and phospholipase D on 1-palmitoyl-2-linoleyl phosphatidylcholine.

CHAPTER **7** **NUCLEOTIDES AND THE BACKBONE STRUCTURE OF NUCLEIC ACIDS**

Nucleotides are the recurring structural units of the nucleic acids and thus participate in the molecular mechanisms by which genetic information is transmitted. Nucleotides also have other functions in the cell, particularly in the transport of chemical energy from energy-yielding to energy-requiring reactions. They also participate in many biosynthetic reactions and some serve as coenzymes.

In this chapter we shall examine the component building blocks and the properties of nucleotides and then consider how nucleotides are linked together in deoxyribonucleic acid (DNA) and in the various forms of ribonucleic acid (RNA). The three-dimensional structure and role of nucleic acids in storage, replication, and transformation of genetic information will be described in Chapters 21–23.

Components of Nucleotides

Nucleotides, also called mononucleotides, contain three characteristic components which are released in free form on complete hydrolysis: (1) a nitrogenous base, (2) a five-carbon sugar, and (3) phosphoric acid.

Pyrimidine and Purine Bases

Two classes of nitrogenous bases are found in mononucleotides derived from nucleic acids. They are derivatives of two parent heterocyclic compounds *pyrimidine*

(Figure 7-1) and purine (Figure 7-2), which are themselves not found in nature. Purine is a derivative of pyrimidine in which an imidazole ring is fused to the pyrimidine ring.

Three pyrimidine bases are common in nucleic acids: uracil, thymine, and cytosine, universally abbreviated as U, T, and C (Figure 7-1). Uracil is in general found only in ribonucleic acid and thymine in deoxyribonucleic acid, whereas cytosine is found in both DNA and RNA. There are two common purine bases, found in both DNA and RNA: adenine (A) and guanine (G) (Figure 7-2).

In addition to the major pyrimidines and purines just listed, nucleic acids also contain small amounts of other pyrimidines and purines, particularly methylated derivatives of the major bases listed in Figures 7-1 and 7-2. Among the minor pyrimidines are 5-methylcytosine, 5-hydroxymethylcytosine, and dihydrouracil. Among the minor purines are 1-methylguanine, 2-methyladenine, and dimethylguanine (Figure 7-3).

The pyrimidine and purine bases are nearly flat molecules which are relatively insoluble in water. They show strong absorption of ultraviolet light at 260 nm, a feature which is useful in quantitative analysis of nucleotides and nucleic acids. The bases are easily separated from each other by chromatography.

The Pentose Components

Two types of five-carbon or pentose sugars are components of nucleotides. D-Ribose is the sugar component in the mononucleotides derived from RNA and 2-deoxy-D-ribose is the sugar in mononucleotides derived from DNA. Their structures were given in Chapter 5.

Nucleosides

Nucleosides consist of a pyrimidine or purine base joined by an N-glycosyl linkage to either D-ribose or deoxyribose. The linkage is between nitrogen atom 1 of the pyrimidine (or nitrogen 9 of the purine) and carbon atom 1 of the pentose, which is in its furanose form. There are two anomeric forms of nucleosides, α and β, because the furanose structure makes carbon atom 1 of the pentose anomeric. All nucleosides derived from nucleic acids are the β anomers.

Nucleosides are derived from nucleotides by hydrolytic cleavage of the phosphoric acid group, either by the action of bases or by the action of specific nucleotidases.

Figure 7-1
Pyrimidine and the major pyrimidine bases.

Pyrimidine

The major pyrimidines

Cytosine
(2-Oxy-4-aminopyrimidine)

Uracil
(2,4-Dioxypyrimidine)

Thymine
(5-Methyl-2,4-dioxypyrimidine)

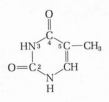

Figure 7-2
Purine and the major purine bases.

Purine

The major purines

Adenine
(6-Aminopurine)

Guanine
(2-Amino-6-oxypurine)

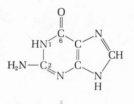

Figure 7-3
Some minor bases.

Two minor pyrimidine bases

5-Methylcytosine

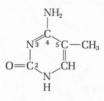

5-Hydroxymethylcytosine

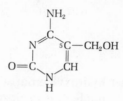

Two minor purine bases

2-Methyladenine

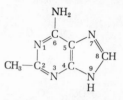

1-Methylguanine

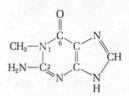

There are two series of nucleosides (Figure 7-4), the *ribonucleosides,* which contain D-ribose as the sugar component, and the *2'-deoxyribonucleosides,* which contain 2-deoxy-D-ribose. The trivial names of the four major ribonucleosides are adenosine, guanosine, cytidine, and uridine, and those of the four major deoxyribonucleosides are 2'-deoxyadenosine, 2'-deoxyguanosine, 2'-deoxycytidine, and 2'-deoxythymidine.

Nucleosides do not occur free in any significant amounts in cells; they are transient intermediates in biological breakdown of the nucleic acids.

Mononucleotides

Mononucleotides are phosphoric acid esters of nucleosides in which the phosphoric acid is esterified to one of the free pentose hydroxyl groups. Nucleotides occur in free form in significant amounts in all cells. They are also formed on partial hydrolysis of nucleic acids, particularly by the action of enzymes called *nucleases.* Nucleotides derived from DNA and thus containing 2-deoxy-D-ribose are *deoxyribonucleotides;* those containing D-ribose are *ribonucleotides.*

Figure 7-4
Two representative nucleosides.

Adenosine   2'-Deoxyadenosine

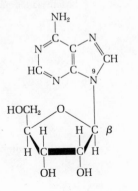   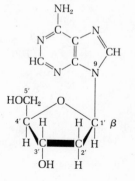

Since there are two or more free hydroxyl groups in nucleosides, the phosphate group of nucleotides can potentially occur in more than one position on the sugar ring. In the case of deoxyribonucleotides, there are only two possible positions in 2-deoxyribose than can be esterified with phosphoric acid, namely, the 3' and 5' positions. Both 3'- and 5'-deoxyribonucleotides occur biologically. In the case of ribonucleotides, the phosphate group may be at the 2', 3', or 5' position; all three types of ribonucleotides have been found as hydrolysis products of RNA, depending on conditions. However, the free nucleotides that predominate in cells have the phosphate group in the 5' position, since the enzymatic reactions normally involved in nucleic acid synthesis and breakdown proceed via the nucleoside 5'-phosphates as intermediates. Figure 7-5 gives the general structure and names of the major ribonucleotides (also called ribonucleoside 5'-phosphates) and deoxyribonucleotides (deoxyribonucleoside 5'-phosphates). The trivial names *adenylic acid, guanylic acid, uridylic acid,* and so on, are ordinarily used. The most abundant 5'-nucleotide occurring in the free state in cells is *adenosine 5'-monophosphate,* or adenylic acid. It participates in the transport of chemical energy in the cell.

The mononucleotides of both series are strong acids since the phosphoric acid component has two strongly dissociated protons. At pH 7.0, the nucleotides exist primarily in the form $R—O—PO_3^{2-}$, where R is the nucleoside group.

Figure 7-5
The major ribonucleotides and
deoxyribonucleotides.

Ribonucleoside
5′-monophosphates

2′-Deoxyribonucleoside
5′-monophosphates

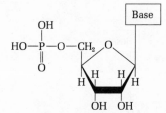

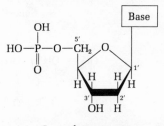

General structure

General structure

Names

Names

Adenosine 5′-phosphoric acid
(adenylic acid; AMP)

Deoxyadenosine 5′-phosphoric acid
(deoxyadenylic acid; dAMP)

Guanosine 5′-phosphoric acid
(guanylic acid; GMP)

Deoxyguanosine 5′-phosphoric acid
(deoxyguanylic acid; dGMP)

Cytidine 5′-phosphoric acid
(cytidylic acid; CMP)

Deoxycytidine 5′-phosphoric acid
(deoxycytidylic acid; dCMP)

Uridine 5′-phosphoric acid
(uridylic acid; UMP)

Deoxythymidine 5′-phosphoric acid
(deoxythymidylic acid; dTMP)

Due to their content of pyrimidine or purine bases, all the mononucleotides show strong light absorption at 260 nm. They are very soluble in water. Hydrolysis of purine nucleotides by dilute acids cleaves the N-glycosyl bond to yield the free purine base plus the pentose phosphate. Hydrolysis of either pyrimidine or purine nucleotides by the action of bacterial alkaline phosphatase yields free nucleosides plus phosphoric acid.

Nucleotides are easily separated from each other by chromatography or by paper electrophoresis.

Nucleoside 5′-Diphosphates (NDPs) and 5′-Triphosphates (NTPs)

In addition to the 5′-monophosphates of the major purine and pyrimidine nucleosides, living tissues also contain the corresponding 5′-diphosphates and 5′-triphosphates. Their names and abbreviations are shown in Figure 7-6. The specific phosphate groups of these compounds are designated by the symbols α, β, and γ. The most abundant and important group are the mono-, di-, and triphosphates of adenosine, namely, *adenosine monophosphate, adenosine diphosphate*, and *adenosine*

Figure 7-6
The general structure and abbreviations of
the NMPs, NDPs, and NTPs.

General structure

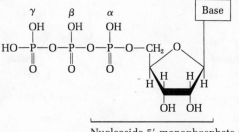

Nucleoside 5'-monophosphate (NMP)

Nucleoside 5'-diphosphate (NDP)

Nucleoside 5'-triphosphate (NTP)

Abbreviations

Ribonucleoside
5'-mono-, di-, and triphosphates

Base	Abbreviations		
Adenine	AMP	ADP	ATP
Guanine	GMP	GDP	GTP
Cytosine	CMP	CDP	CTP
Uracil	UMP	UDP	UTP

Deoxyribonucleoside
5'-mono-, di-, and triphosphates

Adenine	dAMP	dADP	dATP
Guanine	dGMP	dGDP	dGTP
Cytosine	dCMP	dCDP	dCTP
Thymine	dTMP	dTDP	dTTP

triphosphate, abbreviated AMP, ADP, and ATP (Figure 7-6). ATP is the primary carrier of chemical energy in the cell, by transfer of phosphate groups from energy-yielding to energy-requiring processes. In energy-requiring processes the terminal phosphate group of ATP is removed; the ADP so formed is rephosphorylated to ATP during energy-yielding respiration. Although the ATP-ADP-AMP system is the primary or "main-line" phosphate-transferring system in the cell, the other NTPs and dNTPs, such as guanosine triphosphate (GTP) uridine triphosphate (UTP), deoxyguanosine triphosphate (dGTP), and deoxyuridine triphosphate (dUTP), also participate in the channeling of phosphate groups and chemical energy into various biosynthetic processes. The different NTPs are also precursors of the nucleotide units in the biosynthesis of nucleic acids.

The NDPs and NTPs dissociate three and four protons, respectively, from their phosphate groups at pH 7.0. ADP and ATP are therefore highly charged anions. The phosphate groups of the NDPs and NTPs form complexes with divalent cations such as Mg^{2+} and Ca^{2+}. Under intracellular conditions, the NDPs and NTPs exist primarily as Mg^{2+} complexes. The significance of the NDPs and NTPs in cellular energy transformations and in intermediary metabolism will be considered in Chapters 11–13.

Other Nucleotides

In addition to the nucleotides discussed above, other nucleotides or nucleotide-like compounds play important roles in metabolism. Some of these are coenzymes, which are described in Chapter 8, such as flavin mononucleotide (FMN), flavin adenine dinucleotide (FAD), and nicotinamide adenine dinucleotide (NAD).

Nucleic Acids (Polynucleotides)

The mononucleotides are the recurring monomeric units of *oligonucleotides*, which contain several mononucleotide units, and of *polynucleotides* or nucleic acids, which contain many. A polynucleotide consisting of a long covalently linked chain of deoxyribonucleotide units is called *deoxyribonucleic acid* (DNA). When the chain consists of ribonucleotides it is a *ribonucleic acid* (RNA) (Figure 7-7). We shall first consider the distribution and general description of the different types of nucleic acids.

DNA

DNA was first isolated and intensively studied by Friedrich Miescher, a Swiss chemist, in a series of remarkable investigations beginning in 1868. He obtained DNA from fish sperm and from pus cells, which are very rich in DNA. However, it required many years of research before the major building-block components of nucleic acids were identified and the manner of their linkage elucidated. Our present picture of DNA structure did not emerge until the early 1950s.

Figure 7-7
*Polynucleotide structure. In DNA and RNA
the phosphodiester bridges link the 3′-hy-
droxyl of one nucleotide to the 5′-hydroxyl
of the next.*

DNA

RNA

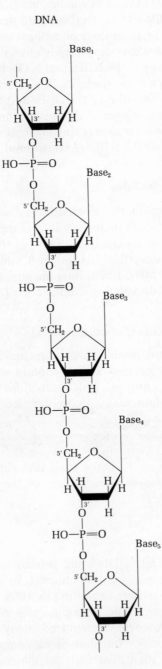

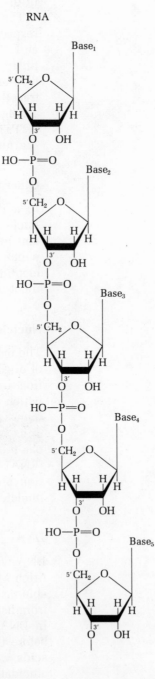

Native DNA molecules have a definite molecular weight, but in most cells they are so long that they are not easily isolated in intact form. In prokaryotic cells, which have only a single chromosome and include the bacteria and blue-green algae, essentially all the DNA is present as a very long, double-stranded molecule exceeding 2,000,000,000 in molecular weight. The DNA makes up about 1 percent of the total weight of such cells. In eukaryotic cells of higher plants and animals, which contain several or many chromosomes, there are, correspondingly, several or many DNA molecules, usually combined with basic proteins, histones or protamines. DNA also occurs in many viruses (see below). DNAs from all types of cells contain four major mononucleotide units, namely, dAMP, dGMP, dTMP, and dCMP, linked in various sequences by 3',5'-phosphodiester bridges. The DNAs isolated from different species of organisms vary in the ratio and sequence of these four mononucleotide units; they also vary in molecular weight. In addition to the major bases—adenine, guanine, thymine, and cytosine—found in all specimens of DNA, small amounts of methylated derivatives of these bases are also present.

RNA

RNA makes up about 5–10 percent of the total weight of the cell. There are three major types of ribonucleic acids, messenger RNA (mRNA), ribosomal RNA (rRNA), and transfer RNA (tRNA). Each has a characteristic molecular weight and base composition (Table 7-1). Each consists of a single polyribonucleotide strand. The three major types of RNA occur in multiple molecular species.

Table 7-1   Properties of E. coli RNAs

Type	Sedimentation coefficient	Mol wt	No. of nucleotide residues	Percent of total cell RNA
mRNA	6S–25S	25,000–1,000,000	75–3,000	~2
tRNA	~4S	23,000–30,000	75–90	16
rRNA	5S	~35,000	~100	
	16S	~550,000	~1,500	82
	23S	~1,100,000	~3,100	

Messenger RNA contains only the four bases A, G, C, and U. It is enzymatically synthesized in the cell nucleus, in such a way that the base sequence of the mRNA molecule is complementary to the base sequence of one of the strands of the DNA molecule (Chapter 22). The mRNA then passes to the cytoplasm, where it serves as the template for the sequential ordering of amino acids during protein synthesis. Each mRNA molecule carries the code for one or more protein molecules. Each cell thus contains hundreds of different mRNA molecules, which are exceedingly difficult to isolate in pure form.

Transfer RNAs are relatively small molecules that act as carriers of specific amino acids during protein synthesis. They have molecular weights in the range of 23,000 to 30,000 and contain from 75 to 90 mononucleotide units (Figure 7-8). Each of the 20 amino acids found in proteins has one or more corresponding tRNAs. Altogether there are 60 or more different tRNAs. The tRNA molecule may exist either in free form or "charged" with its specific amino acid, esterified to the 3'-hydroxyl group of the terminal nucleotide at one end of the polynucleotide chain. Transfer RNAs characteristically contain a rather large number of minor bases, up to 10 percent of the total, in addition to the major bases A, G, C, and U. The minor bases are largely methylated forms of the normal major bases or their derivatives. Transfer RNAs also contain some unusual mononucleotides such as pseudouridylic and ribothymidylic acids (Figure 7-8). Nearly all transfer RNA molecules have a guanylic acid residue at one end; all have the trinucleotide sequence -C-C-A at the other. The 3'-hydroxyl group of the terminal adenylic acid residue is the acceptor for the specific amino acid carried by the tRNA. Another distinctive feature of each tRNA molecule is that it has a specific trinucleotide sequence, called its *anticodon*, which is complementary to a trinucleotide sequence of mRNA specifying a given amino acid, called a *codon*. Chapter 23 considers the function of tRNAs.

Many tRNAs have been isolated in homogeneous form and some have been crystallized. In 1965, R. Holley and his colleagues succeeded in deducing the entire base sequence of yeast alanine tRNA (Figure 7-8). The sequences of a number of other tRNA molecules have since been solved.

Ribosomal RNA (rRNA) is the most abundant type of RNA. It constitutes up to 65 percent of the weight of ribosomes, ribonucleoprotein particles in cells (see below). Ribosomal RNA occurs in three major forms: 5S,

Figure 7-8
The base sequence in a yeast alanine tRNA.
In addition to A, G, U, and C; the following
symbols are used: ψ = pseudouridylic acid,
T = ribothymidylic acid, Uh = dihydro-
uridylic acid, Gm = methylguanylic acid,
GD = dimethylguanylic acid, I = inosinic
acid, Im = methylinosinic acid. The
significance of the "cloverleaf" structure
shown here will be discussed in Chapter 23.
Some tRNAs have an extra arm (Figure 23-4).

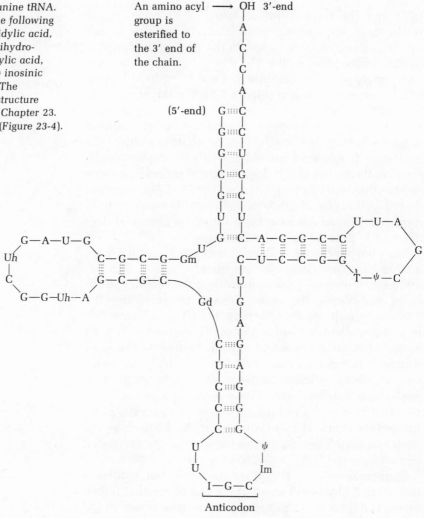

Anticodon

16S, and 23S. The letter S stands for the *svedberg*, the basic unit of measure of the *sedimentation coefficient*, which expresses the rate at which particles sediment in a centrifugal field; it is a composite function of their weight, shape, and density. As is evident in Table 7-1, the sedimentation coefficient is not proportional to molecular weight. Ribosomal RNA contains the four major bases A, G, C, and U; a few of the bases are methylated. Presumably ribosomal RNA plays an important role in the structure and biological function of ribosomes. The structure and function of RNAs are further developed in Chapters 21–23.

The Covalent Backbone of Nucleic Acids

DNA and the three different forms of RNA are linear polymers of successive mononucleotide units, in which one mononucleotide is linked to the next by a phosphodiester bridge between the 3'-hydroxyl group of the pentose moiety of one nucleotide and the 5'-hydroxyl group of the pentose of the next (Figure 7-7). The covalent backbone of nucleic acids thus consist of alternating pentose and phosphoric acid groups. The purine and pyrimidine bases are side chains attached to the pentose units of the backbone. It has been proved that 3'- to 5'-phosphodiester bonds are the sole linkage between successive mononucleotide units in both DNA and RNA. For one thing, partial hydrolysis of DNA or RNA yields many simple dinucleotides, which have been found by chemical degradation to possess only 3',5' linkages.

The phosphodiester bridges of both DNA and RNA are attacked by two classes of enzymes, a and b. For example, an a enzyme present in rattlesnake venom specifically hydrolyzes the ester linkage between the 3'-hydroxyl group and phosphoric acid (site a in Figure 7-9) to yield nucleoside 5'-phosphates. The b enzymes, such as the phosphodiesterase of spleen, hydrolyze the ester linkage between the phosphoric acid and the 5'-hydroxyl end of phosphodiester bridges (site b), to yield only nucleoside 3'-phosphates. These enzymes require a free terminal hydroxyl and proceed stepwise along the polynucleotide chain. Because they start their attack at the ends of polynucleotide chains these enzymes are called _exonucleases_.

Endonucleases, on the other hand, do not require a free 3'- or 5'-hydroxyl group at the end of the chain; they attack certain a or b linkages wherever they occur in the polynucleotide chain. An important example is _ribonuclease_ of bovine pancreas, which attacks b linkages of RNA whose a linkage is to a pyrimidine nucleotide.

These enzymes are extremely important tools in analyzing the base sequence of nucleic acids, just as trypsin and chymotrypsin are important in sequence analysis of proteins.

Hydrolysis of Nucleic Acids by Bases

DNA is not hydrolyzed by bases, whereas RNA is. This finding indicates that the 2'-hydroxyl group of the pentose, which is lacking in DNA, is required for alkaline hydrolysis. Dilute NaOH produces from RNA a mix-

Figure 7-9
Site of action of nucleases. In both DNA and RNA snake venom phosphodiesterase hydrolyzes only a linkages starting from the end having a free 3'-hydroxyl group, whereas bovine spleen phosphodiesterase hydrolyzes only b linkages starting from the end having a free 5'-hydroxyl group.

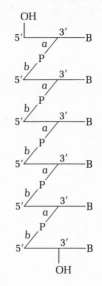

Figure 7-10
Structure of 2',3'-cyclic AMP.

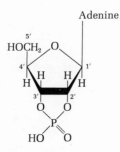

Figure 7-11
Subunit structure of the 70S ribosomes of E. coli.

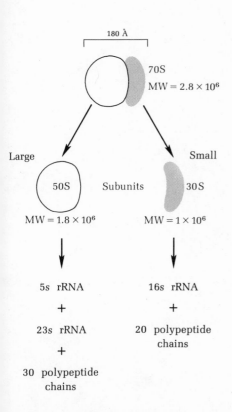

ture of nucleoside 2',3'-cyclic monophosphates (Figure 7-10). The latter are obligatory intermediates in the hydrolysis of RNA by NaOH or by ribonuclease. Alkali cleaves the nucleoside 2',3'-monophosphate at either P—O bond to give a mixture of the 2'- and 3'-phosphates, whereas ribonuclease cleaves only at the 2'-bond to yield the 3'-phosphates.

Conformation of Nucleic Acids

Like the proteins, nucleic acids have characteristic three-dimensional structures in their native, biologically occurring form. Because the conformation of nucleic acids, particularly DNA, is closely related to their biological activity in storage and transfer of genetic information, this subject is discussed in Part IV of this book (Chapters 21–23).

Nucleic Acid-Protein Complexes

Some nucleic acids are characteristically associated with specific proteins to form systems having very complex structures and functional activities. Chief among these are the ribosomes and viruses.

Ribosomes

Ribosomes are complexes of RNA and proteins; they are the sites of protein synthesis. In prokaryotic cells they consist of 60 to 65 percent rRNA and about 35 to 40 percent protein; in eukaryotic cells, they contain about 50 percent rRNA and 50 percent protein. In E. coli cells, ribosomes occur in the cytoplasm and may make up about 25 percent of the total cell weight. E. coli ribosomes have a particle weight of about 2,800,000 daltons and a diameter of 180 Å. The ribosomes in eukaryotic cells are substantially larger, and many of them are associated with the endoplasmic reticulum. Ribosomes may also be found associated in beadlike strings called polyribosomes or, more simply, polysomes, which are formed by the attachment of a number of ribosomes to a single molecule of messenger RNA.

Ribosomes of E. coli, which have a sedimentation coefficient of 70S, consist of two subunits of unequal size (Figure 7-11), the larger having a sedimentation coefficient of 50S and the smaller of 30S (particle weight 1.8 million and 1.0 million, respectively). The 50S subunit

133

contains one molecule of 23S rRNA and the 30S subunit contains a molecule of 16S rRNA. Both subunits contain a large number of separate polypeptide chains, over 30 in the 50S subunit and 20 in the 30S subunit. Ribosomal structure and function are considered in more detail in Chapter 23.

Viruses

Viruses, which have aptly been described as structures "at the threshold of life," are stable complexes containing a nucleic acid molecule and many polypeptide subunits organized into a characteristic three-dimensional arrangement. Although viruses have extremely large particle weights, many have been isolated in homogeneous form and some have been crystallized. Virus particles (called virions) have no power to reproduce themselves in the test tube. However, when a viral particle gains access to the interior of a specific host cell, it has the capacity to direct its own replication. The viral nucleic acid, which is the infective part of a virion, can "monopolize" the biosynthetic machinery of the host cell, forcing it to synthesize the molecular components of virus molecules rather than the normal host cell components. The RNA of RNA viruses and the DNA of DNA viruses can directly or indirectly act as templates specifying the synthesis of viral proteins and nucleic acids.

Viruses vary considerably in size, shape, and chemical composition. The tobacco mosaic virus, a rod-like RNA virus which causes a disease in leaves of the tobacco plant, contains a single RNA molecule and 2200 identical polypeptide chains. Its particle weight is about 40,000,000 daltons. The E. coli bacteriophage T$_4$ virus contains a DNA molecule and many different protein subunits; it has a particle weight exceeding 200,000,000 and has a very complex structure. Figure 7-12 shows the structure of these viruses.

All plant viruses contain RNA and are either rod-like helices, as in the case of the tobacco mosaic virus, or eicosahedral (20-sided), as in tomato bushy stunt virus. Most bacterial viruses (called bacteriophages) contain DNA, but some are RNA viruses. Bacterial viruses are convenient to study because they replicate in large numbers in bacterial suspensions and are easily isolated. The most important bacterial viruses are those of E. coli cells, such as the bacteriophages T$_2$, T$_4$, T$_6$, ϕX-174, and λ. Many viruses infecting human tissues are also known, such as the poliomyelitis virus.

Figure 7-12
Electron micrographs of viruses.

Tobacco mosaic virus

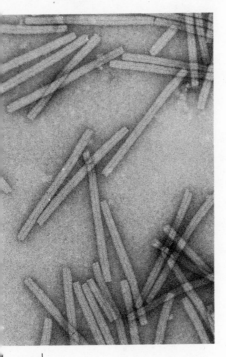

0.1 μ

Bacteriophage T₄

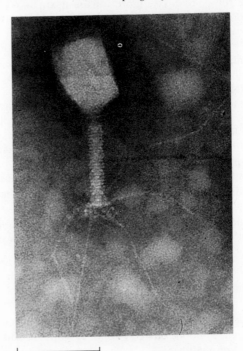

0.1 μ

Summary

Nucleotides, the monomeric structural units of nucleic acids, contain one molecule of a purine or pyrimidine base, one of D-ribose or 2-deoxy-D-ribose, and one of phosphoric acid. The common purine bases are adenine and guanine and the common pyrimidine bases are cytosine, thymine, and uracil. The purine or pyrimidine bases are covalently bonded in β-glycosyl linkage to carbon atom 1 of D-ribose or of 2-deoxy-D-ribose to form ribonucleosides or deoxyribonucleosides, respectively. The most common nucleotides are phosphorylated at the 5'-hydroxyl group of the pentose. Ribonucleotides contain one of the pyrimidines cytosine and uracil or one of the purines adenine and guanine; deoxyribonucleotides contain one of the pyrimidines cytosine and thymine or one of the purines adenine and guanine. All ribonucleotides and deoxyribonucleotides also occur in the form of pyrophosphoric and triphosphoric acid esters, which are called nucleoside (or deoxyribonucleoside) 5'-diphosphates and 5'-triphosphates. Nucleotides have other functions, particularly as coenzymes.

Nucleic acids are chains of nucleotides joined by phosphodiester bridges between the 3'-hydroxyl group of one nucleotide and the 5'-hydroxyl group of the next. Deoxyribonucleic acids contain only deoxyribonucleotides; they serve to store genetic information. Most have extremely high molecular weight; from one species of organism to another they vary in base composition and sequence. There are three major types of ribonucleic acids: messenger RNA, transfer RNA, and ribosomal RNA.

The internucleotide linkages of DNA and RNA may be hydrolyzed by enzymes or bases. Some enzymes hydrolyze either DNA or RNA in such a way as to yield only 5'-phosphorylated nucleotides. Others cleave the phosphodiester linkage to yield only 3'-phosphorylated nucleotides. DNA is not hydrolyzed by bases, but RNA is. Cyclic nucleoside 2',3'-monophosphates are intermediates in the hydrolysis of RNA by either base or ribonuclease.

Nucleic acids are frequently associated with specific proteins to form high-molecular-weight supramolecular complexes, such as ribosomes and viruses.

References

DAVIDSON, J. N.: The Biochemistry of the Nucleic Acids, Methuen, London, 6th ed., 1969. A popular and useful introduction (350 pages).

LEHNINGER, A. L.: Biochemistry, Worth Publishers, New York, 1970. Chapters 12 and 28–31 provide further general text material.

Problems

1. Write the complete structures of the following:
   (a) 2'-deoxycytidine 5'-diphosphate
   (b) thymidylic acid
   (c) uridine 3'-monophosphate
   (d) 2',3'-cyclic GMP
   (e) 3',5'-cyclic adenylic acid
   (f) the trinucleotide (5')AGC(3')

2. What are the products of the action of:
   (a) dilute acid on GMP
   (b) ribonuclease on (5')UCGU(3')
   (c) alkaline phosphatase on cytidylic acid
   (d) snake venom phosphodiesterase on (5')CCGUAU(3')

3. Predict the effects of (a) rattlesnake venom phosphodiesterase and (b) spleen phosphodiesterase on a segment of DNA having the sequence (5')dGdAdGdTdAdAdT(3')

All animals must take in relatively large amounts of organic nutrients, largely in the form of carbohydrates and lipids, to furnish energy and to serve as sources of carbon. They also require amino acids in their diet for the synthesis of their proteins and other nitrogenous components. Carbohydrates, lipids, and amino acids are thus the bulk organic nutrients required for life.

There is another class of organic nutrients, which are also absolutely essential to proper growth, function, and reproduction of animals, the *vitamins*. In contrast to the bulk nutrients, the vitamins are required in only trace amounts in the diet. When one or another of these substances is lacking, the result is a nutritional deficiency disease, which may be prevented or cured by feeding of the missing trace nutrient.

In this chapter we shall consider the chemical nature of the vitamins. We shall give particular attention to the water-soluble vitamins, which serve as vital components of certain coenzymes or enzyme prosthetic groups and thus participate in the catalysis of essential metabolic reactions.

General Nature of Vitamin Function

Nutritional deficiency diseases were first recognized in the human over 200 years ago. The disease *scurvy*, once common among sailors subsisting on diets lacking in fresh fruit or vegetables, was found by British navy physicians to be prevented or cured by adding the juice of limes to the diet. This is why British sailors have been called "limeys." *Vitamin C*, or *ascorbic acid*, first isolated in 1930, is the nutrient whose deficiency causes the symptoms of scurvy. Today all of the vitamins essential in the nutrition of man are known (Table 8-1). Actually, most of the vitamins are present and perform the same function in all animal and plant cells. However, not all these vitamins are required in the diet of all species. For example, vitamin C is required in the diet of only man, monkeys, the guinea pig, and the Indian fruit bat. Other animal species do not require it because they are genetically capable of manufacturing vitamin C from simple precursors. Nevertheless, cells of all higher species contain vitamin C and utilize it in certain vital cell processes. The term "vitamin" is thus more generally defined as a vital trace substance required in normal cell function which some species are unable to synthesize and must obtain from exogenous sources. The name vitamin derives from *vita*, life, and the chemical term amine. As it happened, the first vitamin to be identified chemically, vitamin B_1, is an amine, but not all vitamins have amino groups.

The biochemical function of some of the vitamins first

Table 8-1   Vitamins and their role in enzyme function

Water-soluble	Active form	Type of reaction promoted
Thiamine	Thiamine pyrophosphate	Decarboxylation of α-keto acids
Riboflavin	Flavin mononucleotide Flavin adenine dinucleotide	Oxidation–reduction
Niacin	Nicotinamide adenine dinucleotide Nicotinamide adenine dinucleotide phosphate	Oxidation–reduction
Pantothenic acid	Coenzyme A	Acyl group transfer
Pyridoxol	Pyridoxal phosphate	Amino group transfer
Biotin	Biocytin	CO_2 transfer
Folic acid	Tetrahydrofolic acid	One-carbon transfer
Vitamin B_{12}	Deoxyadenosyl cobalamin	Alkyl group transfer
Lipoic acid	Lipoamide	Hydrogen and acyl group transfer
Ascorbic acid		Cofactor in hydroxylation reactions
Fat-soluble		
Vitamin A		Visual cycle
Vitamin D		Calcium transport
Vitamin E		Lipid antioxidant
Vitamin K		Prothrombin synthesis

Nicotinic acid   Nicotinamide

became clear in the late 1930s through the confluence of two lines of research. The German biochemist O. Warburg had succeeded in isolating and identifying the structure of a coenzyme now called nicotinamide adenine dinucleotide, necessary in certain vital enzyme-catalyzed oxidation–reduction reactions in the cell. He found that one of the components of this coenzyme was the substance nicotinamide, the amide of nicotinic acid (Figure 8-1). Simultaneously, the American biochemists C. Elvehjem and D. Woolley were studying a deficiency disease of dogs called "blacktongue," whose counterpart in the human is *pellagra,* once common in poor farm laborers in the southeastern United States. They finally identified nicotinic acid as the substance whose lack caused this disease. These discoveries established that nicotinic acid is essential to animals in trace amounts because it is an important component of a coenzyme required in enzymatic catalysis of certain vital metabolic reactions. Today we know that most of the vitamins play similar roles as components of enzymes or coenzymes. Since coenzymes are catalytic in function and occur in very small concentrations in cells, the nutritional requirements for vitamins can be satisfied by very small amounts. For example, only a fraction of a microgram (less than one-millionth of a gram) of the vitamin biotin is required daily by an albino rat.

Classification of the Vitamins

Vitamins are divided into two classes, *water-soluble* and *fat-soluble* (Table 8-1). The water-soluble vitamins include thiamine (vitamin B_1), riboflavin (vitamin B_2), nicotinic acid, pantothenic acid, pyridoxol (vitamin B_6), biotin, folic acid, lipoic acid, vitamin B_{12}, and ascorbic acid (vitamin C). The coenzyme function of all of these is known, with the possible exception of ascorbic acid.

The fat-soluble vitamins include vitamins A, D, E, and K. Their functions in animal tissues are fairly well understood, but it is not known whether they serve as components of coenzymes.

The Water-Soluble Vitamins and Their Coenzyme Forms

The nature and properties of each vitamin and its corresponding coenzyme will now be described, as well as the general nature of the enzymatic reactions in which each is concerned (Table 8-1).

Thiamine (Vitamin B₁)

Vitamin B_1 is necessary in the nutrition of most vertebrates and in some microbial species. Its deficiency in the diet of man causes the disease <u>beri-beri</u>, which was once prevalent among Japanese naval seamen due to their diet of refined or polished rice. Rice husks, which are removed in the refining of rice, are rich in this vitamin, as are other cereal grains. The structure of thiamine was solved in the early 1930s. It contains two ring systems, a pyrimidine and a thiazole (Figure 8-2). In

Figure 8-2
Thiamine and its coenzyme forms.

Thiamine
(vitamin B₁)

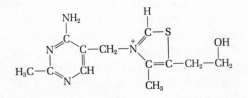

Thiamine
pyrophosphate

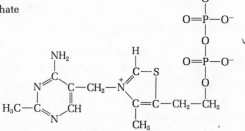

α-Hydroxyethylthiamine pyrophosphate, the transitory form in which an aldehyde group (color) is transferred.

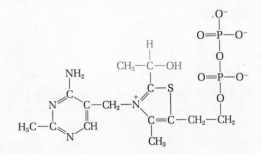

Figure 8-3
The reaction catalyzed by pyruvate decarboxylase, which requires thiamine pyrophosphate as coenzyme.

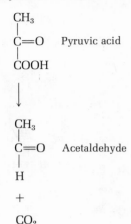

CH₃
|
C=O   Pyruvic acid
|
COOH

↓

CH₃
|
C=O   Acetaldehyde
|
H

+

CO₂

animal tissues, thiamine obtained from the diet is converted into its coenzyme form *thiamine pyrophosphate* (abbreviated TPP), also called *cocarboxylase.*

Thiamine pyrophosphate serves as the coenzyme for a class of enzymes catalyzing the removal or transfer of aldehyde groups. In these reactions the thiamine pyrophosphate, which is tightly bound to the enzyme, serves as a transitory intermediate carrier of an aldehyde group. The most important enzymes for which thiamine pyrophosphate serves as the coenzyme are (1) α-keto acid decarboxylases, (2) α-keto acid dehydrogenases, and (3) transketolases. A well-known example is the enzyme *pyruvate decarboxylase* of yeast, which catalyzes the decarboxylation of pyruvic acid to form acetaldehyde and CO_2, an important step in alcoholic fermentation (Figure 8-3).

Riboflavin (Vitamin B₂)

Riboflavin, yellow in color, consists of a complex isoalloxazine ring to which is attached the five-carbon alcohol ribitol (Figure 8-3). It is required for the growth of most vertebrates. Riboflavin occurs largely in combined form as the two coenzymes *flavin mononucleotide,* abbreviated FMN, and *flavin adenine dinucleotide* (FAD). These coenzymes, often referred to as the flavin coenzymes or flavin nucleotides, are required for the function of certain enzymes catalyzing oxidation–reduction reactions. They are usually tightly bound to the enzyme protein and undergo reversible oxidation and reduction. Their oxidized forms, which are yellow in color, undergo bleaching when they are converted to their reduced forms (symbolized $FMNH_2$ and $FADH_2$) by addition of hydrogen atoms enzymatically transferred from a substrate (Figure 8-4). *Succinate dehydrogenase* is an example of a flavoenzyme. It catalyzes the removal of a pair of hydrogen atoms from succinate and their transfer to the FAD prosthetic group (Figure 8-4).

Nicotinic Acid

Both nicotinic acid and nicotinamide are active in preventing the symptoms of pellagra or blacktongue. Nicotinic acid has been given the alternative name of *niacin.* Although nicotinic acid was first found and named as a derivative of nicotine, an alkaloid of tobacco, it is widely distributed in plant and animal tissues. Meat is especially rich in nicotinic acid. There are two coenzyme

Figure 8-4
*Riboflavin, the riboflavin coenzymes, and the flavin-linked succinate
dehydrogenase reaction.*

Riboflavin (vitamin B$_2$)

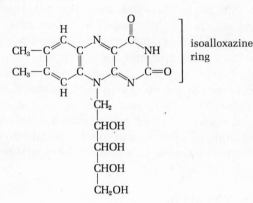

isoalloxazine
ring

Flavin adenine dinucleotide (FAD). In the
reduced form the isoalloxazine ring accepts
hydrogen atoms exactly as shown for FMN
at lower left.

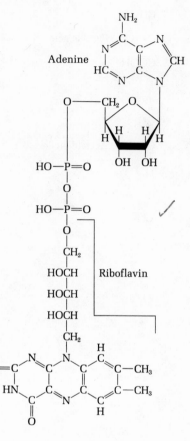

Flavin mononucleotide (FMN)
(oxidized form)

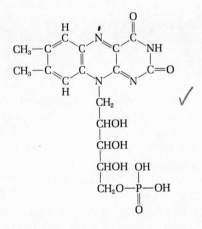

The reduced form of FMN. Only the isoalloxazine ring is shown.
The hydrogen atoms accepted from the substrate are in color.

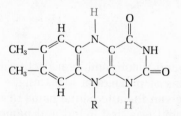

The succinate dehydrogenase reaction,
a typical flavin-linked oxidation.

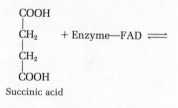

Succinic acid

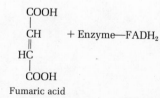

Fumaric acid

Figure 8-5
Nicotinamide adenine dinucleotide (NAD⁺) and nicotinamide adenine
dinucleotide phosphate (NADP⁺). Below is shown the reaction
catalyzed by malate dehydrogenase, a typical NAD-linked
dehydrogenase.

Oxidized form of NAD.

Reduced form. One hydrogen atom from the substrate goes to the
nicotinamide ring. The other becomes a hydrogen ion. The mechanism
is given in Chapter 11.

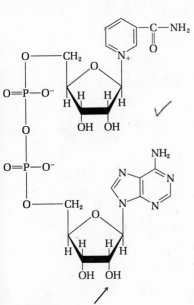

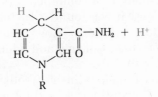

In NADP⁺ this
hydroxyl group is
esterified with phosphate.

The malate dehydrogenase reaction

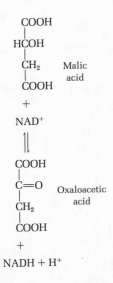

forms of this vitamin: *nicotinamide adenine dinucleo-*
tide (NAD) and its phosphorylated derivative *nico-*
tinamide adenine dinucleotide phosphate (NADP). They
are also known by their older names *diphosphopyridine*
nucleotide and *triphosphopyridine nucleotide*, respectively (Figure 8-5). These two coenzymes are referred to
as *pyridine coenzymes* or *pyridine nucleotides*, since
nicotinamide is a derivative of pyridine. The pyridine
nucleotides serve as coenzymes for a large class of oxido-
reductases called pyridine-linked dehydrogenases. They
are relatively loosely bound to the enzyme protein. They
can undergo reversible oxidation and reduction and thus
function as intermediate carriers of electrons from the
substrate to other systems capable of accepting electrons
during biological oxidation. It is the nicotinamide portion of the oxidized forms of these coenzymes, abbreviated NAD⁺ and NADP⁺, which participates in electron
transfer. The reduced forms are usually abbreviated
NADH and NADPH, respectively.

An example of a reaction catalyzed by a pyridine-
linked dehydrogenase is the oxidation of malate to oxaloacetate by *malate dehydrogenase,* an important step in
the biological oxidation of sugar (Figure 8-5). Malic dehydrogenase from the liver is normally specific for NAD
as coenzyme and will not accept NADP. Other pyridine-
linked dehydrogenases are specific for NADP, such as

143

glucose 6-phosphate dehydrogenase, and a few, such as glutamate dehydrogenase, will function with either NAD or NADP.

Pantothenic Acid

Pantothenic acid, a water-soluble vitamin, is required for the growth of mammals. Its deficiency results in degeneration of the cortex of the adrenal gland and numerous metabolic abnormalities. In animal tissues pantothenic acid occurs largely in a combined form as _coenzyme A,_ so named because it was found to be important in biological acetylation reactions. Coenzyme A (abbreviated CoA—SH) was first isolated and identified in the late 1940s (Figure 8-6).

Coenzyme A functions in metabolism as a carrier for acyl groups. The sulfhydryl or thiol group of the coenzyme A molecule becomes acylated to form a thioester, designated CoA—S—CO—R. The acyl group of the acyl CoA ester may then be transferred to some acceptor molecule by another enzyme. Figure 8-6 illustrates how CoA—SH functions in the utilization of the acetyl group of pyruvic acid to form citric acid from oxaloacetic acid.

Vitamin B_6

There are three forms of vitamin B_6: pyridoxol (also known as pyridoxine), pyridoxal, and pyridoxamine (Figure 8-7). This vitamin is especially abundant in cereal grains. Deficiency of vitamin B_6 causes skin lesions and disturbances in the central nervous system. The phosphorylated derivatives of pyridoxal and pyridoxamine (Figure 8-7), namely, _pyridoxal phosphate_ and _pyridoxamine phosphate_, function as coenzymes. They participate in many different enzymatic reactions in amino acid metabolism, particularly those in which amino groups are transferred. For example, in the class of enzymes called _transaminases_ or _aminotransferases_, which catalyze the general reaction shown in Figure 8-6, pyridoxal phosphate functions as a tightly bound coenzyme to which the amino group of an amino acid becomes transiently attached.

Biotin

Biotin was first recognized by its ability to prevent a deficiency disease that can be induced by feeding of large amounts of raw egg whites to animals or man. Egg white contains a protein called _avidin_ which can specifically

Figure 8-6
Pantothenic acid and coenzyme A.

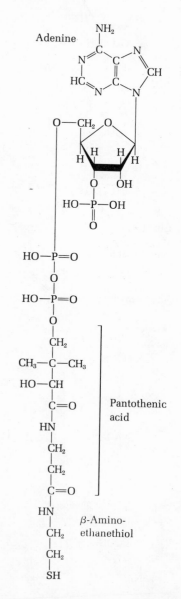

Figure 8-6 (continued)
The role of coenzyme A in the pyruvate dehydrogenase and citrate
synthase reactions.

Formation of acetyl CoA

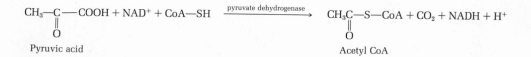

Utilization of acetyl CoA

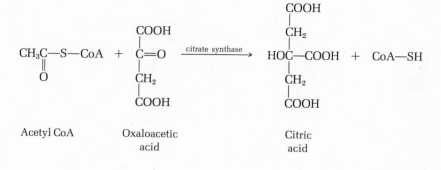

bind biotin very tightly and prevent its absorption from the gastrointestinal tract. Biotin is especially abundant in liver and yeast. It occurs in the cell covalently combined with a specific lysine residue of certain enzymes concerned in the formation or utilization of carbon dioxide; the biotinyl derivative of lysine is called biocytin. The fused ring system of the biotin molecule serves as an intermediate carrier of the CO_2 molecule, in the form of a carboxybiotin derivative (Figure 8-8).

Folic Acid

Folic acid, also known as pteroyl-L-glutamic acid, is required for the growth of chicks and some microorganisms. Its deficiency in chicks results in anemia, a condition in which the number of red blood cells is reduced.

Folic acid itself has no coenzyme activity, but its reduced form, tetrahydrofolic acid, is the active coenzyme (Figure 8-9). Tetrahydrofolic acid, often symbolized FH_4, functions as an intermediate carrier of certain one-carbon

Figure 8-7
Vitamin B₆, its coenzyme forms, and the transaminase reaction.

Pyridoxol

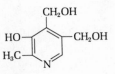

Pyridoxal

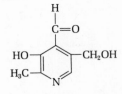

Pyridoxamine

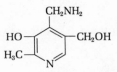

Pyridoxal phosphate the amino group
acceptor form.

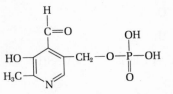

Pyridoxamine phosphate, the amino donor form.
The amino group is in color.

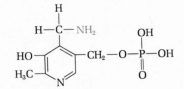

The transaminase reaction

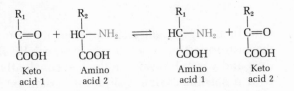

groups, particularly the formyl group, which is required
in the synthesis of purines. There are three related one-
carbon compounds of tetrahydrofolate—formyl FH_4,
methenyl FH_4, and methylene FH_4—each of which can
donate its single carbon atom in one or another specific
enzymatic reaction.

Vitamin B₁₂

Vitamin B_{12}, also called *cyanocobalamin* (Figure 8-10), is
present only in animals and some microorganisms, but
apparently not in higher plants. It is especially abundant
in meat and liver. The disease *pernicious anemia* in
humans results from the failure of vitamin B_{12} to be ab-

Figure 8-8
Biotin, biocytin and the pyruvate carboxylase reaction.

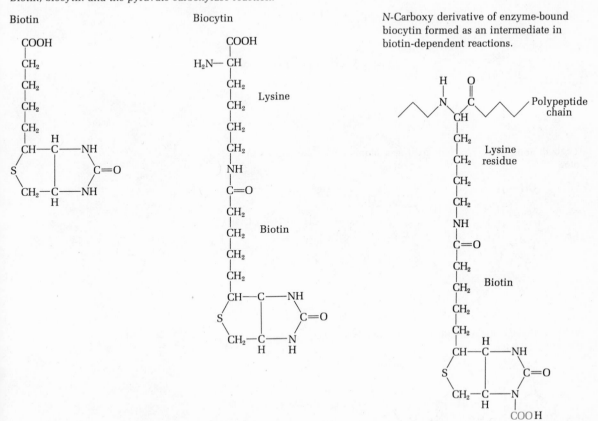

The pyruvate carboxylase reaction. The N-carboxybiocytin prosthetic group of the enzyme serves to transfer the CO_2 to pyruvic acid

$$ATP + CH_3-\underset{\underset{O}{\|}}{C}-COOH + CO_2 \rightleftharpoons ADP + P_i + \underset{\substack{| \\ C=O \\ | \\ COOH}}{\overset{COOH}{\underset{}{CH_2}}}$$

Pyruvic acid   Oxaloacetic acid

sorbed from the intestine. In the coenzyme form of vitamin B_{12}, which is called *deoxyadenosylcobalamin*, the cyanide group is replaced by 5'-deoxyadenosine, which is attached via the 5'-carbon atom of the ribose portion. The vitamin B_{12} coenzyme functions as a transitory carrier of alkyl and substituted alkyl groups, particularly in those reactions in which such groups are transferred

Figure 8-9
Folic acid, tetrahydrofolic acid and N⁵,N¹⁰-methylenetetrahydrofolate.

Folic acid

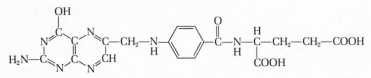

Tetrahydrofolic acid. The four hydrogen atoms added are shown in color.

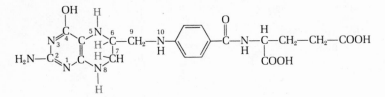

N⁵,N¹⁰-methylenetetrahydrofolate. The methylene group undergoing transfer is in color.

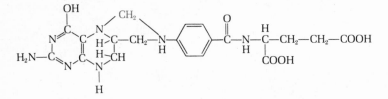

from one carbon atom to an adjacent one in the substrate. An example is the enzyme *methylaspartate mutase*, in which the —CH(NH₂)COOH group is transferred from one carbon atom of methylaspartic acid to the next to form glutamic acid (Figure 8-10).

Lipoic Acid

Lipoic acid, a growth factor which is required by many microorganisms, is a derivative of octanoic acid, a fatty acid (Figure 8-11). In its coenzyme form, sometimes called lipoamide, the lipoic acid is covalently bonded to a specific lysine residue of the enzyme protein. Lipoic acid in its reduced form has two sulfhydryl groups; its oxidized form is a cyclic disulfide. Lipoic acid is an es-

Figure 8-10
Vitamin B₁₂, its coenzyme form and the methylaspartate mutase reaction.

In Vitamin B_{12} (cobalamin), the R group is
$-CN$

In dexoxyadenosylcobalamin, the coenzyme form, the R group is 5′-deoxyadenosyl

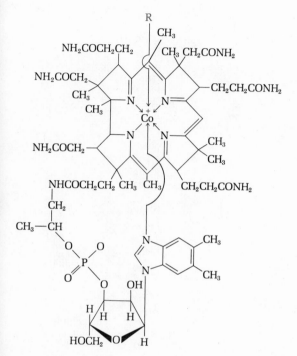

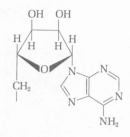

The methylaspartate mutase reaction

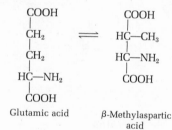

| Glutamic acid | β-Methylaspartic acid |

sential component of α-keto acid dehydrogenases, in which its sulfhydryl groups serve to carry hydrogen atoms and also acyl groups (Chapter 12).

Vitamin C (Ascorbic Acid)

Vitamin C is very simple in structure. It is a lactone of a sugar acid (Figure 8-12), which is readily oxidized to dehydroascorbic acid. Citrus fruits and tomatoes are especially rich in ascorbic acid. It is, however, quickly destroyed on exposure to air. Ascorbic acid appears to act as a necessary cofactor in certain hydroxylation reactions. However, it is believed to have other important functions which are still obscure.

Figure 8-11
Lipoic acid and its active coenzyme form.

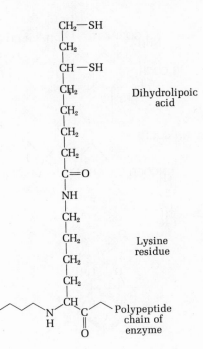

Lipoic acid
(oxidized form)

Dihydrolipoic acid
(reduced form)

Coenzyme form (lipoamide). The lipoic acid is covalently bonded to a lysine residue of the enzyme protein. The reduced form is shown.

Dihydrolipoic acid

Lysine residue

Polypeptide chain of enzyme

Fat-Soluble Vitamins

The fat-soluble vitamins (A, D, E, and K) may also be classified as lipids, since they are insoluble in water and are extracted from tissues with nonpolar fat solvents such as chloroform or ether. They are nonsaponifiable (Chapter 6).

The fat-soluble vitamins share a common denominator in that they are all ultimately built from isoprenoid building blocks, as will be shown in their structural formulas.

Vitamin A

Vitamin A is required for normal growth and function of higher animals. Its deficiency results in a scaly, dry skin and in night blindness (in humans). Vitamin A is especially abundant in fish livers, butter, and eggs. It is not found in plants, but the latter contain carotenes (Chapter 6), which are direct precursors of vitamin A. The structures of vitamin A and β-carotene, an important precursor, are shown in Figure 8-13. Their double bonds are

Figure 8-12
Vitamin C.

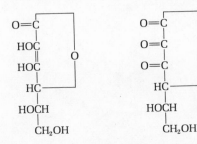

L-Ascorbic acid   L-Dehydroascorbic acid

all in the *trans* configuration. Vitamin A is formed by animals from ingested β-carotene by cleavage of its central double bond.

In man an early symptom of vitamin A deficiency is night blindness, a condition in which the light receptors of the rod cells of the retina fail to respond normally. The light-receptor molecule is rhodopsin, a complex of a protein and an altered form of vitamin A, retinal, in which the double bond at position 11 is *cis*. When rhodopsin absorbs light energy, the retinal undergoes a structural change so that the double bond at position 11 goes into the *trans* configuration. This change excites the nerve

Figure 8-13
Vitamin A₁ and its precursor β-carotene. Their isoprene structural units are set off by colored lines. Cleavage of β-carotene yields two molecules of vitamin A₁.

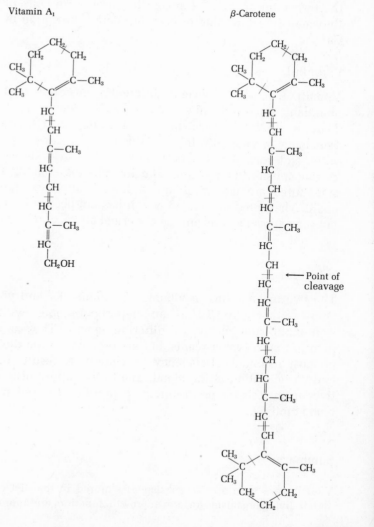

Vitamin A₁

β-Carotene

cells of the retinal rods. The all-*trans*-retinal then reverts back to the 11-*cis* form.

Vitamin D

Vitamin D consists of a group of closely related steroids, of which the most important are vitamin D_2 or *calciferol* and vitamin D_3 (Figure 8-14). Deficiency of vitamin D leads to abnormalities of calcium and phosphorus metabolism, causing the disease rickets in children.

Vitamin D_2 is produced in the skin of animals by irradiation of the ingested plant sterol *ergosterol* with sunlight. Vitamin D_3 is produced from the precursor 7-dehydrocholesterol in the liver of various fishes. Fish liver oils are thus rich sources of both vitamin A and vitamin D. Vitamin D is believed to function by stimulating the formation of a Ca^{2+}-transporting protein in cells of the small intestine; this protein promotes absorption of Ca^{2+}.

Vitamin E

Vitamin E exists in several different forms; the most abundant are α-, β-, and γ-tocopherols (Figure 8-15). They are found in vegetable oils and in wheat germ. Deficiency of vitamin E in rats leads to a scaly skin, sterility, and muscular weakness; vitamin E is also known as the antisterility vitamin. The tocopherols contain a substituted aromatic ring and a long isoprenoid side chain. Their precise mode of action has not been solved, but they appear to prevent the abnormal attack of oxygen on lipids.

Vitamin K

The two major forms of vitamin K, vitamin K_1 and vitamin K_2 (Figure 8-16), are naphthoquinones with isoprenoid side chains of differing length. They are found in most higher plants, but are required in the diet of most mammals. Deficiency of vitamin K results in faulty coagulation of the blood, due to the failure of the liver to synthesize *prothrombin*, a protein required in blood clotting.

Summary

Vitamins are trace organic substances required in the diet of certain living organisms for proper growth, function, and repro-

Figure 8-14
Forms of vitamin D.

Vitamin D_2 (calciferol)

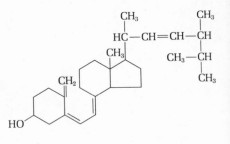

Vitamin D_3

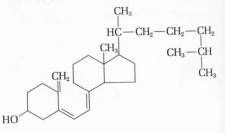

Figure 8-15
Vitamin E (α-tocopherol).

The isoprene units of the side chain are indicated by colored bars.

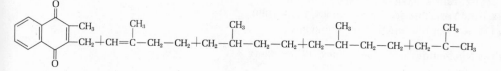

Figure 8-16
Forms of vitamin K.

Vitamin K$_1$

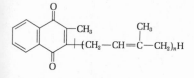

Vitamin K$_2$
(n=6,7, or 9)

duction. Most are present in and perform the same function in all animal and plant cells. However, not all of them are required in the diet of all species since some organisms can manufacture certain vitamins from simple precursors.

The water-soluble vitamins function as molecular components of certain specific coenzymes or prosthetic groups of certain enzymes. The coenzyme form of vitamin B$_1$ (thiamine) is thiamine pyrophosphate, which functions as a transitory carrier of aldehyde groups in certain enzymatic reactions. Vitamin B$_2$ (riboflavin) is the active portion of the oxidation-reduction coenzymes flavin mononucleotide and flavin adenine dinucleotide, which are reversibly oxidized and reduced. Nicotinic acid is the active, reversibly oxidized and reduced group in the oxidation-reduction coenzyme nicotinamide adenine dinucleotide and its phosphate derivative. Pantothenic acid is an essential component of Coenzyme A, which functions as a transitory carrier of acyl groups in metabolism. Vitamin B$_6$ in the form of pyridoxal phosphate functions as an intermediate carrier of amino groups in enzymatic reactions involving amino group transfer. Biotin functions as a carrier of CO_2 in enzymes catalyzing certain carboxylation and decarboxylation reactions. Tetrahydrofolic acid, a reduced form of the vitamin folic acid, participates in one-carbon group transfer reactions. Vitamin B_{12} in its coenzyme form functions to carry alkyl groups in certain enzymatic reactions.

Coenzyme functions for the fat-soluble vitamins are not yet known. Vitamin A and its precursor β-carotene is involved in light perception in the retina, vitamin D functions in the regulation of Ca^{2+} metabolism, vitamin E in the prevention of abnormal oxidations, and vitamin K in the synthesis of prothrombin of the blood.

References

BOYER, P. D., H. LARDY, and K. MYRBÄCK: (eds.) *The Enzymes*, Academic Press, New York, 2nd ed., four volumes, 1959–1960. Comprehensive reviews of the function of coenzymes.

HUTCHINSON, D. W.: *Nucleotides and Coenzymes*, Wiley, New York, 1964.

LEHNINGER, A. L.: *Biochemistry*, Worth Publishers, New York, 1970. See particularly Chapters 15–20, 22, 24, 25.

PART II CATABOLISM AND THE GENERATION
OF PHOSPHATE-BOND ENERGY

PART II CATABOLISM AND THE GENERATION
OF PHOSPHATE-BOND ENERGY

In Part I of this book we examined the structure and properties of the cell macromolecules: the nucleic acids, proteins, polysaccharides, and also the lipids, as well as their component building blocks. Now we shall see how these molecules interact together in what we have called the molecular logic of the living state (Introduction). A central feature of this logic is that cells function as chemical engines, extracting energy and raw materials from their environment and converting these into new cell components.

We have seen in the Introduction to this book that metabolism consists of many consecutive organic reactions, each catalyzed by an enzyme. Metabolism actually consists of two major networks of reactions. The first of these networks, which we shall now examine in Part II of this book, functions to produce the chemical energy of adenosine triphosphate (ATP) from the energy yielded by degradation of fuel molecules or from trapped sunlight.

The second great network of metabolism, which is to be the subject of Parts III and IV of this book, harnesses the chemical energy of the ATP for the purpose of synthesizing new cell components, for active transport of mineral ions and metabolites, for muscular contraction, and for the self-replication of the organism.

Throughout our discussion of these two networks we must also keep in mind that these metabolic networks are self-regulating, always acting on the principle of maximum economy.

CHAPTER **9** A **SURVEY OF INTERMEDIARY METABOLISM**

Metabolism is often briefly defined as the sum total of all the enzymatic reactions occurring in the cell. However, this definition is incomplete since it does not indicate that metabolism is a highly integrated and purposeful activity in which many sets of multienzyme systems participate. There are four specific functions of metabolism: (1) to extract chemical energy from organic nutrients or from sunlight, (2) to convert nutrients from the environment into the building blocks or precursors of the macromolecular components of cells, (3) to assemble the building blocks into proteins, nucleic acids, lipids, polysaccharides, and other characteristic cell components, and (4) to form and degrade those biomolecules required in specialized functions of cells.

Although intermediary metabolism involves hundreds of different enzyme-catalyzed reactions, the central metabolic pathways have a simple organizational plan and are easy to understand. Moreover, they are identical in most forms of life. In this chapter we shall survey the sources of matter and energy for metabolism, the central pathways by which cell components are synthesized and degraded, and the mechanisms by which chemical energy is transferred during metabolism. We shall also survey various experimental approaches to the study of metabolism and the relationships between cell structure and metabolic activities.

Sources of Carbon and Energy for Cellular Life

Cells can be divided into two large groups on the basis of the chemical form of carbon they require from the environment. *Autotrophic* ("self-feeding") cells can utilize carbon dioxide as the sole source of carbon and construct from it all their carbon-containing biomolecules. *Heterotrophic* cells ("feeding on others"), on the other hand, cannot utilize carbon dioxide and must obtain carbon from their environment in a relatively complex, reduced form, such as glucose. Autotrophs are relatively self-sufficient cells, whereas heterotrophs, with their requirement for carbon in a fancier form, must subsist on the products formed by other cells. Most autotrophic organisms obtain their energy from sunlight and are thus photosynthetic, whereas the cells of higher animals and most microorganisms are heterotrophic and obtain their energy from the degradation of organic nutrients such as glucose.

Heterotrophic organisms can in turn be divided into two major classes. *Aerobes* live in air and use molecular oxygen to oxidize their organic nutrient molecules. *Anaerobes* live in the absence of oxygen and degrade their nutrients in ways not requiring oxygen. Many cells can live either aerobically or anaerobically; such organisms are called *facultative*. Anaerobes that cannot utilize oxygen at all are called *strict anaerobes*; in fact, many strict anaerobes are poisoned by oxygen. Most heterotrophic cells, particularly those of higher organisms, are facultative, and if oxygen is available to them, they prefer to use it.

Not all cells of a given organism are of the same class. For example, in higher plants the green chlorophyll-containing cells of leaves are photosynthetic autotrophs, whereas the root cells are heterotrophs. Moreover, green leaf cells function as photosynthetic autotrophs in the sunlight but as heterotrophs in the dark.

Catabolism and Anabolism

Metabolism takes place via sequences of consecutive enzyme-catalyzed reactions and many chemical intermediates (Figure 9-1), hence the term *intermediary metabolism* to describe the pathways of metabolism. Metabolic intermediates are also called *metabolites*.

Metabolism is divided into *catabolism* and *anabolism*. Catabolism refers to the *degradative* phase of metabolism. Organic nutrient molecules, such as carbohydrates, lipids, and proteins, coming either from the environment

Figure 9-1
Sequential reactions connected by common intermediates. E_1, E_2, E_3, etc., are enzymes.

Written as equations

$$A \xrightarrow{E_1} B$$

$$B + C \xrightarrow{E_2} D$$

$$D \xrightarrow{E_3} E$$

$$E \xrightarrow{E_4} F + G$$

$$F \xrightarrow{E_5} H$$

$$H \xrightarrow{E_6} I$$

Written schematically

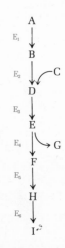

or from the cell's own nutrient storage depots, may be degraded, usually by oxidative reactions, into smaller, simpler end products, such as lactic acid, acetic acid, CO_2, ammonia, or urea. Catabolism is accompanied by release of the energy inherent in the complex structure of large organic molecules. This energy is conserved in the form of adenosine triphosphate (ATP).

Anabolism is the building-up or synthetic phase of metabolism; it is also called *biosynthesis*. In anabolism simple, small precursor molecules are built up into relatively large molecular components of cells, such as polysaccharides, nucleic acids, proteins, and lipids. Since biosynthesis results in increased size and complexity of structure, it requires input of free energy, which is furnished by the breakdown of ATP. Catabolism and anabolism take place concurrently in cells.

The Stages of Catabolism

Metabolism is organized into three major stages, as is shown in Figure 9-2. We shall first consider the stages of catabolism.

The enzymatic degradation of each of the bulk nutrients of cells, namely, carbohydrates, lipids, and proteins, proceeds in a stepwise manner through a number of consecutive enzymatic reactions. The enzymes catalyzing these steps and the various chemical intermediates that are formed en route to the end products are for the most part rather well understood. In stage I of catabolism (Figure 9-2) large nutrient molecules are degraded to their major building blocks. Thus polysaccharides are degraded to hexoses or pentoses; lipids are degraded to fatty acids, glycerol, and other components; and proteins are hydrolyzed to their 20 component amino acids. In stage II, the various products formed in stage I are collected and then converted into a smaller number of simpler molecules. Thus, the hexoses, pentoses, and glycerol from stage I are degraded into the three-carbon phosphorylated sugar, glyceraldehyde 3-phosphate, and then to a single two-carbon species, the acetyl group of acetyl coenzyme A (Chapter 8). Similarly, in stage II the various fatty acids and amino acids are broken down to form acetyl CoA as a common end product. Acetyl CoA is thus the major end product of stage II of catabolism. In stage III the acetyl group of acetyl CoA is fed into the tricarboxylic acid cycle, the final common pathway in which the acetyl group is ultimately oxidized to carbon dioxide, the end product of catabolism.

Figure 9-2

*The three stages of catabolism and anabolism. The catabolic
pathways are shown in black arrows and the anabolic pathways in
color. Stage III is amphibolic. It not only is the final pathway for
degradation of foodstuffs to CO_2, but also can furnish small molecular
weight precursors for anabolism.*

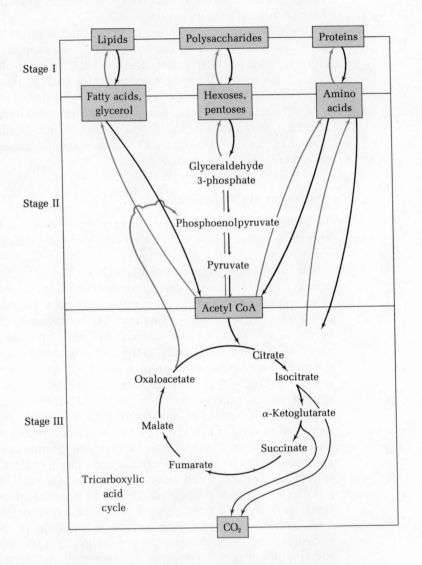

It is important to note that the pathways of catabolism
converge toward a final common pathway. During catabolism in stage I, dozens or even hundreds of different
proteins are degraded to 20 amino acids, in stage II, the
20 amino acids are degraded to acetyl CoA and only a
few other products, and in stage III, the acetyl groups are

oxidized to CO_2 and H_2O. Similarly, in stage I many different polysaccharides and disaccharides are degraded to a few simple sugars, which are all converted into acetyl CoA and CO_2 in stage II. The central pathways of catabolism thus resemble main trunk lines, into which many "feeder" lines flow (Figure 9-3).

The Diverging Pathways of Anabolism

Anabolism also takes place in three stages, beginning with small building blocks originating from stage III of metabolism. Protein synthesis begins in stage III with the formation of α-keto acids and other precursors of amino acids. In stage II the α-keto acids are aminated by amino group donors to form α-amino acids and in stage I the amino acids are assembled into the polypeptide chains of many different proteins (Figure 9-2) Similarly, acetyl groups from stage III are assembled into fatty acids in stage II, and these in turn are assembled to form various lipids. Just as catabolism is converging, anabolism is diverging. The pathways of anabolism thus have many branches leading to the thousands of different cell components (Figure 9-3).

Figure 9-3
Converging pathways of catabolism and diverging pathways of anabolism. Each arrow represents an enzyme-catalyzed step.

Catabolic pathways
converge into a final
common pathway

Biosynthetic pathways diverge, after
starting from a few
precursors

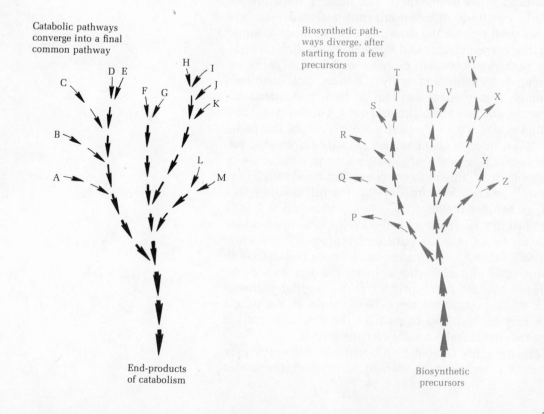

End-products
of catabolism

Biosynthetic
precursors

Each stage in the catabolism or anabolism of a given biomolecule is catalyzed by a multienzyme system. The sequential chemical changes taking place in these central routes of metabolism are largely identical in all forms of life. For example, the conversion of glycogen into pyruvic acid is accomplished through the same chemical intermediates and through the same number of reactions in most living organisms.

The Distinction between Catabolic and Anabolic Pathways

The catabolic and corresponding but opposite anabolic pathways between a given precursor and a given product are usually not identical with respect to their intermediary metabolites or to the enzymes catalyzing the intermediate steps. For example, there is a sequence of 11 specific enzymes catalyzing the successive steps in the degradation of glucose into lactic acid. Although it would seem logical and economical for the synthesis of glucose from lactic acid to occur by simple reversal of the same enzymatic steps used in the catabolic pathway, biosynthesis of glucose utilizes only 9 of the 11 enzymatic steps used in its degradation. The other 2 steps are replaced by entirely different enzyme-catalyzed reactions that are used only in the direction of biosynthesis. Similarly, the corresponding and opposite catabolic and anabolic pathways between proteins and amino acids or between fatty acids and acetyl CoA are not identical. Although it may seem wasteful to have two metabolic pathways, one for catabolism and one for anabolism, such parallel routes are an absolute necessity, since the pathway taken in catabolism is energetically impossible for anabolism. Degradation of a complex organic molecule is energetically a "downhill" process and its synthesis an "uphill" process. We can develop the hill analogy further. A boulder dislodged at the top of a hill will roll downhill in a fairly direct pathway. To haul the boulder back up to its original position by precisely the same pathway taken in its descent may be quite impossible for a tractor of a given horsepower, but the tractor can haul the boulder back up if it follows another pathway with a less precipitous slope. Some parts of the uphill path may be identical to parts of the downhill path if they are energetically feasible for the tractor.

Corresponding catabolic and anabolic pathways may differ in a second way, namely, in their intracellular location. For example, the oxidation of fatty acids in liver cells to the stage of acetyl CoA takes place by the action of a set of enzymes localized in the mitochondria, whereas the synthesis of fatty acids from acetate takes place by another set of different enzymes located in the extramitochondrial cytoplasm. Separate compartmentation of parallel catabolic and anabolic pathways in the cells of higher organisms allows them to take place independently and simultaneously.

There is a third way in which corresponding catabolic and anabolic routes differ: They are independently regulated by different regulatory enzymes. Independent regulation of the rates of the parallel but opposite flows of metabolites between a specific nutrient and its product(s) provides the cell with great metabolic flexibility.

Although the pathways of catabolism and anabolism are not identical, stage III of metabolism (Figure 9-2) constitutes a central meeting ground or pathway which is accessible to both. This is the *amphibolic* stage of metabolism; it has a dual function (*amphi*, both). Stage III can be used catabolically, to bring about completion of the degradation of small molecules derived from stage II of catabolism, and anabolically, to furnish small molecules as precursors for biosynthesis of cell macromolecules.

Nearly all the reactions of metabolism are ultimately linked to each other because the product of one enzymatic reaction becomes the substrate of the next in consecutive sequences. Such sequences are made possible by enzymatic reactions in which specific functional groups of metabolite molecules are removed and transferred to acceptor molecules. Most of the consecutive reactions of intermediary metabolism involve the sequential transfer of amino, acetyl, phosphate, methyl, formyl, or carboxyl groups, or of hydrogen atoms.

The Energy Cycle in Cells

Complex organic molecules such as glucose contain much potential energy because of their high degree of structural order. When the glucose molecule is degraded by oxidation to form CO_2 and H_2O it yields free energy, that is, the form of energy capable of doing work under conditions of constant temperature and pressure. The free energy of glucose so released is harnessed by the cell to do work.

Biological oxidations are in essence flameless, or low-temperature, combustions. Heat cannot be used as an energy source by living organisms, which are essentially isothermal, that is, they have just about the same temperature throughout. Heat can do work only when it can flow from a warmer to a cooler body. Instead, the energy of cellular fuels that is set free by oxidation is conserved in the form of chemical energy, which can do work in isothermal conditions. The chemical energy of fuels is recovered as newly formed adenosine triphosphate (ATP), generated from adenosine diphosphate (ADP) and inorganic phosphate in enzymatic reactions that are coupled to specific oxidation steps during catabolism. The ATP so formed can diffuse to those sites in the cell where its energy is required; ATP is really a transport form of chemical energy. The chemical energy of ATP is then released by transfer of its terminal phosphate group(s) to certain specific acceptor molecules, which become "energized" and can then be assembled to yield larger molecules during anabolism. The ATP–ADP energy cycle is shown in Figure 9-4.

A second way of carrying chemical energy from reactions of catabolism to the energy-requiring reactions of synthesis is in the form of hydrogen atoms or electrons, which are required for the reduction of double bonds to single bonds during biosynthesis of fatty acids. They are transported enzymatically by means of specific coenzymes, the most important of which is nicotinamide adenine dinucleotide phosphate (Chapter 8). NADP thus serves as a carrier of energy-rich electrons from catabolic reactions to electron-requiring anabolic reactions (Figure 9-5), just as ATP is a carrier of energy-rich phosphate groups from reactions of catabolism to those of anabolism.

Regulation of Intermediary Metabolism

The principle of maximum economy pervades all aspects of cellular metabolism, which is under constant regulation. In general, the rate of catabolism is controlled not by the concentration of the available fuels, but rather by the second-to-second needs for energy in the form of ATP. Cells burn their fuels at a rate just sufficient to meet the rate of energy utilization at any particular time. Similarly, the rate of biosynthesis of cell components is also adjusted to immediate needs. For example, growing cells synthesize amino acids at such a rate as to just suffice for the synthesis of new proteins.

The regulation of metabolic pathways is brought about by three different types of mechanisms. The first and most immediately responsive is through the action of

Figure 9-4
The ATP-ADP cycle. The chemical energy inherent in the ATP molecule is used in coupled enzymatic reactions for carrying out energy-requiring mechanical work, transport, and biosynthetic activities, during which inorganic phosphate is released from ATP. The ADP so formed is rephosphorylated to ATP during energy-yielding reactions of catabolism.

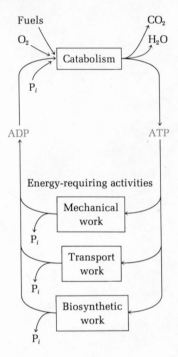

Figure 9-5

Transfer of reducing power via the nicotinamide adenine dinucleotide phosphate (NADP) cycle. Other electron-carrying coenzymes, such as flavin nucleotides, also participate in reductive biosynthesis.

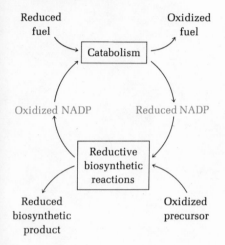

Figure 9-6

Regulation of a metabolic pathway by feedback or end-product inhibition. The first enzyme in the pathway (E_1) is an allosteric enzyme. It is inhibited by the end-product of the sequence. Allosteric inhibition will be indicated in subsequent illustrations by colored dotted lines leading from the inhibitory metabolite to the colored bar across the reaction catalyzed by the allosteric enzyme.

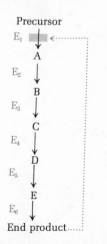

specific allosteric or regulatory enzymes, described in Chapter 4. These enzymes, which usually are located at or near the beginning of a multienzyme sequence, catalyze the rate-limiting step of the sequence (Figure 9-6). In catabolic pathways that lead to the generation of ATP from ADP, the end-product ATP often functions as an allosteric inhibitor, whereas in anabolic pathways it is the biosynthetic end product that usually functions as an allosteric inhibitor. Some regulatory enzymes are activated or stimulated by specific positive modulators; thus, an allosteric enzyme regulating a catabolic sequence may be stimulated by ADP. Some allosteric enzymes are multivalent, that is, they can respond to two or more specific modulators which may be the products of two or more different metabolic sequences. In this way the rates of two or more enzyme systems may be integrated.

The second level at which metabolic regulation is exerted is through control of the *concentration* of a given enzyme in the cell. The concentration of an enzyme at any given time is the result of a balance between the rate of its synthesis and the rate of its degradation. The rate of synthesis of enzymes may vary widely, depending on conditions. Enzymes that are always present in nearly constant amounts in a given cell are called constitutive. Others that are synthesized only in response to the presence of certain substrates are called *adaptive* or *induced* enzymes. The genes specifying the synthesis of induced enzymes are usually repressed and come into play (that is, undergo derepression) only in response to the presence of the specific substrates of such enzymes.

Metabolic control is exerted at a third level in those higher multicellular organisms that possess endocrine systems. Hormones elaborated by various endocrine glands are chemical messengers that stimulate or inhibit specific metabolic activities in other tissues or organs. For example, deficiency of insulin secretion by the pancreas results in impaired transport of glucose into muscle and liver cells, which leads to a number of secondary metabolic effects, such as a decrease in the biosynthesis of fatty acids from glucose and an excessive formation of metabolites called *ketone bodies* by the liver. Administration of trace amounts of insulin repairs these defects. How hormones control metabolic activity is not yet known in detail. Some are known to function by stimulating the conversion of an inactive form of a key enzyme to its active form. Others function by stimulating the synthesis of a specific enzyme or group of enzymes.

Experimental Approaches to Intermediary Metabolism

There are two major goals in research on metabolic pathways: (1) the identification of the chemical stoichiometry and mechanism of each reaction step in the sequence, which requires the isolation of each enzyme; (2) the identification of the mechanisms by which the rate of the pathway is regulated. Three general approaches have been used to approach these goals.

Cell-free systems

The first step in direct elucidation of a metabolic sequence is to obtain extracts of cells or tissues capable of catalyzing the overall process. For example, our modern knowledge of the details of the conversion of glucose to ethanol and carbon dioxide during alcoholic fermentation arose from the discovery of E. Buchner in 1892 that cell-free extracts of yeast catalyze alcoholic fermentation. Similarly, it was shown later that conversion of glucose to lactic acid occurs in cell-free extracts of muscle. Once such preparations were available, various metabolic intermediates in these pathways were made to accumulate by use of specific enzyme inhibitors, by inactivation of specific enzymes, or by removal or inactivation of essential coenzymes. Chemical identification of intermediates accumulating after such treatments ultimately permitted identification and isolation of the enzymes forming or utilizing them. In many cases metabolic pathways have been reconstructed _in vitro_ ("in glass") starting from highly purified enzymes, coenzymes, and the requisite metal ions.

Genetic Defects in Metabolism; Auxotrophic Mutants

An important experimental approach to the study of intermediary metabolism is afforded by genetic mutations of organisms in which a given enzyme fails to be synthesized in active form. Such a defect, if it is not lethal, may result in the accumulation and excretion of the substrate of the defective enzyme; in the normal organism, of course, this intermediate would not accumulate. One such genetic defect in human patients leads to the excretion of _homogentisic acid_ in the urine. Since the excretion of this acid is increased by feeding phenylalanine and tyrosine and is decreased by withholding these amino acids from the diet, it was concluded that

Figure 9-7

Auxotrophic mutants of Neurospora crassa with defective enzymes (color) at different points in the biosynthesis of arginine (Arg) from precursor A.

WILD-TYPE $A \xrightarrow{E_1} B \xrightarrow{E_2} C \xrightarrow{E_3} D \xrightarrow{E_4}$ Arg

MUTANT I $A \xrightarrow{E_1} B \xrightarrow{E_2} C \xrightarrow{E_3} D \xrightarrow{E_4}$ Arg

MUTANT II $A \xrightarrow{E_1} B \xrightarrow{E_2} C \xrightarrow{E_3} D \xrightarrow{E_4}$ Arg

MUTANT III $A \xrightarrow{E_1} B \xrightarrow{E_2} C \xrightarrow{E_3} D \xrightarrow{E_4}$ Arg

Table 9-1 Some isotopes useful as tracers†

Iso-tope	Relative natural abundance, %	Type of radiation	Half-life
$^{2}_{1}H$	0.0154		Stable
$^{3}_{1}H$		β^-	12.1 years
$^{13}_{6}C$	1.1		Stable
$^{14}_{6}C$		β^-	5,700 years
$^{15}_{7}N$	0.365		Stable
$^{18}_{8}O$	0.204		Stable
$^{24}_{11}Na$		β^-,γ	15 hours
$^{32}_{15}P$		β^-	14.3 days
$^{35}_{16}S$		β^-	87.1 days
$^{36}_{17}Cl$		β^-	3.1×10^5 years
$^{42}_{19}K$		β^-	12.5 hours
$^{45}_{20}Ca$		β^-	152 days
$^{59}_{26}Fe$		β^-,γ	45 days
$^{131}_{53}I$		β^-,γ	8 days

† The superscript before the symbol of the element designates the mass number, the subscript the atomic number. β^- radiation is due to negative electrons. Radioactivity is expressed in terms of the *curie*, the quantity undergoing the same number of disintegrations per second as 1.0 gram of radium (3.7 × 10^{10} sec^{-1}); the millicurie and microcurie are more convenient units. Stable isotopes are measured in atoms percent excess over the natural abundance.

homogentisic acid is formed as an intermediate during oxidative catabolism of phenylalanine and tyrosine. Several other genetic defects in human amino acid metabolism are known. Such nonlethal genetic defects involving major metabolic pathways are relatively rare in mammals and not easily recognized. However, genetic defects in metabolism can be produced at will in microorganisms and they are powerful tools for studying metabolic pathways.

Wild-type cells of the mold *Neurospora crassa*, that is, normal or unmutated cells, can grow on a simple medium containing glucose as the sole carbon source and ammonia as the sole nitrogen source. However, if *Neurospora* spores are treated with mutagenic agents, such as x-rays, mutant cells arise which are no longer capable of growing on this simple medium. Such mutants often grow if the medium is supplemented with one or another specific metabolite. For example, certain *Neurospora* mutants will grow normally if they are supplemented with the amino acid arginine. In such mutants one of the enzymes required in the synthesis of arginine from ammonia is genetically defective; for lack of arginine these cells cannot manufacture proteins containing arginine, and thus fail to grow. But when arginine is supplied in the culture medium, the mutant cells grow readily. Mutants of *Neurospora* defective in the capacity to make arginine are not all identical; they may differ with respect to the specific step in arginine biosynthesis which is defective (Figure 9-7). Such mutants defective in a biosynthetic pathway, whose growth can be restored by providing them with the normal product of the pathway, are called *auxotrophic mutants*.

Auxotrophic mutants can also be used to analyze catabolic pathways. For example, wild-type *E. coli* cells are able to obtain the carbon they need for growth from lactose, glucose, or galactose. In some *E. coli* mutants, the capacity to grow on lactose is lost, but the cells are still able to grow on glucose. The genetic defect in this case involves the enzyme that can hydrolyze lactose to glucose and galactose. Many intermediate steps in metabolism have been established by the use of auxotrophic mutants, particularly in the biosynthesis of amino acids, purines, and pyrimidines.

Isotope Tracer Method

Another powerful method for determining the pathways of metabolism is the use of an isotopic form of an element (Table 9-1) to label a given metabolite. The labeled molecule is chemically indistinguishable from normal metabolite molecules, but can easily be detected and measured by physical methods, through its abnormal content of the isotopic element. For example, acetic acid may be synthesized in the laboratory in such a way that its carbon atoms are highly enriched in the radioactive carbon isotope ^{14}C, which otherwise occurs in only very small and constant amounts in the normal carbon compounds of the biosphere. When such a sample of radioactively labeled acetic acid is fed to an animal its metabolic fate can be readily traced. If the long-chain fatty acid, palmitic acid, is subsequently isolated from the liver lipids of the animal, it will be found to contain large amounts of isotopic carbon, indicating that acetic acid is a biosynthetic precursor of palmitic acid. An extraordinary range of important observations have been made with the isotope tracer technique applied to intact animals, among them the discovery that the carbon atoms of acetic acid are metabolic precursors of all the carbon atoms of cholesterol and that glycine is a precursor in the synthesis of purines.

The isotope tracer method can also be used to determine the rate of metabolic processes in intact organisms and whether a given metabolic pathway postulated by study of isolated enzymes in the test tube actually occurs in the intact cell.

One of the most significant advances made with the isotope tracer method is the discovery that certain molecular components of cells and tissues undergo _metabolic turnover_, that is, they exist in a dynamic steady state in the cell, in which the rate of synthesis is exactly counterbalanced by the rate of degradation. For example, the proteins of rat liver have a half-life of about 5 to 6 days, whereas the liver cell itself has a much longer lifetime of several months (Table 9-2). On the other hand, most of the proteins and lipids of muscle tissue or the brain do not turn over rapidly, nor do those of bacteria grown on a medium of constant composition.

Cell Structure and the Compartmentation of Metabolism

The two great classes of cells, _prokaryotic_ and _eukaryotic_, differ greatly in their size, internal structure, and their genetic and metabolic organization. Prokaryotic cells, which include all the different species of bacteria and blue-green algae, are comparatively primitive; they are very small and simple cells. Prokaryotic cells have

Table 9-2 Turnover of some components of rat tissues

Tissue	Half-life, days
Liver:	
Total protein	5.0–6.0
Glycogen	0.5–1.0
Phosphoglycerides	1–2
Triacylglycerols	1–2
Cholesterol	5–7
Mitochondrial proteins	9.7
Muscle:	
Total protein	30
Glycogen	0.5–1.0
Brain:	
Triacylglycerols	10–15
Phospholipid	200
Cholesterol	>100

but a single chromosome (Chapter 21) and possess only a single membrane system, that surrounding the cell. They contain no membrane-surrounded internal organelles, such as a nucleus, mitochondria, or chloroplasts. They do not undergo mitosis and reproduce for the most part by simple asexual division. A typical and well-studied member of the class of prokaryotic cells is the bacterium *Escherichia coli* (see introduction to Part 1). An electron micrograph of the structure of an *E. coli* cell is shown in Figure 9-8.

Figure 9-8
Electron micrograph of the bacterium Escherichia coli. This organism is typically found in the intestinal tract of man. The mature cell is about 2 μ long and 1 μ in diameter; it weighs about 2×10^{-12} gram. E. coli cells grow rapidly on a simple medium containing only glucose as carbon source and ammonia as nitrogen source. Because of their rapid growth and short generation time E. coli and other bacteria are widely used for biochemical and genetic study.

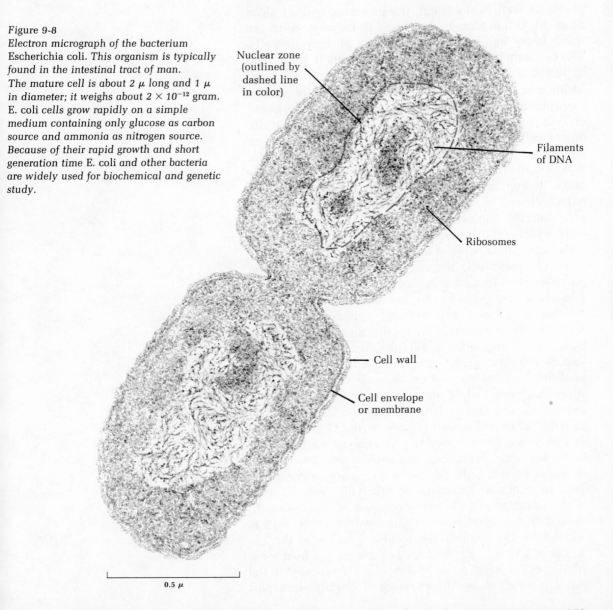

Nuclear zone (outlined by dashed line in color)

Filaments of DNA

Ribosomes

Cell wall

Cell envelope or membrane

0.5 μ

Within prokaryotic cells there are no compartments separated by internal membranes. Thus the metabolism of the bacterial cell proceeds in a single membrane-surrounded chamber. Although there is no compartmentation of metabolism in bacterial cells, there is some degree of segregation of certain enzyme systems. For example, most of the enzymes participating in biosynthesis of proteins are located in the ribosomes of bacteria and many of the enzymes active in synthesis of phospholipids are located in the cell membrane (Figure 9-8).

Eukaryotic cells include those of higher animals and plants, as well as the fungi, the protozoa, and all algae other than the blue-green algae. Eukaryotic cells are much larger and more complex than prokaryotic cells. The cell volume of typical eukaryotic cells is from 1000 to 10,000 times greater than that of prokaryotic cells. In addition to the cell envelope, all eukaryotic cells possess a distinct membrane-surrounded nucleus containing several or many chromosomes, which undergo mitosis during cell division. Eukaryotic cells also contain other membranous internal organelles, such as _mitochondria,_ _endoplasmic reticulum, Golgi bodies,_ and in green plant cells, _chloroplasts._ The electron micrograph of a typical eukaryotic cell in Figure 9-9 shows the exceedingly complex internal compartmentation of this great class of cells, which are believed to have arisen from prokaryotic cells during the course of biological evolution.

In eukaryotic cells the enzymes catalyzing metabolic pathways are localized in specific organelles or compartments. This conclusion has been reached from a number of different experimental approaches, particularly by direct examination of the enzyme content of isolated cell organelles or subfractions. Animal or plant tissues are first gently disrupted in isotonic sucrose medium in order to rupture the cell membranes. The subcellular organelles, such as the nuclei and the mitochondria, can then be isolated from the homogenate by differential centrifugation (Figure 9-10). These fractions can be tested for their capacity to catalyze a given metabolic sequence. From this approach it has been found that specific metabolic pathways take place in characteristic intracellular locations in eukaryotic cells (Figure 9-11). For example, the entire complex of enzymes concerned in the conversion of glucose into lactic acid is located in the _cytosol,_ the soluble portion of the cytoplasm, whereas all the enzymes concerned in the oxidation of the acetyl group of acetyl CoA are located in the _mitochondria,_ as are the enzymes of electron transport

Figure 9-9
Electron micrograph of a thin section of a single rat-liver cell.
Liver cells are polyhedral and about 20 μ in diameter. They are metabolically versatile cells whose most important function is the biochemical processing and distribution of foodstuff molecules brought to the liver from the intestinal tract. They store glucose as glycogen, prepare nitrogenous wastes for excretion, and synthesize blood plasma proteins and lipids. They are perhaps the most thoroughly studied animal cells because of their ready availability and the ease with which nuclei, mitochondria, endoplasmic reticulum (the "microsome" fraction), and other subcellular fractions can be recovered by differential centrifugation of sucrose homogenates of liver (see Figure 9-10).

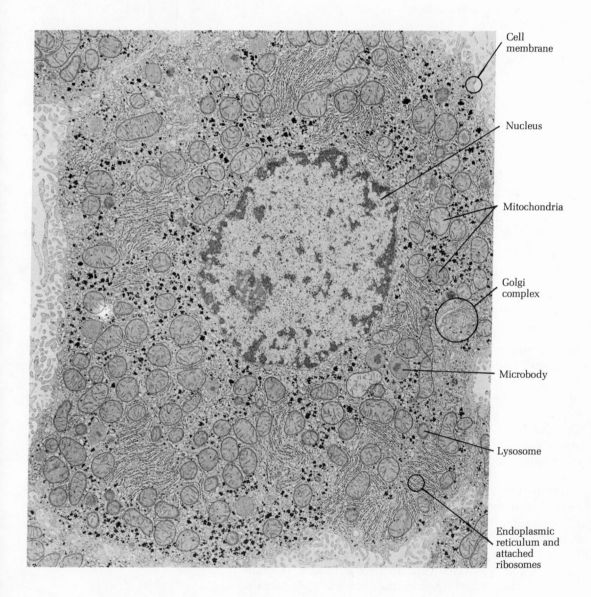

Cell membrane

Nucleus

Mitochondria

Golgi complex

Microbody

Lysosome

Endoplasmic reticulum and attached ribosomes

Figure 9-10
Isolation of intracellular structures by differential centrifugation. The cell membrane is ruptured by the shearing forces developed by the rotating homogenizer pestle. Following removal of connective tissue and fragments of blood vessels and bile ducts by a stainless steel sieve, the cell extract is centrifuged at a series of increasing rotor speeds.

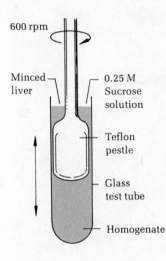

600 rpm

Minced liver

0.25 *M* Sucrose solution

Teflon pestle

Glass test tube

Homogenate

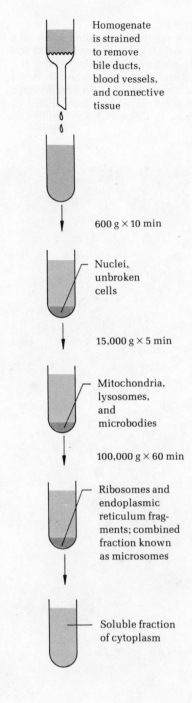

Homogenate is strained to remove bile ducts, blood vessels, and connective tissue

600 g × 10 min

Nuclei, unbroken cells

15,000 g × 5 min

Mitochondria, lysosomes, and microbodies

100,000 g × 60 min

Ribosomes and endoplasmic reticulum fragments; combined fraction known as microsomes

Soluble fraction of cytoplasm

and phosphorylation of ADP. It appears probable that some multienzyme systems are so located in the structure of an organelle that each enzyme is physically oriented adjacent to the next in the pathway, so that the entire sequence can function with a high degree of coordination. This is true of the enzymes of electron transport, which are located in the inner membrane of mitochondria, and the enzymes concerned in protein synthesis, most of which are located in the ribosomes.

Compartmentation of enzyme systems also permits control and integration of some intracellular activities. For example, the synthesis of glucose from pyruvate involves a complex interplay of a series of enzymes, some located in the mitochondria and some in the soluble phase of the cytoplasm. Still another advantage of compartmentation is that it segregates chemically incompatible reactions. At one and the same time a cell may be oxidizing long-chain fatty acids to the stage of acetic acid and carrying out the reverse process, the reduction of

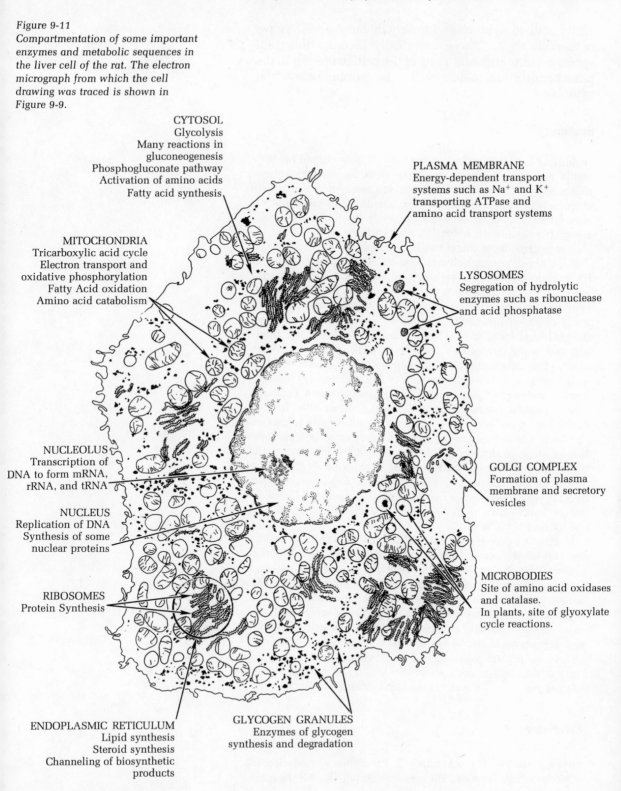

Figure 9-11
Compartmentation of some important enzymes and metabolic sequences in the liver cell of the rat. The electron micrograph from which the cell drawing was traced is shown in Figure 9-9.

CYTOSOL
Glycolysis
Many reactions in gluconeogenesis
Phosphogluconate pathway
Activation of amino acids
Fatty acid synthesis

PLASMA MEMBRANE
Energy-dependent transport systems such as Na^+ and K^+ transporting ATPase and amino acid transport systems

MITOCHONDRIA
Tricarboxylic acid cycle
Electron transport and oxidative phosphorylation
Fatty Acid oxidation
Amino acid catabolism

LYSOSOMES
Segregation of hydrolytic enzymes such as ribonuclease and acid phosphatase

NUCLEOLUS
Transcription of DNA to form mRNA, rRNA, and tRNA

GOLGI COMPLEX
Formation of plasma membrane and secretory vesicles

NUCLEUS
Replication of DNA
Synthesis of some nuclear proteins

RIBOSOMES
Protein Synthesis

MICROBODIES
Site of amino acid oxidases and catalase.
In plants, site of glyoxylate cycle reactions.

ENDOPLASMIC RETICULUM
Lipid synthesis
Steroid synthesis
Channeling of biosynthetic products

GLYCOGEN GRANULES
Enzymes of glycogen synthesis and degradation

173

acetic acid to synthesize long-chain fatty acids, as we have seen above. These chemically incompatible processes occur in different parts of the cell: oxidation in the mitochondria and reduction in the extramitochondrial cytoplasm.

Summary

Organisms can be classified on the basis of their carbon requirements. Autotrophs can utilize carbon dioxide, whereas heterotrophs require carbon in a more complex reduced form, such as glucose. Many autotrophic cells, such as those of green plants, obtain their energy from sunlight, while heterotrophs obtain energy from oxidation-reduction reactions.

Intermediary metabolism can be divided into catabolism, the degradation of energy-rich nutrient molecules, and anabolism, the biosynthesis of new cellular components. Catabolism and anabolism occur in three major stages. In the first stage of catabolism polysaccharides, lipids, and proteins are enzymatically degraded into their building blocks, in the second stage the building blocks are oxidized to acetyl CoA as major product, and in the third stage, the acetyl groups of acetyl CoA are oxidized to carbon dioxide. Catabolic pathways converge into a final common pathway, whereas anabolic pathways diverge from a common origin. Corresponding anabolic and catabolic pathways are not enzymatically identical, they are often located in different parts of the cell, and they are differently regulated. Catabolism of nutrient molecules is accompanied by conservation of some of the energy of the nutrient in the form of the phosphate-bond energy of adenosine triphosphate (ATP). The energy for anabolic pathways is provided by dephosphorylation of ATP, which serves as a carrier of chemical energy. Chemical energy is also carried from catabolic to anabolic pathways in the form of NADPH.

Metabolic pathways are studied in extracts of cells or tissues, from which the component enzymes may be isolated. Microorganisms which are genetically defective in a given pathway (auxotrophs) provide a powerful tool for analysis of metabolic pathways, as does the isotope tracer technique. Metabolism is regulated (1) through the intrinsic properties of enzymes, (2) by regulatory or allosteric enzymes, (3) through genetic repression and depression of enzyme synthesis, and (4) by endocrine control. In eukaryotic cells, the enzymes catalyzing various pathways of metabolism are compartmented in different organelles such as nuclei, mitochondria, and endoplasmic reticulum.

References

CHASE, G. D. and J. L. RABINOWITZ: *Principles of Radioisotope Methodology*, 2nd ed., Burgess Publishing Co., Minneapolis,

1962. Application of isotopic approaches to biochemical problems.

LEHNINGER, A. L.: *Bioenergetics*, 2nd ed., W. A. Benjamin, Menlo Park, Cal., 1971. Elementary treatment of the molecular basis of energy transformations.

ROODYN, D. B. (ed.): *Enzyme Cytology*, Academic Press, New York, 1967. Intracellular location of enzyme systems.

ATP was first isolated in 1929 by C. Fiske and Y. Subbarow, who found it in acid extracts of muscle. From its discovery it was suspected to play a role in cellular energy transfer, but it was not until some ten years later that Lipmann proposed the hypothesis that ATP serves as the principal carrier of chemical energy in the cell.

In this chapter we shall examine the chemical principles underlying the function of ATP in carrying chemical energy from energy-yielding to energy-requiring reactions of metabolism.

Occurrence and Properties of ATP, ADP, and AMP

ATP, ADP, and AMP (Figure 10-1) are present in all cells. The sum of their concentrations is relatively constant; for most cells it is in the range 5–15 mM. However, the relative proportions of ATP, ADP, and AMP can vary considerably, depending on the metabolic state of the cell. In normal respiring cells ATP makes up at least 75 percent of the total soluble adenine nucleotides.

At pH 7.0, both ATP and ADP are highly charged anions; ATP has four negative charges and ADP three. In the intact cell ATP and ADP exist largely as the $MgATP^{2-}$ and $MgADP^-$ complexes, because of the high affinity of the pyrophosphate groups for divalent cations and the high concentration of Mg^{2+} in intracellular fluid. In many enzymatic reactions in which ATP participates as phosphate donor, its active form is the $MgATP^{2-}$ complex (Figure 10-1).

Figure 10-1
ATP, ADP, and AMP at pH 7.0. The high-
energy bonds are shown by the symbol ~

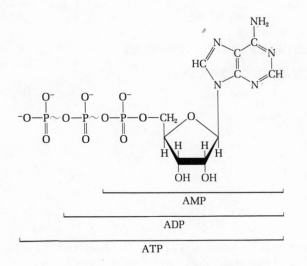

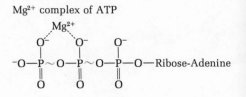

Mg²⁺ complex of ATP

Principles of Chemical Thermodynamics

A description of the chemical basis of the function of
ATP in the energy cycle of the cell requires a brief review
of some principles of thermodynamics, the science
dealing with energy exchanges. To analyze energy exchanges we must first identify the *system*, the collection
of matter under study. All other matter in the universe
apart from the specified system is called the *sur-*
roundings. During any chemical or physical change in
the system energy may pass from the system to the surroundings or from the surroundings to the system. It is
the working approach of thermodynamics to measure the
kinds and amount of energy exchanged as the system and
surroundings undergo a physical or chemical change
from their initial state to their final state of equilibrium.
In this way an energy balance sheet is prepared. The
amount of energy exchanged by any given system as it
undergoes a physical or chemical process is independent
of the rate of the process; it depends only on the initial
and final states of the system.

Two basic laws govern energy exchanges. *The first law*
of thermodynamics is the principle of the conservation
of energy. It states that in any process the *total* energy of
the system plus the surroundings remains constant.

Although energy is neither created nor destroyed during chemical or physical processes, it may undergo transformation from one form to another, such as heat, light, electrical, mechanical, and chemical energy.

The *second law of thermodynamics* states that all processes tend to proceed in such a direction that the *entropy* of the system plus surroundings, that is, of the *universe*, always increases until ultimately an equilibrium is reached. At this point the entropy is the maximum that can be attained by the system plus surroundings under the prevailing conditions of temperature and pressure. Entropy may be defined qualitatively as the degree of disorder or randomness in the universe. An *equilibrium* is defined as a state in which no further net chemical or physical change is taking place and in which temperature, pressure, and concentration are uniform throughout the system. The second law thus says that once a process has occurred with an increase in entropy and has attained a condition of equilibrium, it cannot by itself reverse and return to its initial state, which would require a decrease in entropy. Put in another way, a randomized system never unrandomizes itself spontaneously (Figure 10-2). As a system proceeds toward equilibrium it can do work, but once it has reached equilibrium it has exhausted its capacity for doing work.

The increase in entropy during a chemical or physical process can be used to calculate how much work the system can do as it proceeds toward equilibrium under a given set of conditions. But entropy is rather difficult to measure in chemical reactions. There is a more convenient way of determining how much useful energy can

Figure 10-2
The increase of entropy or randomness in two physical systems. Such flows never reverse spontaneously.

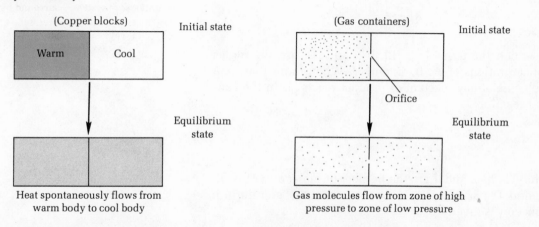

(Copper blocks)   Initial state

| Warm | Cool |

Equilibrium state

Heat spontaneously flows from
warm body to cool body

(Gas containers)   Initial state

Orifice

Equilibrium state

Gas molecules flow from zone of high
pressure to zone of low pressure

be delivered by a chemical reaction as it proceeds toward equilibrium. Under conditions of constant temperature and pressure, such as exist in the cell, the entropy change is related to the total energy change by an equation that combines the first and second laws:

$$\Delta G = \Delta E - T \, \Delta S$$

In this equation ΔE is the change in total energy of the system under the conditions described above, ΔS is the change in entropy, T is the absolute temperature, and ΔG is the change in free energy, the kind of energy that can do useful work at constant temperature and pressure. The free energy change can be defined as *useful* energy, that portion of the total energy change (ΔE) which is available to do work as a system proceeds to equilibrium at constant temperature and pressure. As the system approaches equilibrium, the entropy of the universe increases to a maximum, but the free energy of the system decreases to a minimum.

Figure 10-3 summarizes the relationship between changes in free energy and entropy in a system and its surroundings under the special conditions when temperature and pressure in the system are constant.

The Standard Free Energy Change of Chemical Reactions

The free energy change occurring during any given chemical reaction under standard conditions of temperature, pressure, and concentration can be calculated from *its equilibrium constant* (Chapter 1). The equilibrium constant for a simple reaction such as $A + B \rightleftharpoons C + D$ is

$$K'_{eq} = \frac{[C][D]}{[A][B]}$$

in which the terms [A], [B], [C], and [D] are the molar concentrations of A, B, C, and D. When more than one molecule of any reactant or product reacts, as in the general reaction

$$aA + bB \longrightarrow cC + dD$$

where a, b, c, and d are the number of molecules of A, B, C, and D participating in the reaction, the equilibrium constant is given by

Figure 10-3
Summary of free energy and entropy relationships between a system and its surroundings when the temperature, pressure, and volume of the system are constant. The entropy of the system + surroundings, or the free energy of the system alone, are the criteria for predicting the direction of chemical reactions under these conditions.

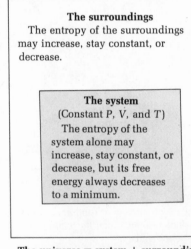

The surroundings
The entropy of the surroundings may increase, stay constant, or decrease.

The system
(Constant P, V, and T)
The entropy of the system alone may increase, stay constant, or decrease, but its free energy always decreases to a minimum.

The universe = system + surroundings
The entropy of the universe always increases to a maximum.

$$K'_{eq} = \frac{[C]^c[D]^d}{[A]^a[B]^b}$$

From the equilibrium constant K'_{eq} of any chemical reaction we can calculate its _standard free energy change_ ($\Delta G°$) with the expression

$$\Delta G° = -2.303\ RT\ \log_{10} K'_{eq}$$

in which R is the gas constant (1.987 calories per mole degree) and T is the absolute temperature. The standard free energy change of a chemical reaction is in reality the difference in the free energy content of the reactants and the free energy content of the products. If the equilibrium constant for a reaction is 1.0, then the standard free energy change is $\Delta G° = 0.0$. If the equilibrium constant is greater than 1.0, the standard free energy change $\Delta G°$ is negative. If the equilibrium constant is less than 1.0, $\Delta G°$ is positive. Chemical reactions with a _negative_ standard free energy change are termed _exergonic;_ starting from 1.0 M concentrations of all components, such reactions proceed further in the direction written. Reactions with a _positive_ standard free energy change are called _endergonic:_ they do not proceed in the direction written starting from 1.0 M concentrations of all reactants and products; in fact, they proceed in the reverse direction. Table 10-1 shows the quantitative relationship between the standard free energy change $\Delta G°$ and the magnitude of the equilibrium constant K_{eq}. In biochemical energetics pH 7.0 is designated as the standard state. The standard free energy change at pH 7.0 is designated by $\Delta G°'$.

Now let us make a sample calculation of the standard free energy change of the reaction catalyzed by the enzyme _phosphoglucomutase,_

Glucose 1-phosphate $\rightleftharpoons$ glucose 6-phosphate

Chemical analysis shows that if we start with 0.020 M glucose 1-phosphate in the presence of excess enzyme and allow the reaction to go in the forward direction or if we start with 0.020 M glucose 6-phosphate and allow the reaction to go in the reverse direction, the final equilibrium mixture in either case will contain 0.001 M glucose 1-phosphate and 0.019 M glucose 6-phosphate at 25°C and pH 7.0. We can then calculate the equilibrium constant

$$K'_{eq} = \frac{[\text{glucose 6-phosphate}]}{[\text{glucose 1-phosphate}]} = \frac{0.019}{0.001} = 19$$

Table 10-1 Relationship between the equilibrium constant and the standard free energy change at 25°C

K'_{eq}	$\Delta G°$, cal
0.001	+4089
0.01	+2726
0.1	+1363
1.0	0
10.0	−1363
100.0	−2726
1,000.0	−4089

From this value of K'_{eq} the standard free energy change $\Delta G^{\circ\prime}$ is calculated

$$\begin{aligned}
\Delta G^{\circ\prime} &= -2.303 \ RT \ \log_{10} K'_{eq} \\
&= -2.303 \times 1.987 \times 298 \ \log 19 \\
&= -1745 \ \text{cal}
\end{aligned}$$

Since the sign of the standard free energy change is negative the conversion of glucose 1-phosphate to glucose 6-phosphate is an exergonic process. In quantitative terms, these data mean that when 1.0 mole of glucose 1-phosphate is converted to 1.0 mole of glucose 6-phosphate at 25°C and pH 7.0, under conditions in which the concentration of each is maintained at 1.0 M, a decline in free energy of 1745 cal occurs. It is usually more convenient to express thermodynamic data in kilocalories; -1745 cal becomes -1.745 kcal.

Table 10-2 gives the standard free-energy changes for a number of biologically important reactions. Note that oxidation reactions proceed with especially large decreases in standard free energy; they are the most important energy-yielding reactions in metabolism.

Table 10-2 Standard free-energy changes of some chemical reactions (pH = 7.0; T = 25°C)

Reaction	$\Delta G^{\circ\prime}$, kcal
Hydrolysis	
Acid anhydrides	
Acetic anhydride + $H_2O \longrightarrow$ 2 acetate	-21.8
Pyrophosphate + $H_2O \longrightarrow$ 2 phosphate	$- 8.0$
Esters	
Ethyl acetate + $H_2O \longrightarrow$ ethanol + acetate	$- 4.7$
Glucose 6-phosphate + $H_2O \longrightarrow$ glucose + phosphate	$- 3.3$
Amides	
Glutamine + $H_2O \longrightarrow$ glutamate + $NH_4{}^+$	$- 3.4$
Glycylglycine + $H_2O \longrightarrow$ 2 glycine	$- 2.2$
Glycosides	
Sucrose + $H_2O \longrightarrow$ glucose + fructose	$- 7.0$
Maltose + $H_2O \longrightarrow$ 2 glucose	$- 4.0$
Rearrangement	
Glucose 1-phosphate $\longrightarrow$ glucose 6-phosphate	$- 1.7$
Fructose 6-phosphate $\longrightarrow$ glucose 6-phosphate	$- 0.4$
Elimination	
Malate $\longrightarrow$ fumarate + H_2O	$+ 0.75$
Oxidation	
Glucose + $6O_2 \longrightarrow 6CO_2 + 6H_2O$	-686
Palmitic acid + $23O_2 \longrightarrow 16CO_2 + 16H_2O$	-2338

The Standard Free Energy of Hydrolysis of ATP

ATP may undergo enzymatic hydrolysis of its terminal phosphate group with the formation of ADP and inorganic phosphate

$$\text{ATP} + \text{HOH} \longrightarrow \text{ADP} + \text{phosphate}$$

From the relationship

$$\Delta G^{\circ\prime} = -2.303RT \log K'_{eq}$$

applied to equilibria of reactions in which ATP loses its terminal phosphate group, it has been found that the standard free energy of hydrolysis of ATP expressed by the above equation is -7.3 kcal. It is important to note that this value assumes pH 7.0, 37°C, an excess of Mg^{2+} ions, and 1.0 M concentrations of reactants and products.

The terminal phosphate group of ADP may also be hydrolyzed according to the equation

$$\text{ADP} + \text{H}_2\text{O} \rightleftharpoons \text{AMP} + \text{P}_i \qquad \Delta G^{\circ\prime} = -7.3 \text{ kcal}$$

The $\Delta G^{\circ\prime}$ of this reaction is thus about the same as for the hydrolysis of the terminal phosphate group of ATP.

Standard free-energy changes have been determined for the hydrolysis of other phosphorylated compounds (Table 10-3). We note that some phosphate compounds yield more and some less energy than ATP when they undergo hydrolysis. For example, for the reaction

$$\text{Glucose 6-phosphate} + \text{H}_2\text{O} \xrightarrow{\text{phosphatase}} \text{glucose} + \text{phosphate}$$

the $\Delta G^{\circ\prime}$ is -3.3 kcal, indicating that this reaction yields much less energy than the hydrolysis of ATP ($\Delta G^{\circ\prime} = -7.3$ kcal).

The Structural Basis of the Free Energy Change during Hydrolysis of ATP

What structural features of the ATP molecule cause it to deliver considerably more free energy on hydrolysis of the terminal phosphate group than the hydrolysis of glucose 6-phosphate? This is tantamount to asking why the equilibrium of hydrolysis lies farther in the direction of completion for ATP than it does for glucose 6-

Table 10-3   Standard free energy of hydrolysis
of some phosphorylated compounds

	$\Delta G^{\circ\prime}$, kcal	Direction of phosphate group transfer
Phosphoenolpyruvate	−14.80	
1,3-Diphosphoglycerate	−11.80	
Phosphocreatine	−10.30	
Acetyl phosphate	−10.10	
Phosphoarginine	− 7.70	
ATP	− 7.30	
Glucose 1-phosphate	− 5.00	
Fructose 6-phosphate	− 3.80	
Glucose 6-phosphate	− 3.30	
Glycerol 1-phosphate	− 2.20	

phosphate. The answer to this question can be found in the properties of both the substrate and the reaction products, since we will recall that the standard free energy of hydrolysis is a measure of the *difference* in free energy of the reactants and products.

One reason is that at pH 7.0 ATP molecules have closely spaced negative charges, which repel each other strongly (Figure 10-1). When the terminal phosphate bond is hydrolyzed, some of this electrical stress is relieved. Both of the resulting products, the anions HPO_4^{2-} and ADP^{3-}, are negatively charged and thus have little tendency to approach each other again because of charge repulsion (like charges repel each other). Thus ADP^{3-} and phosphate^{2-} do not readily recombine to form ATP. In contrast, when glucose 6-phosphate undergoes hydrolysis, one product, glucose, has no net charge. Since glucose and free phosphate do not repel each other electrostatically, they have a greater tendency to recombine.

The second major feature contributing to the more negative $\Delta G^{\circ\prime}$ of ATP hydrolysis is the fact that the two products ADP^{3-} and HPO_4^{2-} are *resonance hybrids*, in which the electrons can sink to lower energy levels than in unhydrolyzed ATP. This fact causes the free ADP^{3-} and HPO_4^{2-} anions to have much less free energy than they would have if they were still combined as ATP^{4-}.

Factors Affecting the Free Energy of Hydrolysis of ATP

The standard free energy of hydrolysis of ATP has the value −7.3 kcal only under arbitrarily set conditions, that is, when the pH is 7.0, the Mg^{2+} concentration is sufficiently high so that ATP, ADP, and phosphate are entirely present as Mg^{2+} complexes, and when the concentration of ATP, ADP and phosphate is 1.0 M. However, the standard value of $\Delta G^{\circ\prime} = -7.3$ kcal is not necessarily that which is applicable under the conditions existing in the intact cell, in which the pH and Mg^{2+} concentration are usually not identical with the standard conditions described above. Moreover, the concentrations of ATP, ADP, and phosphate in intact cells are far from the arbitrary standard concentrations of 1.0 M. If appropriate corrections are made for the pH, Mg^{2+} concentration, and the steady-state concentrations of ATP, ADP, and phosphate that actually exist in the intracellular water, the free energy of hydrolysis of ATP in the cell is about -12.5 kcal. However, the free energy of hydrolysis of ATP is not necessarily constant in the cell; it can vary from time to time or even from place to place in the cell, depending on the pH, Mg^{2+} concentration, and the local concentration of ATP, ADP, and phosphate. However, for consistency, we must carry out thermodynamic calculation of biological energy exchanges assuming the arbitrarily defined standard conditions.

The Scale of Phosphate Group Transfer Potentials

Each of the many phosphorylated compounds found in cells has a characteristic standard free energy of hydrolysis; some data are shown in Table 10-3. We see that some phosphate compounds release more energy on hydrolysis than ATP, and are called "high-energy" compounds, and some yield less ("low-energy" compounds). Actually, ATP is distinctive because it has an *intermediate* value in this thermodynamic scale, although it itself is considered to be a high-energy phosphate. It is the whole function of the ATP–ADP system to serve as an obligatory intermediate carrier of phosphate groups originating from compounds high on the scale, that is, those releasing more energy on hydrolysis than ATP, to acceptor molecules that are low on this scale, those releasing less energy than ATP under standard conditions. The term *phosphate group transfer potential* is often used to indicate the relative tendency of phosphate groups in different phosphorylated compounds to be transferred to acceptor molecules.

The Common-Intermediate Principle in Energy Transfers

We have seen that the reactions of intermediary metabolism take place via chains of consecutive or sequential reactions linked by common intermediates. Thus the reactions

$$A + B \longrightarrow C + D$$
$$D + E \longrightarrow F + G$$

are linked by the common intermediate D. The only way chemical energy can be transferred from the first reaction to the second at constant temperature is for the two reactions to have such a common intermediate. In the cell, ATP functions as a common intermediate linking the reactions delivering energy and the reactions requiring energy. During catabolism high-energy phosphate compounds are generated at the expense of energy released on degradation of cell nutrients. These compounds donate their phosphate groups and thus their energy to ADP. The ATP so formed then donates its terminal phosphate group and thus its energy to an acceptor molecule, which then increases in energy content. This pattern is shown in the following equations, in which $X-P$ designates a donor of a high-energy phosphate group to ADP, Y an acceptor of a phosphate group from ATP, and E_1 and E_2 are specific phosphate-transferring enzymes:

$$X-P + ADP \xrightarrow{E_1} X + ATP$$

$$ATP + Y \xrightarrow{E_2} ADP + Y-P$$

Now let us examine the enzymatic transfer of phosphate from high-energy donors to ADP.

Transfer of High-Energy Phosphate Groups to ADP

Among the phosphorylated compounds having a standard free energy of hydrolysis substantially more negative than that of ATP are those generated during enzymatic breakdown of fuel molecules.

The two most important are _1,3-diphosphoglycerate_ and _phosphoenolpyruvate_ (Figure 10-4), which are formed during the energy-yielding anaerobic degradation of glucose to lactic acid. In these compounds is conserved much of the energy released on degradation of

Figure 10-4
Structures of high-energy phosphate compounds formed during anaerobic glucose breakdown.

Phosphoenolpyruvate

1,3-Diphosphoglycerate

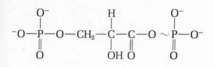

glucose (Chapter 11). In the intact cell, 1,3-diphosphoglycerate does not undergo hydrolysis; instead its 1-phosphate group is transferred by the action of a specific enzyme to ADP to form ATP and 3-phosphoglycerate

1,3-Diphosphoglycerate + ADP $\rightleftharpoons$

3-phosphoglycerate + ATP

This reaction proceeds far to the right because 1,3-diphosphoglyceric acid has a much more negative $\Delta G^{\circ\prime}$ value (-11.8 kcal) and thus a greater phosphate group "pressure" than ATP ($\Delta G^{\circ\prime} = -7.3$ kcal). Thus the phosphate group will tend to move from diphosphoglycerate to ADP when the necessary enzyme is present.

Similarly, phosphoenolpyruvate formed during breakdown of glucose donates its phosphate group to ADP in an enzymatic reaction

Phosphoenolpyruvate + ADP $\rightleftharpoons$ pyruvate + ATP

again, because the $\Delta G^{\circ\prime}$ for its hydrolysis (-14.8 kcal) is much larger than that for ATP. Thus both 1,3-diphosphoglycerate and phosphoenolpyruvate are high-energy phosphate-group donors which carry the energy of anaerobic glucose breakdown and donate it to ADP to yield ATP.

High-energy phosphate bonds are universally symbolized by the squiggle $\sim$ and high-energy phosphate groups as $\sim$ P. However, the term high-energy phosphate bond can be misleading since it wrongly suggests that the bond itself contains the energy. This is not the case: it actually requires energy to break chemical bonds. The free energy released by hydrolysis of phosphate esters comes not from the bond that is broken but is the result of the fact that the products have a smaller total free energy content than the reactants.

Transfer of High-Energy Phosphate Groups from ATP

Some examples of the large number of low-energy phosphate compounds found in cells are shown in Table 10-3. Most low-energy phosphate compounds are phosphoric acid esters of organic alcohols. Many enzymes are known that catalyze transfer of a phosphate group from ATP to specific phosphate acceptors to form low-energy phosphate compounds, among them the enzymes *glycerol kinase* and *hexokinase,* which catalyze phosphate transfer from ATP to glycerol and from ATP to D-glucose, respectively

$$\text{ATP} + \text{glycerol} \xrightarrow{\text{glycerol kinase}} \text{ADP} + \text{L-glycerol 3-phosphate}$$

$$\text{ATP} + \text{D-glucose} \xrightarrow{\text{hexokinase}} \text{ADP} + \text{D-glucose 6-phosphate}$$

Since the $\Delta G^{\circ\prime}$ for hydrolysis of glycerol 3-phosphate ($\Delta G^{\circ\prime} = -2.2$ kcal per mole) and glucose 6-phosphate ($\Delta G^{\circ\prime} = -3.3$ kcal per mole) is less negative than for ATP ($\Delta G^{\circ\prime} = -7.3$ kcal per mole) these reactions tend to go to the right as written. Note from these equations that this standard free energy charge represents the difference between the standard free energy of hydrolysis of the reactant and products. Thus, for the first reaction the standard free energy change $\Delta G^{\circ\prime}$ is -5.1 kcal, the difference between the $\Delta G^{\circ\prime}$ for hydrolysis of ATP (-7.3 kcal) and that for glycerol phosphate (-2.2 kcal).

The products of these reactions, glycerol phosphate and glucose 6-phosphate, have a higher energy content than their unphosphorylated forms. They are now "activated" or "energized" and can undergo enzymatic reactions in which they can serve as building blocks for synthesis of larger molecules. For example, glycerol phosphate is an activated building block in lipid synthesis and glucose 6-phosphate is a precursor in glycogen synthesis.

Thus the chemical energy originally transferred from degradation of glucose can be transmitted to glycerol, glucose, or other phosphate donors, with ATP serving as the common intermediate between energy-donating and energy-requiring reactions.

The Enzymatic Pathways of Phosphate Transfers

Figure 10-5 shows a flow sheet of enzymatic phosphate transfer reactions in the cell. An important feature is that virtually all high-energy phosphate groups must pass to low-energy phosphate acceptors via the ATP system, catalyzed in two steps by specific phosphate-transferring enzymes. Direct phosphate-group transfer from a high-energy donor to a low-energy acceptor ordinarily does not occur in the mainstreams of intermediary metabolism.

Figure 10-5 also shows the reservoir role of the high-energy phosphate compound *phosphocreatine* (Figure 10-6), which is formed by direct and reversible enzymatic transfer of a phosphate group from ATP to creatine at the expense of ATP. There is no other pathway for

Figure 10-5
Flow of phosphate groups from high-energy phosphate donors to low-energy acceptors via the ATP–ADP system.

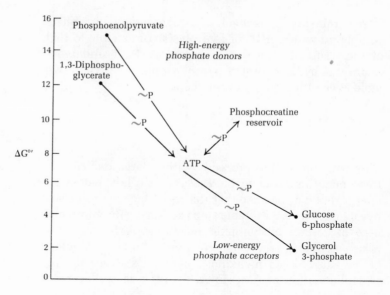

Figure 10-6
Phosphocreatine

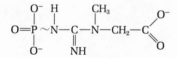

the formation and breakdown of phosphocreatine. Phosphocreatine is a reservoir of high-energy phosphate groups. It is important in various muscles and some nerves, but is not found in liver, kidney, or other mammalian tissues nor in bacteria.

The Role of AMP and Pyrophosphate

Although ADP is the product of many ATP-utilizing reactions in the cell and ADP is the direct phosphate acceptor in energy-yielding reactions, there are some ATP-utilizing reactions in the cell in which the two terminal phosphate groups of ATP are removed in one piece as pyrophosphate (symbolized PP_i), leaving adenylic acid (AMP) as the product. An example is the enzymatic activation of a fatty acid to form its coenzyme A ester (Chapter 14)

$$\text{ATP} + \text{RCOOH} + \text{CoA—SH} \rightleftharpoons \text{AMP} + \text{PP}_i + \text{RCO—S—CoA}$$
Fatty acid   Fatty acyl CoA

This reaction proceeds by what is termed a *pyrophosphate cleavage* of ATP, in contrast to the usual *orthophosphate cleavage*, in which ATP loses a single orthophosphate group, as in the reaction

$$\text{ATP} + \text{D-glucose} \rightleftharpoons \text{ADP} + \text{D-glucose 6-phosphate}$$

189

Inorganic pyrophosphate is a high-energy phosphate compound with a $\Delta G^{\circ\prime}$ of hydrolysis comparable to that of the terminal phosphate bond of ATP. It undergoes hydrolysis by the action of *pyrophosphatase* to yield two molecules of inorganic orthophosphate

$$PP_i + H_2O \longrightarrow 2P_i \qquad\qquad \Delta G^{\circ\prime} = -7.2 \text{ kcal}$$

This reaction, which proceeds with a large decrease in free energy, appears to be wasteful of phosphate-bond energy, but we shall see that the secondary hydrolysis of pyrophosphate is a valuable step in assuring the completeness of certain biosynthetic reactions. AMP can then return to the energy cycle by the action of *adenylate kinase*, also called *myokinase*, which catalyzes the rephosphorylation of AMP to ADP in the reversible reaction

$$ATP + AMP \rightleftharpoons ADP + ADP$$

Other Energy-Rich Nucleoside 5′-Triphosphates

We shall see in subsequent chapters that the 5′-triphosphates of guanosine, uridine, and cytidine (GTP, UTP, and CTP) (Chapter 7) also participate in intermediary metabolism. Their terminal phosphate groups are also high-energy groups with about the same $\Delta G^{\circ\prime}$ for hydrolysis as that of ATP. Similarly, the corresponding deoxyribonucleoside 5′-triphosphates (Chapter 7), such as dATP, dGTP, dCTP, and dTTP, are also high-energy compounds. Although ATP is the mainstream carrier of phosphate groups in the cell, the other types of triphosphates function in special roles. They acquire their terminal phosphate groups from ATP in reactions catalyzed by *nucleoside diphosphokinase*

$$ATP + UDP \rightleftharpoons ADP + UTP$$
$$ATP + GDP \rightleftharpoons ADP + GTP$$
$$ATP + CDP \rightleftharpoons ADP + CTP$$
$$GTP + UDP \rightleftharpoons GDP + UTP$$
$$ATP + dCDP \rightleftharpoons ADP + dCTP$$
$$GTP + dADP \rightleftharpoons GDP + dATP$$

Summary

Energy changes of chemical reactions can be analyzed quantitatively in terms of the first and second laws of thermodynamics, which are combined into the equation $\Delta G = \Delta H - T\,\Delta S$. Under conditions in which biological reactions occur, i.e., at constant temperature and pressure, chemical reactions proceed in such a direction that at equilibrium the entropy S of the system plus surroundings is at a maximum. Every chemical reaction has a characteristic standard free-energy change ($\Delta G^{\circ\prime}$) at standard temperature and pressure, with all reactants and products at 1 M concentration and pH = 7.0. It can be calculated from the equilibrium constant for the reaction by the equation

$$\Delta G^{\circ\prime} = -2.303\ RT\ \log_{10} K'_{eq}$$

The $\Delta G^{\circ\prime}$ of hydrolysis of ATP to ADP and phosphate is -7.30 kcal at pH 7.0 and 37° in the presence of excess Mg^{2+}. However, ΔG_{ATP} varies with pH, Mg^{2+} concentration, and ionic strength, as well as temperature. Under intracellular conditions it is approximately -12.5 kcal. Some phosphorylated compounds, such as phosphoenolpyruvate and 1,3-diphosphoglycerate, which are generated in glycolysis, have much more negative $\Delta G^{\circ\prime}$ values for their hydrolysis than ATP, whereas others (for example, glucose 6-phosphate) have more positive values. The intermediate position of ATP in the thermodynamic scale of phosphate-bond energy and the specificity of the phosphate-transferring enzymes for ADP or ATP as phosphate acceptor or donor, respectively, mean that the ADP-ATP system is the obligatory common intermediate carrying phosphate groups from high-energy phosphate compounds generated during catabolism to low-energy phosphate acceptors, which thus become energized. Phosphocreatine is a reservoir of high-energy phosphate groups.

ATP may undergo loss of either an orthophosphate or a pyrophosphate group during its utilization in biosynthetic reactions to form ADP or AMP, respectively. AMP is rephosphorylated to ADP by the adenylate kinase reaction. Other nucleoside 5′-triphosphates such as GTP, UTP, CTP, dATP, dTTP, etc., also participate as carriers of high-energy phosphate groups, which they channel into specific biosynthetic routes.

References

FLORKIN, M., and E. H. STOTZ (eds.): *Bioenergetics*, Vol. 22 of *Comprehensive Biochemistry*, American Elsevier Publishing Company, New York, 1967. Comprehensive articles and reviews.

KALCKAR, H. M.: *Biological Phosphorylations. Development of Concepts*, Prentice-Hall, Englewood Cliffs, N.J., 1969. A collection of reprinted papers describing classical investigations in bioenergetics, with an accompanying narrative.

LEHNINGER, A. L.: *Bioenergetics*, W. A. Benjamin, Menlo Park, Cal., 2nd ed., 1971. Elementary paperback stressing biochemical aspects.

LEHNINGER, A. L.: *Biochemistry*, Worth Publishers, New York, 1970. Chapter 13 provides a more detailed textbook treatment.

Problems

1. Calculate the equilibrium constants for the following reactions at pH = 7.0 and T = 25°C, using the $\Delta G^{\circ\prime}$ values of Table 10-2:
   (a) Glucose 6-phosphate + $H_2O \rightleftharpoons$ glucose + phosphate
   (b) Sucrose + $H_2O \rightleftharpoons$ glucose + fructose
   (c) Malate $\rightleftharpoons$ fumarate + H_2O

2. Calculate the standard free-energy changes of the following reactions at 25°C from the equilibrium constants given (pH 7.0):
   (a) Glutamate + oxaloacetate $\rightleftharpoons$

   aspartate + α-ketoglutarate $K'_{eq} = 6.8$
   (b) Isopropanol + $NAD^+ \rightarrow$

   acetone + NADH + H^+ $K'_{eq} = 7.2 \times 10^{-9}$ M
   (c) Malate + $NAD^+ \rightleftharpoons$

   oxaloacetate + NADH + H^+ $K'_{eq} = 7.5 \times 10^{-13}$ M

3. Glucose 1-phosphate is converted to fructose 6-phosphate in two successive reactions

   Glucose 1-phosphate $\rightleftharpoons$ glucose 6-phosphate
   Glucose 6-phosphate $\rightleftharpoons$ fructose 6-phosphate

   Using the $\Delta G^{\circ\prime}$ values of Table 10-2 calculate the equilibrium constant K'_{eq} for the sum of the two reactions.

4. From data in Table 10-3 calculate the $\Delta G^{\circ\prime}$ value for the reactions
   (a) Phosphocreatine + ADP $\rightleftharpoons$ creatine + ATP
   (b) ATP + fructose $\rightarrow$ ADP + D-fructose 6-phosphate

Glycolysis ("splitting of sugar") is the process by which organisms break glucose down into lactic acid in the absence of molecular oxygen for the purpose of obtaining chemical energy. It is one of several pathways used by different species of organisms to degrade glucose anaerobically; such anaerobic pathways are known generically as *fermentations*. Glycolysis is the type of anaerobic glucose fermentation occurring in the cells of higher animals and in many plants and microorganisms. Another type of glucose fermentation is the *alcoholic fermentation* of yeasts, in which glucose is degraded to ethanol and carbon dioxide. As we shall see, glycolysis and alcoholic fermentation have nearly identical pathways.

The Relationship between Fermentation and Respiration

We have seen that there are two classes of anaerobic organisms. The *strict anaerobes*, which cannot utilize oxygen at all, consist of various species of bacteria and invertebrates that live in environments having no oxygen, as in deep soils, deep sea waters, and marine mud. The *facultative anaerobes*, on the other hand, can obtain energy either from anaerobic fermentation or by aerobic respiration. When they live in the absence of oxygen, they degrade glucose by the same type of anaerobic processes as strict anaerobes. When they live aerobically, they usually continue to degrade their fuels by their anaerobic pathway, but then oxidize the products of the anaerobic pathway at the expense of molecular oxygen. In facultative cells, therefore, the anaerobic pathway of glucose breakdown is the obligatory first stage before the aerobic phase of respiration which follows (Figure 11-1). This pattern is characteristic not only of many bacteria, yeasts, and fungi, but also of the aerobic cells of most higher animals and plants. It is also consistent with what we know about the evolution of cells. The first living organisms arose at a time when the earth's atmosphere lacked oxygen; they had to develop anaerobic mechanisms for obtaining energy from the breakdown of glucose and other organic nutrients. But most present-day aerobic organisms have retained the capacity to degrade glucose by means of this ancient anaerobic pathway, which has become the preparatory stage for the aerobic phase of glucose oxidation.

The Equation of Glycolysis

The most commonly employed fuels for anaerobic fermentations are the six-carbon sugars, particularly D-glucose. In glycolysis (also called lactic fermentation) the six-carbon glucose molecule is degraded to two molecules of the three-carbon lactic acid as sole end product

$$C_6H_{12}O_6 \longrightarrow 2CH_3CHOHCOOH$$
Glucose   Lactic acid

In alcoholic fermentation glucose is broken down into two molecules of the two-carbon compound ethanol (C_2H_5OH) and two molecules of CO_2

$$C_6H_{12}O_6 \longrightarrow 2C_2H_5OH + 2CO_2$$
Glucose   Ethanol

However, these equations are incomplete statements; they describe only the fate of the carbon skeleton of glucose during glycolysis and alcoholic fermentation. Actually, ATP is formed from ADP and phosphate during both these processes. The complete balanced equations for glycolysis and alcoholic fermentation are

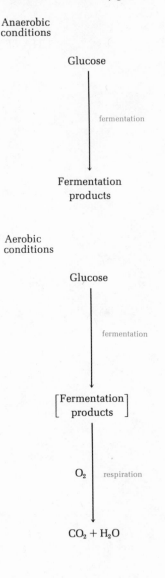

Figure 11-1
Glucose utilization in facultative organisms. The fermentation pathway is common to both the aerobic and anaerobic utilization of glucose.

Anaerobic conditions

Glucose

fermentation

Fermentation products

Aerobic conditions

Glucose

fermentation

[Fermentation products]

O_2 respiration

$CO_2 + H_2O$

Glycolysis:

$$C_6H_{12}O_6 + 2P_i + 2ADP \longrightarrow 2CH_3CHOHCOOH + 2ATP + 2H_2O$$
$$\text{Lactic acid}$$

Alcoholic fermentation:

$$C_6H_{12}O_6 + 2P_i + 2ADP \longrightarrow$$

$$2CH_3CH_2OH + 2CO_2 + 2ATP + 2H_2O$$
$$\text{Ethanol}$$

These equations do not show the various reactants and products in the ionization state in which they exist in the intact cell, whose contents are near pH 7.0. In the cell, phosphate, ADP, and lactic acid exist as their anions and CO_2 as the bicarbonate ion (HCO_3^-). In this and subsequent chapters enzyme-catalyzed reactions will usually be written with components shown or assumed to be the form existing at pH 7.0.

It is a major purpose of fermentation reactions to yield useful energy. We can analyze the energetics of glycolysis by breaking down its overall equation into two processes, the conversion of glucose to lactate, which is exergonic, and the formation of ATP from ADP and phosphate, which is endergonic

Exergonic process:

$$\text{Glucose} \longrightarrow 2 \text{ lactate} \qquad \Delta G_1^{\circ\prime} = -47.0 \text{ kcal}$$

Endergonic process:

$$2P_i + 2ADP \longrightarrow 2ATP + 2H_2O$$
$$\Delta G_2^{\circ\prime} = 2 \times 7.30 = +14.6 \text{ kcal}$$

From the standard free energy changes of these two reactions it is clear that the breakdown of glucose to lactate ($\Delta G^{\circ\prime} = -47$ kcal) provides more than sufficient energy to cause the phosphorylation of two molecules of ADP to ATP ($\Delta G^{\circ\prime} = 2 \times 7.3 = +14.6$ kcal). We can also calculate that $14.6/47.0 \times 100 = 31$ percent of the free energy decrease during breakdown of glucose to lactate is conserved as the phosphate-bond energy of ATP. Actually, if such calculations are adjusted to take account of the actual intracellular concentrations of the reactants and products of glycolysis, the true efficiency of energy recovery during glycolysis turns out to be closer to 50 percent. The two molecules of ATP generated by the degradation of one molecule of glucose are the cell's free energy profit; the remainder of the free energy decrease during glycolysis is lost as heat.

Some Historical Landmarks

Glycolysis is catalyzed by the consecutive action of a group of 11 enzymes, most of which have been crystallized and thoroughly studied. So far as is known today, the chemical pathway of glycolysis is identical in all forms of life, with only relatively minor differences in the structure and mechanism of action of the individual enzymes catalyzing the sequence. The glycolytic enzymes are easily extracted in soluble form from cells and thus are believed to be localized in the soluble portion of the cytoplasm.

The mechanism of glycolysis was elucidated over the course of many years of research. Some important historical landmarks will illustrate the experimental approaches used in the investigation of this fundamental metabolic pathway. E. Büchner first discovered in 1892 that an extract of yeast, freed of intact cells by filtration, retained the ability to ferment glucose to ethanol. This observation demonstrated that the enzymes of fermentation can function independently of intact cell structure.

Some years later Harden and Young in England discovered that alcoholic fermentation in yeast extracts requires a supply of phosphate. As the phosphate is utilized, a phosphorylated sugar accumulates in the fermenting mixture. This sugar was later identified as fructose 1,6-diphosphate, and it was postulated to be an intermediate in the conversion of glucose to ethanol and CO_2. This view was later verified when it was found that fructose 1,6-diphosphate added to yeast extract instead of glucose was also fermented with the formation of ethanol and CO_2.

Another important set of observations came from the use of substances later identified as inhibitors of specific enzymes of glycolysis. Sodium fluoride added to fermenting yeast extracts was found to produce an accumulation of two phosphate esters, 3-phosphoglycerate and 2-phosphoglycerate. This effect was later found to be caused by the inhibition of *enolase*, one of the enzymes of fermentation. On the other hand, the inhibitor iodoacetic acid (ICH_2COOH) caused an accumulation of both fructose 1,6-diphosphate and the triose phosphates. Once these intermediates were identified, it became possible to deduce and arrange in the proper sequence the several enzymatic reactions by which they were formed and utilized. Among the most important contributors to the elucidation of glycolysis in the 1930s were the German biochemists G. Embden and O. Meyerhof; the glycolytic sequence is often called the *Embden-Meyerhof pathway*.

The Stages of Glycolysis

The first stage of glycolysis (Figure 11-2) serves as a preparatory or collection stage, which a number of different hexoses may enter following their phosphorylation at the expense of ATP. They are all converted into fructose 1,6-diphosphate, which is then cleaved to yield glyceraldehyde 3-phosphate.

The second stage of glycolysis is the final common pathway for all sugars. In it occur the oxidoreduction steps and the energy-conserving mechanisms by which ADP is phosphorylated to ATP.

Three different types of chemical transformation take place during glycolysis; their pathways are interconnected: (1) the sequence of reactions by which the carbon skeleton of glucose is degraded to form the lactic acid end product, that is, the pathway of the carbon atoms; (2) the sequence of reactions by which inorganic phosphate becomes the terminal phosphate group of ATP (the pathway of phosphate); and (3) the sequence of oxidoreductions (the pathway of hydrogen atoms or electrons). We shall recapitulate these steps later.

The Enzymatic Steps in the First Stage of Glycolysis

Phosphorylation of Glucose

In the first step of the first stage of glycolysis, the D-glucose molecule is made ready or "primed" for the subsequent steps by its phosphorylation at the 6-position to *glucose 6-phosphate* (Figure 11-3) at the expense of ATP. This reaction, which is irreversible under intracellular conditions, is catalyzed by the broadly distributed enzyme *hexokinase*

$$\text{ATP} + \alpha\text{-D-glucose} \xrightarrow{\text{Mg}^{2+}}$$

$$\text{ADP} + \alpha\text{-D-glucose 6-phosphate} \qquad \Delta G^{o\prime} = -4.0 \text{ kcal}$$

Hexokinase catalyzes the phosphorylation not only of D-glucose, but also of certain other hexoses, such as D-fructose, D-mannose, and D-glucosamine; it has a higher affinity for aldohexoses than for ketohexoses. The hexokinase of yeast has been crystallized (mol wt 96,000). Hexokinases of all cells require Mg^{2+} (or Mn^{2+}), which first combines with ATP to form the true substrate of the

Figure 11-2
Stages of glycolysis.

Stage 1
Collection of
simple sugars
and conversion
to the common
product glyceralde-
hyde 3-phosphate

Stage 2
Conversion of
glyceraldehyde
3-phosphate
to lactate and
coupled formation
of ATP

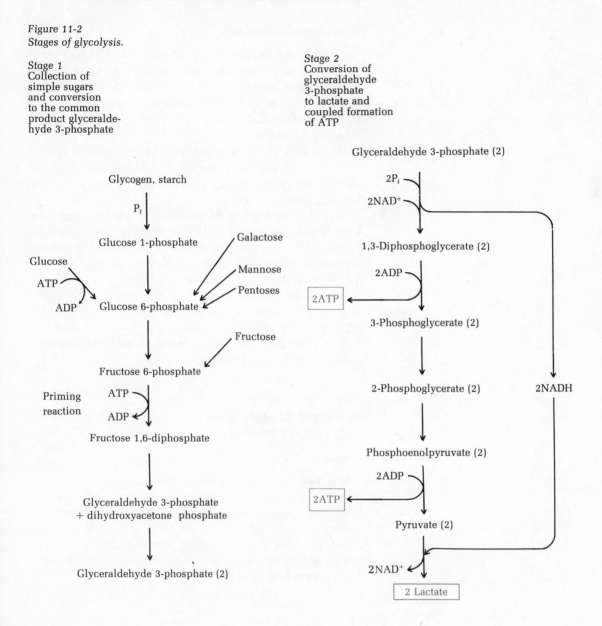

enzyme, the $MgATP^{2-}$ complex. Hexokinase is inhibited
by certain sulfhydryl reagents, especially arsenicals. It
is also strongly inhibited by its own product, glucose
6-phosphate. In fact, hexokinase is a regulatory enzyme
which sets the rate at which free glucose enters the
glycolytic scheme. Whenever an excess of glucose 6-phos-
phate accumulates in the cell, it temporarily inhibits hexokinase until the excess glucose 6-phosphate is utilized.

Figure 11-3
α-D-Glucose 6-phosphate.

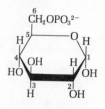

In the liver there is a second enzyme capable of catalyzing the phosphorylation of glucose, namely, _glucokinase._ This enzyme differs from hexokinase in being specific for glucose. It has a lower affinity for glucose than hexokinase and only comes into play when the glucose concentration in the blood becomes excessively high, particularly in the disease _diabetes mellitus,_ in which there is a defect in the utilization of blood glucose.

Conversion of Glucose 6-Phosphate to Fructose 6-Phosphate

Figure 11-4
α-D-Fructose 6-phosphate.

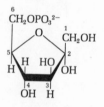

Phosphoglucoisomerase, which has been isolated in highly purified form from muscle tissue, catalyzes the isomerization of glucose 6-phosphate to fructose 6-phosphate (Figure 11-4)

α-D-Glucose 6-phosphate $\rightleftharpoons$
      α-D-fructose 6-phosphate $\Delta G^{\circ\prime} = +0.4$ kcal

The reaction proceeds readily in either direction, as is shown by the relatively small standard free energy change. Phosphoglucoisomerase requires Mg^{2+} or Mn^{2+} and is specific for glucose 6-phosphate and fructose 6-phosphate.

Phosphorylation of Fructose 6-Phosphate to Fructose 1,6-Diphosphate

Figure 11-5
Fructose 1,6-diphosphate.

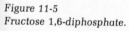
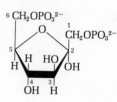

In this, the second of the two "priming" reactions of glycolysis, a second molecule of ATP is invested to phosphorylate D-fructose 6-phosphate in the 1 position to yield fructose 1,6-diphosphate (Figure 11-5) by action of _phosphofructokinase_

ATP + D-fructose 6-phosphate $\longrightarrow$
    ADP + D-fructose 1,6-diphosphate $\Delta G^{\circ\prime} = -3.40$ kcal

Mg^{2+} is required, presumably because the true substrate is $MgATP^{2-}$. The strongly negative $\Delta G^{\circ\prime}$ value for the phosphofructokinase reaction indicates that it is essentially irreversible in the cell.

The phosphorylation of fructose 6-phosphate is an important control point in the glycolytic sequence; phosphofructokinase is an allosteric or regulatory enzyme

(Chapter 4). Like most allosteric enzymes, it has a rather high molecular weight (360,000). The positive or stimulating modulators for phosphofructokinase activity are ADP or AMP, as well as the substrate itself, fructose 6-phosphate. The most important negative modulator is ATP (Figure 11-6). Thus the activity of phosphofructokinase is accelerated whenever the cell's supply of ATP is depleted and ADP predominates, or whenever there is an excess of fructose 6-phosphate. On the other hand, the enzyme is inhibited whenever the cell has ample ATP. In many cell types the phosphofructokinase reaction is the rate-limiting step of glycolysis. The regulation of sugar degradation will be summarized further below.

Cleavage of Fructose 1,6-Diphosphate

This reaction is catalyzed by the enzyme _aldolase,_ which is easily isolated in crystalline form from rabbit muscle extracts. The reaction catalyzed is a reversible aldol condensation which yields two triose phosphates, glyceraldehyde 3-phosphate and dihydroxyacetone phosphate

D-Fructose 1,6-diphosphate $\rightleftharpoons$
dihydroxyacetone phosphate + D-glyceraldehyde 3-phosphate
$$\Delta G^{\circ\prime} = +5.73 \text{ kcal}$$

The structures of the products are shown in Figure 11-7. Although this reaction has a strongly positive $\Delta G^{\circ\prime}$, at physiological concentrations a large fraction of fructose 1,6-diphosphate is cleaved. Moreover, the reaction products are quickly removed, thus pulling the reaction farther to the right.

Skeletal muscle aldolase has a molecular weight of 150,000 and contains four major subunits. It cleaves not only fructose 1,6-diphosphate, but also a number of different ketose 1-phosphates. The aldolases found in bacteria, yeasts, and fungi are different from those in animal tissues and higher plants; they require specific divalent metal ions, usually Zn^{2+}, Ca^{2+}, or Fe^{2+}, as well as K^+. Their molecular weight is about 65,000, or one-half that of the animal aldolases. Aldolases of animal tissues require no metal ion as cofactor.

Interconversion of the Triose Phosphates

Only one of the two triose phosphates, namely, _glyceraldehyde 3-phosphate,_ can be directly degraded in

Figure 11-6
Regulation of phosphofructokinase activity by positive and negative allosteric modulators. Inhibition by negative allosteric modulators is shown by the bar notation described in Figure 9-6. Stimulation by a positive allosteric modulator is shown by a colored arrow parallel to the reaction arrow.

ATP + fructose 6-phosphate (F6P)

Positive modulators
ADP
AMP
F6P

Negative modulators
ATP, citrate

ADP + fructose 1,6-diphosphate

Figure 11-7
The triose phosphates. The carbon atoms are numbered to correspond to those of fructose 1,6-diphosphate.

Glyceraldehyde 3-phosphate

$(6)\ CH_2OPO_3^{2-}$
|
$(5)\ H - C - OH$
|
$(4)\ CH$
‖
O

Dihydroxyacetone phosphate

$(3)\ CH_2OH$
|
$(2)\ C = O$
|
$(1)\ CH_2OPO_3^{2-}$

the further reactions of glycolysis. The other, *dihydroxy-acetone phosphate*, is reversibly converted to glyceraldehyde 3-phosphate by the enzyme *triose phosphate isomerase*

Dihydroxyacetone phosphate $\rightleftharpoons$
  D-glyceraldehyde 3-phosphate $\Delta G^{\circ\prime} = +1.83$ kcal

Note that by this reaction carbon atoms 1, 2, and 3 of the starting glucose now become indistinguishable from carbon atoms 6, 5, and 4, respectively (Figure 11-8). Dihydroxyacetone phosphate constitutes over 95 percent of the equilibrium mixture of the two triose phosphates.

This reaction completes the first stage of glycolysis, in which the glucose molecule has been prepared for the second stage by two phosphorylation steps and cleavage to two molecules of the three-carbon glyceraldehyde 3-phosphate.

The Second Stage of Glycolysis

The second stage of glycolysis (see Figure 11-2) includes the oxidoreductions and the phosphorylation steps in which ATP is generated. Since one molecule of glucose forms two of glyceraldehyde 3-phosphate, both halves of the glucose molecule follow the same pathway.

Oxidation of Glyceraldehyde 3-Phosphate
to 1,3-Diphosphoglycerate

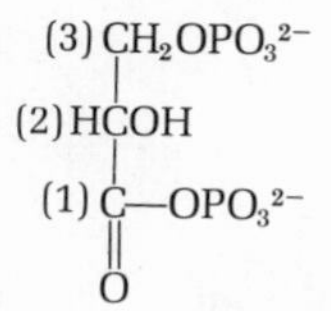

This is one of the most important steps of the glycolytic sequence, since it conserves the energy of oxidation of the aldehyde group of glyceraldehyde 3-phosphate in the form of the high-energy phosphate compound formed as the oxidation product, 1,3-diphosphoglycerate (Figure 11-9).

The enzyme catalyzing this step, *glyceraldehyde 3-phosphate dehydrogenase*, has been isolated in crystalline form from rabbit muscle. It has a molecular weight of 140,000 and contains four identical subunits, each consisting of a single polypeptide chain of some 330 residues, the amino acid sequence of which has been deduced. The overall reaction catalyzed by the enzyme is

D-Glyceraldehyde 3-phosphate + NAD^+ + P_i $\longrightarrow$
    1,3-diphosphoglycerate + NADH + H^+ $\Delta G^{\circ\prime} = +1.5$ kcal

In this reaction the aldehyde group of D-glyceraldehyde

Figure 11-8
The triose phosphate isomerase reaction.

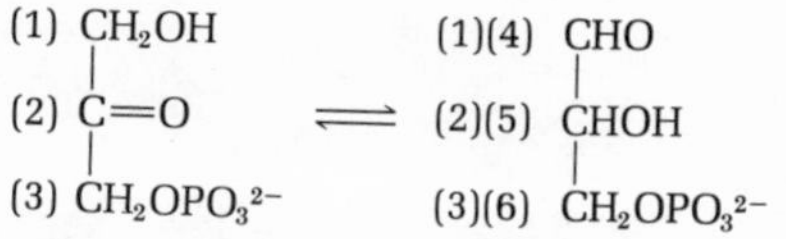

(1) CH_2OH
(2) $C{=}O$ $\rightleftharpoons$ (1)(4) CHO
(3) $CH_2OPO_3{}^{2-}$ (2)(5) CHOH
                        (3)(6) $CH_2OPO_3{}^{2-}$

Dihydroxy-   Glyceraldehyde
acetone   3-phosphate
phosphate

Figure 11-9
Phosphoglycerates.

3-Phosphoglycerate

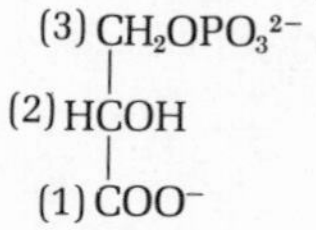

(3) $CH_2OPO_3{}^{2-}$
(2) HCOH
(1) COO^-

,3-Diphosphoglycerate

(3) $CH_2OPO_3{}^{2-}$
(2) HCOH
(1) $C{-}OPO_3{}^{2-}$
      $\|$
      O

3-phosphate is oxidized to the oxidation level of a carboxyl group. However, instead of a free carboxylic acid,
the reaction yields a mixed anhydride of phosphoric acid
and the carboxyl group of 3-phosphoglyceric acid,
namely, 1,3-diphosphoglycerate, which, as we have seen,
is a high-energy phosphate compound having a more negative standard free energy of hydrolysis than ATP
(Chapter 10). The other important component of this
reaction is the coenzyme (Figure 11-10) abbreviated as
NAD$^+$ (nicotinamide adenine dinucleotide, Chapter 8),
which accepts hydrogen atoms from the aldehyde group
of D-glyceraldehyde 3-phosphate to yield its reduced
form NADH. When NAD$^+$ is reduced (Figure 11-10) a
hydride ion (:H$^-$) is transferred from the substrate molecule to the 4-position of the nicotinamide ring, ultimately leading to the reduction of the latter at positions
1 and 4. The other hydrogen atom of the substrate is lost
to the medium as an H$^+$ ion. For this reason the enzymatic reduction of NAD$^+$ is written to include the H$^+$
ion formed

Figure 11-10
Nicotinamide adenine dinucleotide.
The nicotinamide moiety, which is the
portion undergoing reversible
reduction, is shown in color.

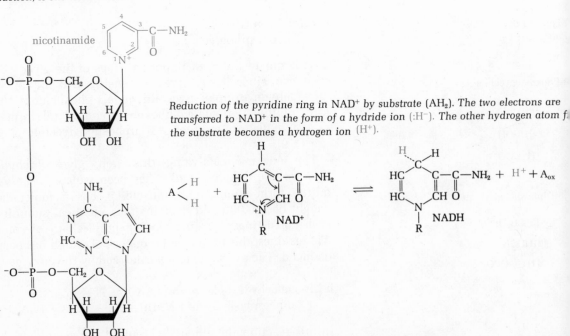

Reduction of the pyridine ring in NAD$^+$ by substrate (AH$_2$). The two electrons are
transferred to NAD$^+$ in the form of a hydride ion (:H$^-$). The other hydrogen atom f.
the substrate becomes a hydrogen ion (H$^+$).

$$\text{Substrate} + \text{NAD}^+ \rightleftharpoons \text{oxidized substrate} + \text{NADH} + \text{H}^+$$

The function of NAD in carrying electrons or hydrogen atoms will be considered in more detail in Chapter 13.

The mechanism of action of glyceraldehyde phosphate dehydrogenase is rather complex. The substrate first combines with an —SH group at the active site of the enzyme. In this form it reacts with a molecule of NAD^+, also bound at the active site. The resulting acyl–enzyme complex then reacts with phosphate to discharge free 1,3-diphosphoglyceric acid. Glyceraldehyde 3-phosphate dehydrogenase is inhibited by iodoacetic acid, which blocks its essential —SH group.

The most important feature of this reaction is that the energy yielded on oxidation of the aldehyde group of glyceraldehyde 3-phosphate is conserved in the form of the high-energy phosphate bond of the product, 1,3-diphosphoglycerate.

Transfer of Phosphate from 1,3-Diphosphoglycerate to ADP

The 1,3-diphosphoglycerate formed in the preceding reaction next reacts enzymatically with ADP, with formation of ATP and 3-phosphoglycerate (Figure 11-9). The transfer of the phosphate group is catalyzed by *phosphoglycerate kinase*

$$\text{1,3-Diphosphoglycerate} + \text{ADP} \rightleftharpoons$$
$$\text{3-phosphoglycerate} + \text{ATP} \qquad \Delta G^{\circ\prime} = -4.50 \text{ kcal}$$

The overall equation for the two reactions, involving oxidation of the aldehyde group of glyceraldehyde 3-phosphate to the carboxyl group of 3-phosphoglycerate, with coupled formation of ATP from ADP and phosphate, is

$$\text{Glyceraldehyde 3-phosphate} + \text{P}_i + \text{ADP} + \text{NAD}^+ \rightleftharpoons$$
$$\text{3-phosphoglycerate} + \text{ATP} + \text{NADH} + \text{H}^+ \qquad \Delta G^{\circ\prime} = -3.0 \text{ kcal}$$

The result of these two reactions is that the energy of oxidation of an aldehyde group to a carboxylate group has been largely conserved as the phosphate-bond energy of ATP.

Conversion of 3-Phosphoglycerate to 2-Phosphoglycerate

This reaction is catalyzed by the enzyme *phosphoglyceromutase*

3-Phosphoglycerate $\rightleftharpoons$

$\qquad$ 2-phosphoglycerate $\qquad$ $\Delta G^{\circ\prime} = +1.06$ kcal

Mg^{2+} is essential for this reaction, which involves transfer of the phosphate group from the 3 to the 2 position of glyceric acid (Figure 11-11). This reaction is freely reversible.

Dehydration of 2-Phosphoglycerate to Phosphoenolpyruvate

This is the second reaction of the glycolytic sequence in which a high-energy phosphate bond is generated; the enzyme _enolase_ catalyzes dehydration of 2-phosphoglycerate to yield the high-energy compound phosphoenolpyruvate (Figure 11-11).

2-Phosphoglycerate $\rightleftharpoons$

$\qquad$ phosphoenolpyruvate $+$ H_2O $\qquad$ $\Delta G^{\circ\prime} = +0.44$ kcal

Enolase has been obtained in pure crystalline form from several sources (mol wt 85,000). It has an absolute requirement for a divalent cation (Mg^{2+} or Mn^{2+}), which makes a complex with the enzyme before the substrate is bound. Despite the relatively small standard free energy change in this reaction, there is a very large difference in the standard free energy of hydrolysis of the phosphate group of the reactant and product, that of 2-phosphoglycerate (a low-energy phosphate) being about -4.2 kcal and that of phosphoenolpyruvate (a high-energy phosphate) about -14.8 kcal (Chapter 10). Evidently there is a large change in the distribution of energy within the 2-phosphoglycerate molecule when it is dehydrated to phosphoenolpyruvate. Enolase is inhibited by fluoride salts.

Transfer of Phosphate from Phosphoenolpyruvate to ADP

The transfer of the high-energy phosphate group from phosphoenolpyruvate to ADP (Figure 11-11) is catalyzed by the enzyme _pyruvate kinase_

Phosphoenolpyruvate $+$ ADP $\rightleftharpoons$

$\qquad$ pyruvate $+$ ATP $\qquad$ $\Delta G^{\circ\prime} = -7.5$ kcal

which has been obtained in pure crystalline form (mol wt 250,000). The enzyme requires K^+ and either Mg^{2+} or

Figure 11-11
Intermediates in the second stage of glycolysis.

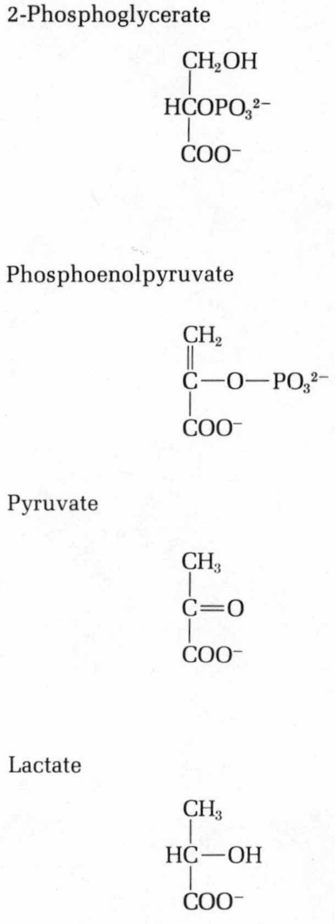

2-Phosphoglycerate

$$CH_2OH$$
$$|$$
$$HCOPO_3{}^{2-}$$
$$|$$
$$COO^-$$

Phosphoenolpyruvate

$$CH_2$$
$$\|$$
$$C\!-\!O\!-\!PO_3{}^{2-}$$
$$|$$
$$COO^-$$

Pyruvate

$$CH_3$$
$$|$$
$$C\!=\!O$$
$$|$$
$$COO^-$$

Lactate

$$CH_3$$
$$|$$
$$HC\!-\!OH$$
$$|$$
$$COO^-$$

Mn^{2+}. This reaction is essentially irreversible under intracellular conditions, as is suggested by its large negative $\Delta G^{\circ\prime}$ value.

Reduction of Pyruvate to Lactate

In the last step of glycolysis, pyruvate is reduced to lactate (Figure 11-11) at the expense of electrons originally donated by glyceraldehyde 3-phosphate to NAD^+ (see above). These electrons are carried to this last step in the form of NADH (Figure 11-10). This reaction is catalyzed by _lactate dehydrogenase_, which has been isolated in crystalline form

$$Pyruvate + NADH + H^+ \rightleftharpoons$$
$$lactate + NAD^+ \qquad \Delta G^{\circ\prime} = -6.0 \text{ kcal}$$

The overall equation of this reaction is far to the right, as shown by the large negative value of $\Delta G^{\circ\prime}$.

We have seen that lactate dehydrogenase exists in at least five different molecular forms or isozymes in the tissues of higher animals. Heart, liver, and kidney tissue are especially rich in the H-type isozymes (Chapter 4), which have a relatively low affinity for pyruvate, whereas skeletal muscle is rich in the M-type isozymes, which have a high affinity for pyruvate and tend to favor formation of lactic acid.

The Overall Balance Sheet

A balance sheet for glycolysis can now be constructed to account for (1) the fate of the carbon skeleton of glucose, (2) the oxidoreduction reactions, and (3) the input and output of phosphate, ADP, and ATP. The left-hand part of the following equation shows all the inputs (consult Figure 11-2), adjusted for the fact that each molecule of glucose yields two molecules of glyceraldehyde 3-phosphate:

$$Glucose + 2ATP + 2NAD^+ + 2P_i + 4ADP$$
$$+ 2NADH + 2H^+ \longrightarrow 2 \text{ lactate} + 2ADP$$
$$+ 2NADH + 2H^+ + 2NAD^+ + 4ATP + 2H_2O$$

By canceling out common terms on both sides of the equation we get

$$Glucose + 2P_i + 2ADP \longrightarrow 2 \text{ lactate} + 2ATP + 2H_2O$$

In the overall process D-glucose is converted to two molecules of lactate (the pathway of carbon); two molecules of ADP and phosphate are converted to ATP (the pathway of phosphate); and four electrons have been transferred from two molecules of glyceraldehyde 3-phosphate to two of pyruvate via two molecules of NAD$^+$ (the pathway of electrons). Although two oxidoreduction steps have taken place in the sequence, there is no *net* change in oxidation–reduction state.

Alcoholic Fermentation

In organisms like brewer's yeast, which ferment glucose to ethanol and CO$_2$ rather than to lactic acid, the fermentation pathway is identical to that described for glycolysis except for the terminal step catalyzed by lactate dehydrogenase. This is replaced by two other enzymatic steps (Figure 11-12), catalyzed by *pyruvate decarboxylase* and *alcohol dehydrogenase*. In the first, pyruvate resulting from the breakdown of glucose is decarboxylated by *pyruvate decarboxylase* to acetaldehyde and CO$_2$

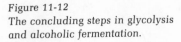

Figure 11-12
The concluding steps in glycolysis and alcoholic fermentation.

$$\text{Pyruvate} \longrightarrow \text{acetaldehyde} + CO_2$$

The decarboxylation of pyruvate to form acetaldehyde and CO$_2$ is essentially irreversible. Pyruvate decarboxylase requires Mg^{2+} and has a tightly bound coenzyme, *thiamine pyrophosphate* (Chapter 8), whose function as a carrier of the acetaldehyde group will be discussed in Chapter 12.

In the final step of alcoholic fermentation, acetaldehyde is reduced to ethanol, with NADH furnishing the reducing power, through the action of the enzyme *alcohol dehydrogenase*.

$$\text{Acetaldehyde} + \text{NADH} + H^+ \rightleftharpoons \text{ethanol} + \text{NAD}^+$$

Ethanol and CO$_2$, instead of lactate, are thus the end products of alcoholic fermentation. The overall equation of alcoholic fermentation can therefore be written

$$\text{Glucose} + 2P_i + 2\text{ADP} \longrightarrow 2\text{ ethanol} + 2CO_2 + 2\text{ATP} + 2H_2O$$

"Feeder" Pathways Leading to the Glycolytic Sequence

In addition to D-glucose, many other carbohydrates may ultimately enter the glycolytic sequence. The most impor-

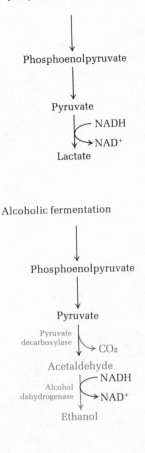

tant are the storage polysaccharides glycogen and starch, the disaccharides maltose, lactose, and sucrose, and the monosaccharides fructose, mannose, and galactose. We shall now consider the pathways by which these sugars may enter the glycolytic sequence.

Glycogen and Starch

The D-glucose units of glycogen and starch gain entrance into the glycolytic pathway through the sequential action of two enzymes, _glycogen phosphorylase_ (or _starch phosphorylase_ in plants) and _phosphoglucomutase_. Glycogen phosphorylase and starch phosphorylase are closely related members of a class of enzymes designated as $\alpha(1 \rightarrow 4)$ _glucan phosphorylases_. Widely distributed in animal, plant, and microbial cells, they catalyze the general reaction below, in which (glucose)$_n$ designates a glucan chain of n D-glucose residues in $\alpha(1 \rightarrow 4)$ linkage and (glucose)$_{n-1}$ designates the glucan chain shortened by one glucose residue (see Chapter 5 for the structure of glycogen and starch).

$$(\text{Glucose})_n + HPO_4^{2-} \rightleftharpoons$$
$$\text{(glucose)}_{n-1} + \text{glucose 1-phosphate} \qquad \Delta G^{\circ\prime} = +0.73 \text{ kcal}$$

Although this reaction is reversible, under intracellular conditions it proceeds only in the direction of degradation to yield glucose 1-phosphate. In this reaction the terminal $\alpha(1 \rightarrow 4)$ glycosidic linkage at the nonreducing end of the glucan chain undergoes _phosphorolysis_, the removal of the terminal glucose residue by attack of phosphate to yield _glucose 1-phosphate_. Left behind is a chain with one less glucose unit (Figure 11-13).

Glycogen phosphorylase acts repetitively on the nonreducing ends of glycogen branches until it meets the $\alpha(1 \rightarrow 6)$ branch points, which it cannot attack. Further degradation can occur only after the action of a "debranching" enzyme, $\alpha(1 \rightarrow 6)$ _glucosidase_, which hydrolyzes the $1 \rightarrow 6$ linkage at the branch point, thus making available another length of the polysaccharide chain to the action of glycogen phosphorylase.

Phosphorylase is situated at a strategic point between the fuel reservoir glycogen and the enzymatic system for utilizing the fuel. Its activity in muscle and liver has been found to be under regulation by an elaborate set of controls. The glycogen phosphorylase of skeletal muscle occurs in two forms, an active phosphorylated form

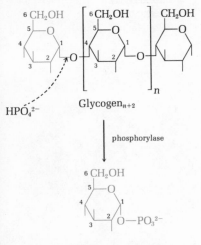

Figure 11-13

Removal of a glucose residue from the nonreducing end of a glycogen chain by phosphorylase.

(phosphorylase a) and a much less active dephosphorylated form (phosphorylase b). Phosphorylase a has been crystallized; it has a molecular weight of 380,000 and is a tetramer, that is, it consists of four identical subunits. Phosphorylase a is itself a regulatory or allosteric enzyme. It is stimulated by the positive modulators AMP and ADP and is inhibited by ATP (Figure 11-14). Thus, whenever the ratio of ATP to ADP in the cell is high, the breakdown of the fuel reservoir glycogen undergoes feedback inhibition.

The activity of phosphorylase is subject to a second set of controls, imposed by the action of certain hormones (Chapter 20). In this second level of regulation the ratio of the active (phosphorylase a) and inactive (phosphorylase b) forms of the enzyme is under control. The active form of phosphorylase a can be attacked by another enzyme, _phosphorylase a phosphatase_, which hydrolyzes specific phosphate groups present in the phosphorylase a molecule (Figure 11-15). This reaction converts phosphorylase a into the dephosphorylated, relatively inactive, phosphorylase b, which exists largely as a dimer, that is, a form having two subunits.

Phosphorylase b is converted back to active phosphorylase a molecules not by simple reversal of the above reaction, which is irreversible, but by an alternate pathway in which four molecules of ATP phosphorylate two molecules of phosphorylase b in the presence of the enzyme _phosphorylase b kinase_ to yield one molecule of phosphorylase a. Thus, by the action of phosphorylase a phosphatase and phosphorylase b kinase the ratio of the active phosphorylase a and the less active phosphorylase b in the cell may be varied, thus varying the rate of breakdown of glycogen into glucose 1-phosphate. The activity of phosphorylase b kinase can itself be controlled, since it in turn occurs in active and inactive forms (see Chapter 20).

Liver also contains active and inactive forms of glycogen phosphorylase. However, in the liver these enzymes have a distinctly different structure than in muscle and they are controlled in a different fashion, as we shall see in Chapter 20.

Glucose 1-phosphate, the end product of the glycogen phosphorylase and starch phosphorylase reactions, is converted into glucose 6-phosphate by phosphoglucomutase, which has been obtained in pure form from many sources. It catalyzes the reaction

$$\text{Glucose 1-phosphate} \rightleftharpoons \text{glucose 6-phosphate}$$

Figure 11-14
Allosteric regulation of the activity of phosphorylase a in glycogen breakdown. The allosteric inhibition of phosphorylase a by ATP is shown in the notation described in Figure 9-6. Allosteric stimulation by AMP and ADP is designated by the parallel colored arrow.

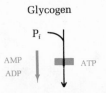

Glycogen

Glucose 1-phosphate

Figure 11-15
Conversion of phosphorylase a to phosphorylase b by phosphorylase a phosphatase and reactivation of phosphorylase b by phosphorylase b kinase.

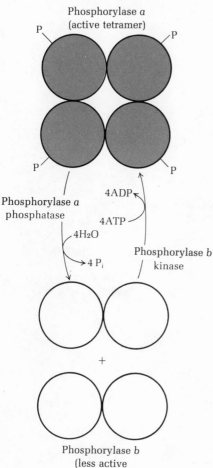

Phosphorylase *a*
(active tetramer)

Phosphorylase *a*
phosphatase

4ADP
4ATP
4H$_2$O
4 P$_i$

Phosphorylase *b*
kinase

+

Phosphorylase *b*
(less active
dimers)

Phosphoglucomutase requires a cofactor, *glucose 1,6-diphosphate*, whose role is indicated by the following sequence of intermediate steps in the action of the enzyme, which cycles between two forms, a phosphorylated and a nonphosphorylated or dephospho form:

Phosphoenzyme + glucose 1-phosphate $\rightleftharpoons$

dephosphoenzyme + glucose 1,6-diphosphate

Dephosphoenzyme + glucose 1,6-diphosphate $\rightleftharpoons$

phosphoenzyme + glucose 6-phosphate

Sum: Glucose 1-phosphate $\rightleftharpoons$ glucose 6-phosphate

Phosphoglucomutase is noteworthy from another point of view. It is one of a group of enzymes containing an essential serine residue at its active site. When phosphoglucomutase becomes phosphorylated by reaction with glucose 1,6-diphosphate, it is this serine hydroxyl group of the enzyme which becomes phosphorylated. The enzymes of the "serine" class are irreversibly inhibited by certain organic phosphates, such as *diisopropyl fluorophosphate* (Chapter 4), which forms an ester with the essential serine hydroxyl group of such enzymes, thus inactivating them.

*Entry of Disaccharides into
the Glycolytic Sequence*

In higher animals the common disaccharides maltose, lactose, and sucrose, do not occur in the blood or tissues. When they are ingested they are enzymatically hydrolyzed in the small intestine to yield their hexose components prior to absorption into the bloodstream

$$\text{Maltose} + \text{H}_2\text{O} \xrightarrow{\text{maltase}} \text{glucose} + \text{glucose}$$

$$\text{Lactose} + \text{H}_2\text{O} \xrightarrow{\text{lactase}} \text{galactose} + \text{glucose}$$

$$\text{Sucrose} + \text{H}_2\text{O} \xrightarrow{\text{invertase}} \text{fructose} + \text{glucose}$$

Like glucose, both galactose and fructose are then phosphorylated, largely in the liver. They are then ultimately converted into one or another intermediate of the glycolytic sequence by pathways now to be described.

Entry of Monosaccharides

D-Fructose and D-mannose may be phosphorylated at the 6-position by hexokinase, which can phosphorylate a number of hexoses.

$$\text{D-Fructose} + \text{ATP} \rightleftharpoons \text{D-fructose 6-phosphate} + \text{ADP}$$

$$\text{D-Mannose} + \text{ATP} \rightleftharpoons \text{D-mannose 6-phosphate} + \text{ADP}$$

D-Mannose 6-phosphate is then isomerized to yield D-fructose 6-phosphate by the action of *phosphomanno-isomerase*

$$\text{D-Mannose 6-phosphate} \rightleftharpoons \text{D-fructose 6-phosphate}$$

In the liver of vertebrates, fructose gains entry into glycolysis by another pathway. The enzyme *fructokinase* catalyzes the phosphorylation of fructose, not at carbon atom 6, but at carbon atom 1 (Figure 11-16)

$$\text{D-Fructose} + \text{ATP} \rightleftharpoons \text{D-fructose 1-phosphate} + \text{ADP}$$

The resulting fructose 1-phosphate is then cleaved into D-glyceraldehyde and dihydroxyacetone phosphate by *fructose 1-phosphate aldolase*, a liver enzyme which resembles aldolase in its action but has a different substrate specificity

$$\text{Fructose 1-phosphate} \rightleftharpoons$$
$$\text{D-glyceraldehyde} + \text{dihydroxyacetone phosphate}$$

One product, dihydroxyacetone phosphate, is an intermediate of glycolysis, and the other, D-glyceraldehyde, is phosphorylated to D-glyceraldehyde 3-phosphate, also an intermediate in glycolysis, by the reaction

$$\text{D-Glyceraldehyde} + \text{ATP} \longrightarrow$$
$$\text{glyceraldehyde 3-phosphate} + \text{ADP}$$

D-Galactose, derived from the D-lactose of milk, is first phosphorylated at the 1-position at the expense of ATP by the enzyme *galactokinase*

$$\text{ATP} + \text{D-galactose} \longrightarrow \text{ADP} + \text{D-galactose 1-phosphate}$$

The resulting product, D-galactose 1-phosphate, is then

Figure 11-16
Fructose 1-phosphate

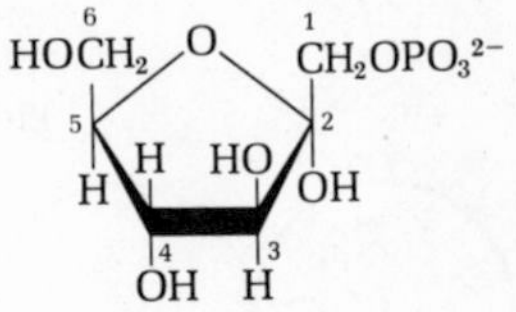

Figure 11-17 (right)
Epimerization of galactose 1-phosphate to glucose 1-phosphate.

$$UTP + \text{Galactose 1-phosphate} \rightleftharpoons \text{UDP-galactose} + PP_i$$

$$\text{UDP-Galactose} \overset{NAD}{\rightleftharpoons} \text{UDP-glucose}$$

$$\text{UDP-Glucose} + PP_i \rightleftharpoons UTP + \text{glucose 1-phosphate}$$

converted into its epimer at carbon atom 4, namely, D-glucose 1-phosphate, by a complex pathway shown in Figure 11-17. In these reactions uridine triphosphate (Chapter 17) plays an important role in the transfer of glycosyl groups. We shall later see other instances of the role of uridine diphosphate sugars as intermediates.

The Regulation of Glycolysis

Figure 11-18 summarizes the regulation of the glycolytic degradation of important carbohydrate fuels in animal cells. Two important points for entry of glucosyl residues into the glycolytic pathway are controlled by regulatory enzymes. One is the entry of free glucose via its phosphorylation by the regulatory enzyme hexokinase and the other is the entry of glucosyl residues from glycogen controlled by the regulatory enzyme glycogen phosphorylase. Once the glucosyl residues have entered the glycolytic scheme, there is another important control point, catalyzed by the regulatory enzyme phosphofructokinase. Each of these three control points is subject to feedback inhibition by an immediate or ultimate end product. Of overriding importance is control of glycolysis by the ADP/ATP ratio. Whenever this ratio is high, the cell's supply of ATP is low; glycolysis is then stimulated to a high rate so as to form more ATP from ADP. On the other hand, whenever the ADP/ATP ratio is low, the cell has ample ATP, which inhibits phosphofructokinase and thus inhibits further ATP formation. Hormonal regulation of this pathway is discussed in Chapter 20.

Figure 11-18
Allosteric regulation of glycolysis. Inhibition by negative modulators is indicated by ▬ *; stimulation by positive modulators by parallel arrows in color.*

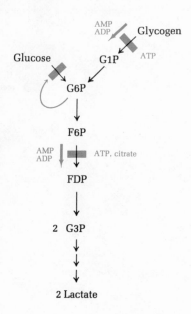

Summary

Anaerobic fermentation is the most primitive pathway for recovery of chemical energy from fuels such as glucose. In facultative cells it is an obligatory first stage in glucose catabolism, which is followed by aerobic oxidation of the fermentation products. The two most common types of fermentation are glycolysis and alcoholic fermentation. Both utilize identical energy-conserving mechanisms, and differ only in their terminal steps. The overall equation for glycolysis is

$$\text{glucose} + 2\text{ADP} + 2\text{P}_i \longrightarrow 2 \text{ lactic acid} + 2\text{ATP} + 2\text{H}_2\text{O}$$

and for alcoholic fermentation is

$$\text{glucose} + 2\text{ADP} + 2\text{P}_i \longrightarrow 2 \text{ ethanol} + 2\text{CO}_2 + 2\text{ATP} + 2\text{H}_2\text{O}$$

Glycolysis is catalyzed by eleven enzymes, acting in sequence. It takes place in two stages. In the first, D-glucose is enzymatically phosphorylated by ATP and ultimately cleaved to yield two molecules of D-glyceraldehyde 3-phosphate. Other hexoses, pentoses, and glycerol are also collected and ultimately converted into glyceraldehyde 3-phosphate following their phosphorylation.

In the second stage of glycolysis, the glyceraldehyde 3-phosphate is oxidized by NAD^+ with uptake of inorganic phosphate to form 1,3-diphosphoglycerate. The latter donates its high-energy phosphate group to ADP to yield ATP and 3-phosphoglycerate, which is then isomerized to 2-phosphoglycerate. After dehydration of the latter by enolase, the phosphoenolpyruvate formed donates its high-energy phosphate group to ADP. The free pyruvate formed is reduced to lactate by NADH from the triose phosphate dehydrogenation. Two molecules of ATP enter the first stage of glycolysis, and four are formed from ADP in the second stage, giving a net yield of two ATP from one molecule of glucose. In alcoholic fermentation, the reaction sequence is identical, but instead of being reduced to lactate, pyruvate is decarboxylated to acetaldehyde, which is in turn reduced to ethanol. Phosphofructokinase, a regulatory enzyme, is rate-limiting for glycolysis. The entry of glucose residues of glycogen and starch into glycolysis is made possible by glycogen or starch phosphorylase and phosphoglucomutase. Glycogen phosphorylase is a regulatory enzyme existing in active (phosphorylase a) and less active (phosphorylase b) forms.

References

DICKENS, F., P. J. RANDLE, and W. J. WHELAN: *Carbohydrate Metabolism and its Disorders*, 2 vols., Academic Press, New York, 1968. A series of reviews on various aspects of carbohydrate metabolism and its regulation.

FLORKIN, M. and E. STOTZ, (eds.): *Carbohydrate Metabolism*, Vol. 17 of *Comprehensive Biochemistry*, American Elsevier Publishing Co., New York, 1967. Comprehensive, detailed articles and reviews.

LEHNINGER, A. L.: *Biochemistry*, Worth Publishers, New York, 1970. Chapter 15 provides more detailed textbook treatment.

Problems

1. Write overall equations for the following, including all phosphorylation steps.

   (a) Conversion of D-mannose to phosphoenolpyruvate

   (b) Conversion of D-galactose to ethanol and CO_2

2. Write series of equations for the conversion of (a) D-fructose, (b) D-galactose, and (c) D-mannose into blood D-glucose in the liver.

3. Calculate the percentage of 3-phosphoglycerate in the equilibrium mixture of the phosphoglyceromutase reaction.

4. Write a series of equations for the quantitative conversion of lactose into D-glyceraldehyde 3-phosphate.

Aerobic cells obtain most of their energy from respiration, the process by which electrons are transferred from organic fuel molecules to molecular oxygen. Respiration is far more complex than glycolysis; it has been said that respiration is to glycolysis what a modern jet turbine is to a one-cylinder reciprocating engine.

In this chapter we shall begin discussion of respiration by examining the Krebs tricarboxylic acid cycle, the final common pathway into which the major fuel molecules of the cell—carbohydrates, fatty acids, and amino acids—ultimately converge during catabolism. We shall also examine the glyoxylate cycle, a variant of the tricarboxylic acid cycle, as well as the phosphogluconate pathway, a secondary pathway for oxidation of glucose which generates reducing power for biosynthetic reactions and also yields D-ribose.

The Energetics of Fermentation and Respiration

The breakdown of glucose to lactic acid during glycolysis releases only a very small fraction of the chemical energy potentially available in the structure of the glucose molecule. Much more energy is released when the glucose molecule is oxidized completely to CO_2 and H_2O, as is shown by comparing the standard free energy changes of these processes

$$\text{Glucose} \longrightarrow \text{2 lactate} \qquad \Delta G^{\circ\prime} = -47.0 \ \text{kcal}$$

$$\text{Glucose} + 6O_2 \longrightarrow 6CO_2 + 6H_2O \qquad \Delta G^{\circ\prime} = -686.0 \ \text{kcal}$$

When cells break down glucose anaerobically the lactic acid formed, which cannot be utilized further and thus leaves the cell, still contains most of the energy of the original glucose molecule. Lactic acid is almost as complex a molecule as glucose and it has the same oxidation state, that is, it retains the same number of hydrogen atoms per carbon as glucose, on the average. On the other hand, carbon dioxide, the product of respiration, is a much simpler and smaller molecule than glucose and its carbon atoms are fully oxidized; it contains only a small fraction of the total energy of the original glucose.

In aerobic cells of animal tissues, the glycolytic breakdown of glucose is a preliminary stage which prepares the fuel for the oxidative process of respiration which follows.

The Flow Sheet of Respiration

As is shown in Figure 12-1, acetyl groups derived from carbohydrates, lipids, and amino acids in stage II of catabolism enter stage III, which consists of the multienzyme system catalyzing the tricarboxylic acid cycle. It is the purpose of this metabolic cycle to dehydrogenate the acetic acid fuel to form CO_2 and hydrogen atoms. The CO_2 leaves the cell as a stable end product of respiration and the hydrogen atoms, or their equivalent electrons, are then fed into a chain of electron carriers. In the ensuing process of *electron transport* (Chapter 13) the electrons pass to molecular oxygen. Large amounts of energy are released during electron transport, much of which is conserved as ATP by the process called *oxidative phosphorylation* (Chapter 13).

The overall reaction catalyzed by the tricarboxylic acid cycle is

$$CH_3COOH + 2H_2O \longrightarrow 2CO_2 + 8H$$

This equation and Figure 12-1 indicate that neither molecular oxygen, inorganic phosphate, nor ATP participate directly in the tricarboxylic cycle.

In each turn of the tricarboxylic acid cycle (Figures 12-1 and 12-2) one molecule of acetic acid, which has two carbon atoms, condenses in the form of acetyl CoA with one molecule of the four-carbon compound oxaloacetic acid to form the six-carbon compound citric acid. The latter then is degraded by a sequence of reactions in such a way as to lose two molecules of CO_2 and to regenerate one molecule of the four-carbon oxaloacetic acid.

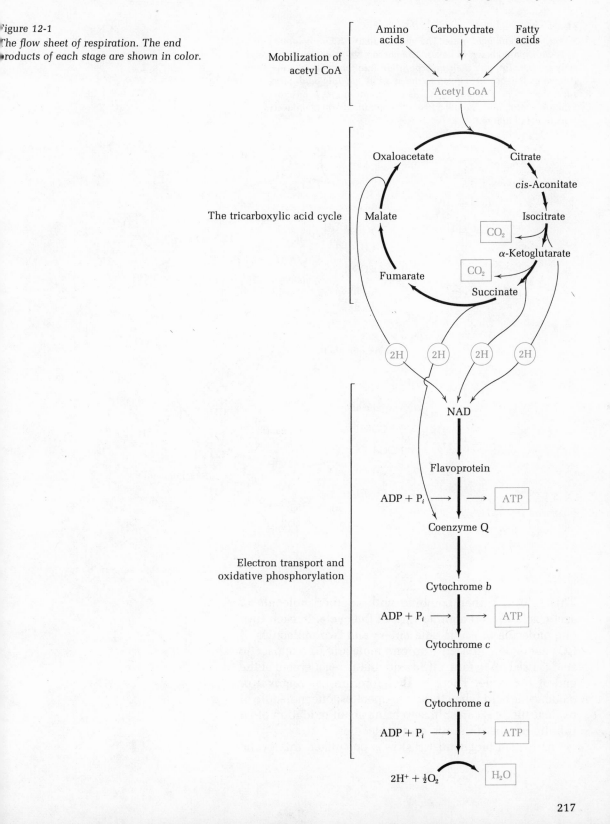

Figure 12-1
The flow sheet of respiration. The end
products of each stage are shown in color.

Figure 12-2
The tricarboxylic acid cycle. The end products (2CO₂ plus four pairs of H atoms) are shown in boxes. The carbon atoms entering as acetyl CoA are in color, and their incorporation into the cycle intermediates (shown as free acids) is given to the stage of succinic acid. Since succinic acid is symmetrical, all its carbon atoms, and those of fumaric, malic, and oxaloacetic acids contain carbon from acetyl CoA in equal amount.

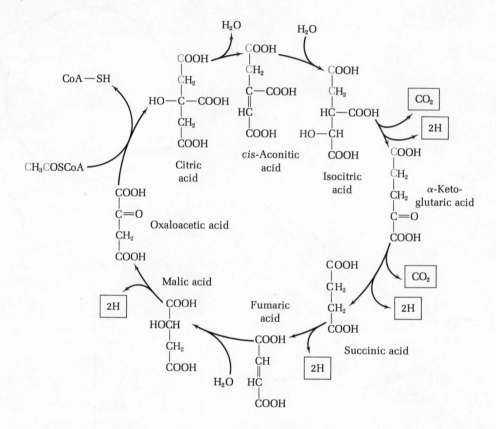

This acid may then combine with another molecule of acetic acid to start another turn of the cycle. In each turn one molecule of acetic acid enters and two molecules of CO_2 come out. In each turn one molecule of oxaloacetic acid is used up to form citric acid, but is regenerated at the end of the cycle. There is, therefore, no net removal of oxaloacetic acid when the cycle operates; one molecule of oxaloacetic acid can suffice to bring about oxidation of an infinite number of acetic acid molecules.

The British biochemist H. Krebs postulated this cyclic

mechanism in 1937 on the basis of his discovery that several specific organic acids known to be present in cells are capable of stimulating in a catalytic manner the consumption of oxygen by tissue suspensions. The compounds having this activity were the six-carbon tricarboxylic acids citric, *cis*-aconitic, and isocitric acids, the five-carbon α-ketoglutaric acid, and the four-carbon dicarboxylic acids succinic, fumaric, malic, and oxaloacetic acids (Figure 12-2). No other organic acids tested showed this activity. Again, we must emphasize that at the pH of the cell, these compounds, as well as most other metabolites, exist largely as anions rather than as free, undissociated acids. Although we shall refer to them in the text either as free acids or as anions (examples: succinic acid or succinate) it will be assumed they occur and function as anions. However, in structural formulas the free acid will sometimes be shown for the sake of simplicity.

Another important observation was that malonic acid, a specific competitive inhibitor of succinate dehydrogenase (Chapter 4), can inhibit the entire oxygen consumption of tissues, regardless of which of the active organic acids was present, indicating that succinate and succinate dehydrogenase must participate in an essential step in a series of reactions involved in respiration. Krebs ultimately established that the various tri- and dicarboxylic acids listed above can be arranged in a metabolic sequence, each step catalyzed by a specific enzyme. Ultimately he demonstrated that this sequence functions in the cyclic manner shown in Figure 12-2, rather than as a linear sequence. From his simple yet elegant experiments and reasoning Krebs postulated the tricarboxylic acid cycle as the main pathway for oxidation of carbohydrate in muscle. Today we know that the tricarboxylic acid cycle is virtually universal in occurrence. It represents the major pathway for oxidation of acetic acid residues in all tissues of higher animals, in most aerobic microorganisms, and in many plant tissues.

The entire set of reactions of the tricarboxylic acid cycle takes place in the mitochondria of higher animal and plant cells; the nuclei, microsomes, and the soluble fraction of the cytoplasm have no activity. The mitochondria not only contain all the enzymes and coenzymes required for the tricarboxylic acid cycle, they also contain all the enzymes required for the last stage of respiration, namely, electron transport and oxidative phosphorylation. For this reason the mitochondria have been called the power plants of the cell.

The Oxidation of Pyruvic Acid to Acetyl CoA

Pyruvic acid formed during glycolysis must first be oxidized, with loss of carbon dioxide, to acetic acid in the form of the acetyl derivative of coenzyme A (Chapter 8) before it can enter the tricarboxylic acid cycle. The overall equation of this reaction is

$$\text{Pyruvate} + NAD^+ + CoA—SH \longrightarrow$$
$$\text{acetyl}—S—CoA + NADH + H^+ + CO_2$$
$$\Delta G^{\circ\prime} = -8.0 \text{ kcal}$$

Because of its large negative standard free energy change, this reaction is essentially irreversible in the intact cell.

The oxidative decarboxylation of pyruvate to acetyl CoA requires three different enzymes (*pyruvate dehydrogenase*, *dihydrolipoyl transacetylase*, and *dihydrolipoyl dehydrogenase*) and five different coenzymes or prosthetic groups (thiamin pyrophosphate, lipoic acid, flavin adenine dinucleotide, coenzyme A, and nicotinamide adenine dinucleotide) organized into a multienzyme complex, the *pyruvate dehydrogenase complex*. Figure 12-3 and 12-4 show how this very complex system functions.

The pyruvate dehydrogenase complex has been isolated in pure form from pig heart mitochondria and from *E. coli*. From the latter source, this enzyme has a particle weight of 4.0 million. It consists of 24 molecules of pyruvate dehydrogenase (mol wt 90,000), each containing one molecule of bound thiamine pyrophosphate, one molecule of dihydrolipoyl transacetylase (mol wt 860,000), containing one molecule of lipoic acid, and 12 molecules of dihydrolipoyl dehydrogenase (mol wt 55,000), each containing one molecule of FAD. The long lipoyllysine side chain of dihydrolipoyl transacetylase has been postulated to serve as a "swinging arm" to transfer acetyl groups and hydrogen atoms (or equivalent electrons) from one enzyme molecule to the next within the pyruvate dehydrogenase complex (Figure 12-4).

The acetyl CoA formed as the end product of the complex pyruvate dehydrogenase system is now ready to enter the tricarboxylic acid cycle.

Figure 12-3
Oxidation of pyruvate to acetyl CoA by the pyruvate dehydrogenase
complex of enzymes. The fate of pyruvate is traced in color.
E_1 = *pyruvate dehydrogenase; TPP = thiamine pyrophosphate;*
E_2 = *dihydrolipoyl transacetylase; E_3 = dihydrolipoyl dehydrogenase.*
Step I in the dehydrogenation of pyruvate to acetyl CoA and CO_2 is
catalyzed by pyruvate dehydrogenase (E_1—TPP). In this reaction the
acetaldehyde resulting from decarboxylation of pyruvate forms a
hydroxyethyl derivative of the thiazole ring of thiamine pyrophos-
phate, which is tightly bound to E_1. In step II the hydroxyethyl group
of the E_1—TPP—CHOH—CH_3 complex is oxidized to an acetyl group
and transferred to the coenzyme lipoic acid (Chapter 8), which is
covalently bound to the second enzyme of the complex, (E_2). Simul-
taneously, a pair of hydrogen atoms from the hydroxyethyl group is
transferred to the disulfide bond of the lipoic acid, which is reduced
to its dithiol form. In step III, the acetyl group is enzymatically trans-
ferred from the lipoyl group to coenzyme A; the acetyl CoA so formed
then leaves the pyruvate dehydrogenase complex in free form. In step
IV, the free dithiol form of dihydrolipoyl transacetylase is reoxidized
to its disulfide form by the third enzyme, dihydrolipoyl dehydrogenase
(E_3—FAD), which contains tightly bound (FAD) as hydrogen acceptor
and becomes reduced to E_3—$FADH_2$. In the final reaction (step V),
E_3—$FADH_2$ is reoxidized by the coenzyme NAD^+ to regenerate the
oxidized form E_3—FAD, with formation of $NADH + H^+$.

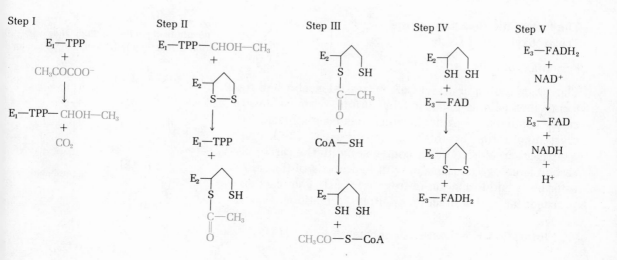

Figure 12-4
The long lipoyllysine side chain of dihydrolipoyl transacetylase (E₂)
serves as a swinging arm to transfer hydrogen atoms from pyruvate
dehydrogenase (E₁) to dihydrolipoyl dehydrogenase (E₃) and acetyl
groups from pyruvate dehydrogenase to coenzyme A.

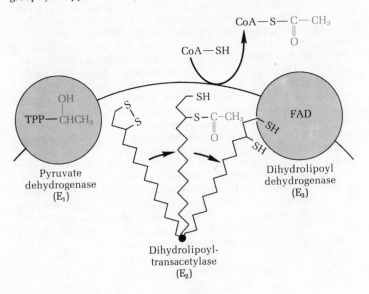

The Enzymatic Reactions of the Tricarboxylic Acid Cycle

Formation of Citric Acid

The condensation of acetyl CoA with oxaloacetic acid to form citric acid is catalyzed by citrate synthase (also called condensing enzyme), which has been isolated in crystalline form. In this reaction (Figure 12-5), the methyl group of acetyl CoA condenses with the carbonyl carbon atom of oxaloacetate, with hydrolysis of the thioester bond and formation of free CoA—SH. The reaction proceeds far in the direction of citrate formation

$$\text{Acetyl CoA} + \text{oxaloacetate} \longrightarrow \text{citrate} + \text{CoA} + H_2O$$

The Aconitase Equilibrium

The enzyme aconitase catalyzes the reversible addition of H_2O to the double bond of cis-aconitic acid in two directions, one leading to citric acid and the other to isocitric acid:

$$\text{Citric acid} \underset{+H_2O}{\overset{-H_2O}{\rightleftharpoons}} \text{cis-aconitic acid} \underset{-H_2O}{\overset{+H_2O}{\rightleftharpoons}} \text{isocitric acid}$$

Figure 12-5
The citrate synthase reaction. The carbon atoms originating from acetate are in color.

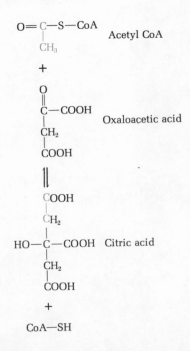

Figure 12-6
The aconitase equilibrium. The carbon atoms arising from the acetyl group of acetyl CoA are in color.

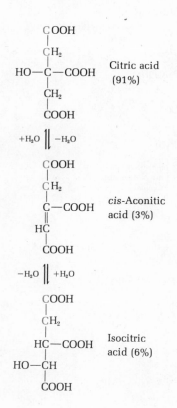

The equilibrium is shown in Figure 12-6. The equilibrium mixture at pH 7.4 and 25°C contains about 91 percent citrate, 6 percent isocitrate, and 3 percent *cis*-aconitate.

Isocitrate Dehydrogenase

Most microorganisms and higher animal and plant tissues contain two types of isocitrate dehydrogenase. One requires NAD^+ as electron acceptor and the other $NADP^+$. The overall reactions catalyzed by the two types of isocitrate dehydrogenase are identical

$$\text{Isocitrate} + NAD^+ \rightleftharpoons \alpha\text{-ketoglutarate} + CO_2 + NADH + H^+$$

$$\text{Isocitrate} + NADP^+ \rightleftharpoons \alpha\text{-ketoglutarate} + CO_2 + NADPH + H^+$$

Both the NAD-linked and NADP-linked isocitrate dehydrogenases have been found in mitochondria of many higher animal tissues, but the former is found only in mitochondria, whereas the latter is also found in the extramitochondrial cytoplasm. It has been established that the NAD-linked isocitrate dehydrogenase is the catalyst normally participating in the tricarboxylic acid cycle, whereas the NADP-linked enzyme is primarily concerned with auxiliary biosynthetic reactions of the cycle.

The NAD-specific isocitrate dehydrogenase of mitochondria requires Mg^{2+} or Mn^{2+} for activity. It oxidizes isocitrate to α-ketoglutarate and CO_2 (Figure 12-7). It is an allosteric or regulatory enzyme, which is stimulated by the positive modulators ADP and NAD^+ and is strongly inhibited by the negative or inhibitory modulators NADH and ATP (see below).

Oxidation of α-Ketoglutarate to Succinate

α-Ketoglutarate formed by isocitrate dehydrogenase undergoes oxidative decarboxylation to form succinyl CoA (Figure 12-8) and CO_2

$$\alpha\text{-Ketoglutarate} + NAD^+ + CoA\text{—}SH \longrightarrow$$
$$\text{succinyl—S—CoA} + CO_2 + NADH + H^+ \qquad \Delta G^{\circ\prime} = -8.0 \text{ kcal}$$

This reaction is comparable to the oxidation of pyruvate to acetyl CoA and CO_2 (see above) and occurs by the same mechanism, with thiamine pyrophosphate, lipoic acid, coenzyme A, NAD^+, and FAD participating as required enzyme-bound cofactors. The α-ketoglutarate dehy-

Fig. 12-7
The isocitrate dehydrogenase reaction.

Figure 12-8
The oxidation of α-ketoglutaric acid.

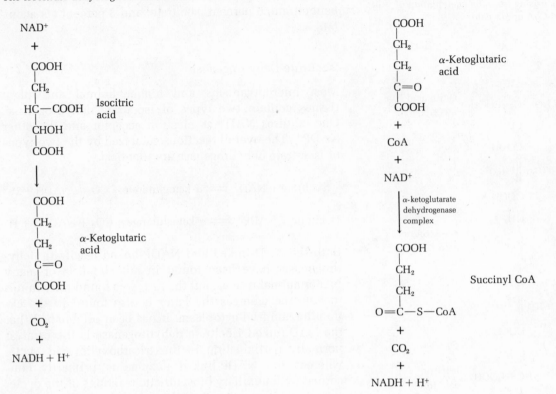

drogenase complex is very similar in structure and properties to the pyruvate dehydrogenase complex. It has been isolated from animal tissues and from *E. coli*, in which it has a particle weight of 2,100,000.

The end product of this reaction is succinyl CoA, a high-energy thioester of succinic acid. Succinyl CoA then undergoes loss of its CoA group, not by simple hydrolysis, but by an energy-conserving reaction with guanosine diphosphate (GDP) and phosphate

Succinyl—S—CoA + P$_i$ + GDP $\rightleftharpoons$
          succinate + GTP + CoA—SH $\Delta G^{o\prime} = -0.7$ kcal

The enzyme catalyzing this reaction, <u>succinyl thiokinase</u>, causes the formation of the high-energy phosphate bond of GTP from GDP and P$_i$ at the expense of the high-energy thioester bond of succinyl CoA. The GTP formed in this reaction then may donate its terminal phosphate

Figure 12-9
The succinate dehydrogenase reaction.

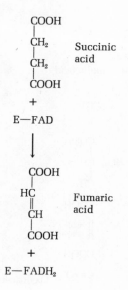

group to ADP to form ATP, by the action of <u>nucleoside</u> <u>diphosphokinase</u>

$$GTP + ADP \rightleftharpoons GDP + ATP$$

The generation of ATP coupled to deacylation of succinyl CoA is often called a <u>substrate-level phosphorylation,</u> to distinguish it from the respiratory chain phosphorylations to be discussed in Chapter 13.

Succinate Dehydrogenase

Succinate is oxidized to fumarate (Figure 12-9) by the flavoprotein *succinate dehydrogenase,* which contains covalently bound flavin adenine dinucleotide. This reducible prosthetic group functions as a hydrogen acceptor in the following reaction:

$$\text{Succinate} + \text{E—FAD} \rightleftharpoons \text{fumarate} + \text{E—FADH}_2$$

Succinate dehydrogenase is tightly bound to the mitochondrial membrane. As isolated from beef heart mitochondria, it has a molecular weight of about 175,000 and contains one molecule of covalently bound FAD, four atoms of iron, and four atoms of sulfur in an unknown chemical form. The iron atoms appear to undergo Fe(II)–Fe(III) valence changes during the action of the enzyme. Since the enzyme contains no heme group, such as is found in hemoglobin or cytochromes (Chapter 13), the iron of this enzyme is spoken of as <u>nonheme iron.</u> The nonheme iron and labile sulfur atoms are present in distinct subunits of succinate dehydrogenase called <u>nonheme-iron proteins.</u>

Succinate dehydrogenase is competitively inhibited by malonate (Chapter 4), which we have seen played an important role in establishing the nature of the tricarboxylic acid cycle. Succinate dehydrogenase is also competitively inhibited by very low concentrations of oxaloacetate. Thus the last dicarboxylic acid of the tricarboxylic acid cycle can inhibit its own formation from succinic acid.

Fumarase

The reversible hydration of fumarate to L-malate (Figure 12-10) is catalyzed by the enzyme *fumarase,* which has been obtained in crystalline form from pig heart

Figure 12-10
The fumarase reaction.

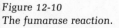

$$\text{Fumarate} + \text{H}_2\text{O} \rightleftharpoons \text{L-malate}$$

Fumarase has a molecular weight of about 200,000 and contains four polypeptide-chain subunits, which are inactive in separated form. It requires no coenzyme.

Oxidation of Malate to Oxaloacetate

In the last reaction of the tricarboxylic acid cycle, the NAD-linked L-*malate dehydrogenase* catalyzes the oxidation of L-malate to oxaloacetate (Figure 12-11)

L-Malate + NAD$^+$ $\rightleftharpoons$

$\qquad$ oxaloacetate + NADH + H$^+$ $\qquad$ $\Delta G°' = 7.1$ kcal

Although the reaction is endergonic as written (Chapter 10), in the cell it readily goes in the direction shown because of the very rapid removal of the reaction products oxaloacetate and NADH in subsequent steps.

We may now sum up the output of one turn around the tricarboxylic acid cycle. Two carbon atoms appear as carbon dioxide, equivalent to, but not identical with, the two carbon atoms of the acetyl group which entered the cycle. Four pairs of hydrogen atoms are yielded by enzymatic dehydrogenation; three pairs have been used to reduce NAD and one pair to reduce the FAD of succinate dehydrogenase. Ultimately these four pairs of hydrogen atoms become H$^+$ ions and their corresponding electrons combine with oxygen, following their transport down the respiratory chain.

Isotopic Tests of the Tricarboxylic Acid Cycle

That the tricarboxylic acid cycle actually takes place in intact cells and can account quantitatively for the oxidation of carbohydrate, fatty acids, and amino acids has been verified by tests with precursors and intermediates in which specific carbon atoms were isotopically labeled with either the ^{13}C or ^{14}C isotopes. As an example, consider Figure 12-2, which traces the fate of the two carbon atoms of acetic acid through the various steps of the tricarboxylic acid cycle. Experiments in which both of the carbon atoms were labeled with the ^{14}C isotope show clearly that the two molecules of carbon dioxide evolved during one revolution of the cycle are not the same two carbon atoms which entered the cycle in the acetic acid molecule. These carbon atoms actually remain in the four-carbon dicarboxylic acids.

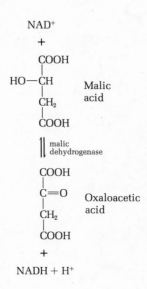

Figure 12-11
Oxidation of malic acid.

Regulation of the Tricarboxylic Acid Cycle

Like virtually every metabolic sequence in the cell, the tricarboxylic acid cycle is regulated by the action of specific allosteric or regulatory enzymes in response to the concentration of certain important metabolites in the cell. The major rate-limiting reaction of the tricarboxylic acid in many tissues is that catalyzed by *citrate synthase* in the first reaction in the cycle. This is an allosteric enzyme. Its negative or inhibitory modulators are ATP, the end product of the energy-conserving reactions of respiration, and NADH, the end product of the dehydrogenation steps of the cycle.

The second regulatory enzyme of the cycle, which may be rate-limiting in certain tissues, is that catalyzed by the NAD-linked isocitrate dehydrogenase. This allosteric enzyme is strongly inhibited by specific negative modulators, particularly ATP and NADH, and is stimulated by the positive modulators ADP and NAD^+.

Whenever the concentration of ATP in the cell exceeds a certain level which satisfies all its energy needs, it acts to inhibit citrate synthase and isocitrate dehydrogenase. Moreover, whenever the concentration of NADH builds up beyond a certain level, it also serves as an inhibitor of these regulatory enzymes. Conversely, whenever the cell has had its ATP concentration lowered, owing to a heavy drain on ATP to carry out some energy-requiring function, then the ensuing high ADP concentration serves to stimulate the overall rate of the tricarboxylic acid cycle, thus replenishing the normal high level of ATP.

Anaplerotic Reactions

We must now recall from Chapter 9 that the tricarboxylic acid cycle is an amphibolic pathway; it functions not only in catabolism but also to generate precursors for anabolic pathways. By means of several important auxiliary reactions, certain intermediates of the tricarboxylic acid cycle, particularly α-ketoglutarate, succinate, and oxaloacetate, can be removed or drained away from the cycle to serve as precursors of amino acids (Chapters 17, 19). When this happens, the rate of the tricarboxylic acid cycle would be expected to decline, because the concentration of the intermediates would be diminished. However, the intermediates of the cycle can be replenished again. Normally the reactions by which cycle intermediates are drained away and by which they are formed are in dynamic balance, so that the concentrations of these intermediates in the mitochondria remain constant.

The special enzymatic mechanisms by which tricarboxylic acid cycle intermediates can be replenished are called *anaplerotic* ("filling up") reactions. The most important is the enzymatic carboxylation of pyruvate to form oxaloacetate (Figure 12-12) by the action of *pyruvate carboxylase*, which catalyzes the reaction

$$\text{Pyruvate} + CO_2 + \text{ATP} \longrightarrow \text{oxaloacetate} + \text{ADP} + P_i \qquad \Delta G^{\circ\prime} = -0.5 \text{ kcal}$$

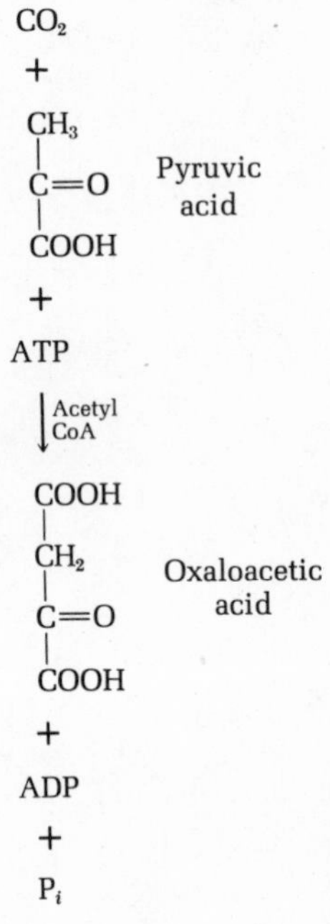

When the tricarboxylic acid cycle is deficient in oxaloacetate or any of the other intermediates, pyruvate is carboxylated to produce more oxaloacetate. Conversely, when oxaloacetate is in excess, it can be decarboxylated, with the formation of pyruvate and CO_2.

Pyruvate carboxylase has a molecular weight of about 650,000 and contains a large number of polypeptide-chain subunits. At 0°C it is inactivated and dissociates into subunits. The native enzyme contains four molecules of the vitamin biotin, which is covalently attached to the enzyme protein through a peptide linkage with the ε-amino group of a specific lysine residue at the active site. A ring nitrogen atom of biotin is the site at which CO_2 combines in an ATP-dependent reaction. In a subsequent reaction at the enzyme-active site, the bound CO_2 is transferred to pyruvate to form oxaloacetate (Figure 12-13).

Pyruvate carboxylase is an allosteric enzyme. The rate of its forward reaction leading to oxaloacetate is very low unless acetyl CoA, its positive modulator, is present. Thus, whenever acetyl CoA, the fuel of the tricarboxylic acid cycle, is present in excess, it stimulates the pyruvic carboxylase reaction to produce more oxaloacetate, thus enabling the cycle to oxidize more acetyl CoA. Although this is the most important anaplerotic reaction in the liver and kidney of higher animals, other reactions appear to perform this function in heart and muscle tissue. One such reaction is that catalyzed by *malic enzyme*

$$\text{Pyruvate} + CO_2 + \text{NADPH} + H^+ \rightleftharpoons \text{L-malate} + \text{NADP}^+$$

Another is that catalyzed by *phosphoenolpyruvate carboxykinase*, an enzyme to be discussed again in Chapter 17.

$$\text{Phosphoenolpyruvate} + CO_2 + \text{GDP} \longrightarrow \text{oxaloacetate} + \text{GTP}$$

Figure 12-12
The pyruvate carboxylase reaction.

CO_2

+

CH_3
|
$C = O$ Pyruvic
| acid
COOH

+

ATP

$\downarrow$ Acetyl CoA

COOH
|
CH_2 Oxaloacetic
| acid
$C = O$
|
COOH

+

ADP

+

P_i

Figure 12-13
Prosthetic group of pyruvate carboxylase. The biotin carboxyl group forms a peptide linkage with the ε-amino group of lysine. During CO_2 transfer an N-carboxy derivative of biotin is reformed, as is shown.

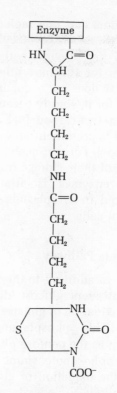

Figure 12-14
The glyoxylate cycle. The overall equation of the cycle is

2 Acetyl CoA $\longrightarrow$ succinate + 2H + 2CoA

The reactions in color are those catalyzed by the auxiliary enzymes isocitratase and malate synthase; all others are reactions of the tricarboxylic acid cycle.

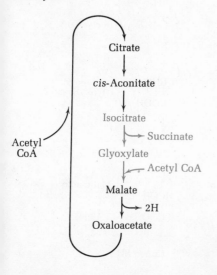

The Glyoxylate Cycle

When acetate must serve both as a source of energy and as a source of various intermediates required to synthesize the carbon skeletons of all the major cellular components (which may occur in higher plants, as well as in microorganisms such as *E. coli*, *Pseudomonas*, and algae), the tricarboxylic acid cycle operates in a modified form called the *glyoxylate cycle*. The overall plan of the glyoxylate cycle is shown in Figure 12-14. Acetyl CoA is the fuel of this cycle; it condenses with oxaloacetate to form citrate. However, the breakdown of isocitrate does not occur via the usual NAD-linked isocitrate dehydrogenase, but via a cleavage catalyzed by the enzyme *isocitratase* to form succinic and glyoxylic acids (Figure 12-15). The glyoxylic acid then condenses with acetyl CoA (Figure 12-16) to yield malate. The malate then is oxidized to oxaloacetate, which can condense with another molecule of acetyl CoA. In each turn of the glyoxylate cycle, two molecules of acetyl CoA enter and one pair of hydrogen atoms and one molecule of succinate are formed. The latter is used for biosynthetic purposes. The glyoxylate cycle thus provides both energy and four-carbon intermediates for the biosynthetic pathways of the cell. This pathway does not occur in higher animals, which have no need for it, simply because they are never forced to survive on two-carbon fuel molecules alone. The glyoxylate cycle is prominent, however, in the seeds of higher plants, which convert acetyl residues derived from the fatty acids of their storage triacylglycerols into carbohydrate. The enzymes isocitratase and malate synthase are localized in cytoplasmic organelles called glyoxysomes in plant cells.

The Phosphogluconate Pathway

Many cells possess, in addition to the tricarboxylic acid cycle, another pathway of glucose degradation whose first reaction is the oxidation of glucose 6-phosphate to 6-phosphogluconate. This phosphogluconate pathway, which is also known as the pentose phosphate pathway, or the hexose monophosphate shunt, is not a "main-line" pathway for the oxidation of glucose. Its primary purpose in most cells is to generate reducing power in the extramitochondrial cytoplasm in the form of NADPH. This function is especially prominent in tissues that actively carry out the synthesis of fatty acids and steroids from small precursors, such as the liver, mammary gland, adipose or fat tissues, and the adrenal cortex. Skeletal muscle, which is not active in synthesizing fatty acids, is virtually lacking in this pathway. A second function of the phosphogluconate pathway is to generate pentoses, particularly D-ribose, which is used in the synthesis of nucleic acids. Another major function is to participate in the formation of glucose from CO_2 in the "dark" reactions of photosynthesis (Chapters 16, 17).

The first reaction of the phosphogluconate pathway is the enzymatic dehydrogenation of glucose 6-phosphate by glucose 6-phosphate dehydrogenase, also known by the German name Zwischenferment, to form 6-phosphogluconate (Figure 12-17). Glucose 6-phosphate dehydrogenase requires $NADP^+$ as electron acceptor. The product is 6-phosphoglucono-δ-lactone, which is hydrolyzed to the free acid by a specific lactonase (Figure 12-17). The overall equilibrium lies far in the direction of formation of NADPH. In the next step 6-phosphogluconate undergoes oxidative decarboxylation by 6-phosphogluconate dehydrogenase to form the ke-

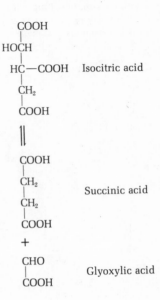

Figure 12-15
The isocitratase reaction.

COOH
|
HOCH
|
HC—COOH   Isocitric acid
|
CH₂
|
COOH

COOH
|
CH₂
| Succinic acid
CH₂
|
COOH

+

CHO
| Glyoxylic acid
COOH

Figure 12-16
The malate synthase reaction.

CH₃
|
C=O   Acetyl CoA
|
S—CoA

+

CHO
| Glyoxylic acid
COOH

COOH
|
CH₂
| Malic acid
HCOH
|
COOH

+

CoA—SH

Figure 12-17
The phosphogluconate pathway.

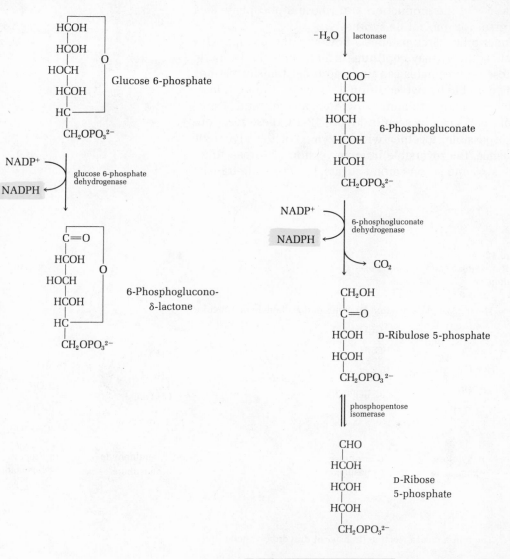

topentose D-*ribulose 5-phosphate* (Figure 12-17), a reaction that generates a second molecule of NADPH. *Phosphopentose isomerase* then converts D-ribulose 5-phosphate into its aldose isomer, D-ribose 5-phosphate (Figure 12-17), which can be used in the synthesis of ribonucleotides and deoxyribonucleotides. In some cells the phosphogluconate pathway ends at this point, and its overall equation is then written

Glucose 6-phosphate + 2NADP$^+$ $\longrightarrow$
$\quad$ D-ribose 5-phosphate + CO_2 + 2NADPH + 2H$^+$

The net result is the production of NADPH for reductive biosynthetic reactions in the extramitochondrial cytoplasm and the production of D-ribose 5-phosphate as a precursor for nucleotide synthesis.

Under other circumstances, however, the phosphogluconate pathway may continue further, since the isomeric pentose 5-phosphates can undergo other transformations made possible by two additional enzymes, _transketolase_, which contains thiamine pyrophosphate, and _transaldolase_. As can be seen in Figure 12-18, these reactions make possible, together with enzymes of the glycolytic sequence, the reversible interconversion of three-, four-, five-, six-, and seven-carbon sugars, by reversible transfer

Figure 12-18
The transketolase and transaldolase reactions.

The transketolase reaction [transfer of glycolaldehyde (in color)]

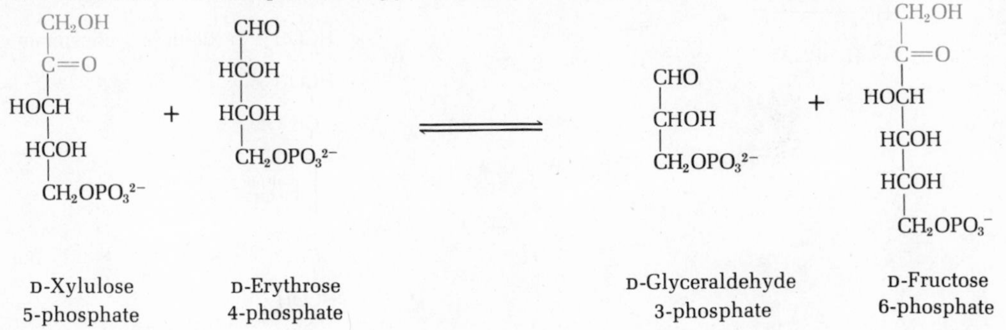

CH₂OH	CHO		CHO	CH₂OH	
C=O	HCOH		CHOH	C=O	
HOCH	HCOH		CH₂OPO₃²⁻	HOCH	
HCOH	CH₂OPO₃²⁻			HCOH	
CH₂OPO₃²⁻				HCOH	
				CH₂OPO₃⁻	
D-Xylulose 5-phosphate	D-Erythrose 4-phosphate		D-Glyceraldehyde 3-phosphate	D-Fructose 6-phosphate	

The transaldolase reaction [transfer of dihydroxyacetone (in color)]

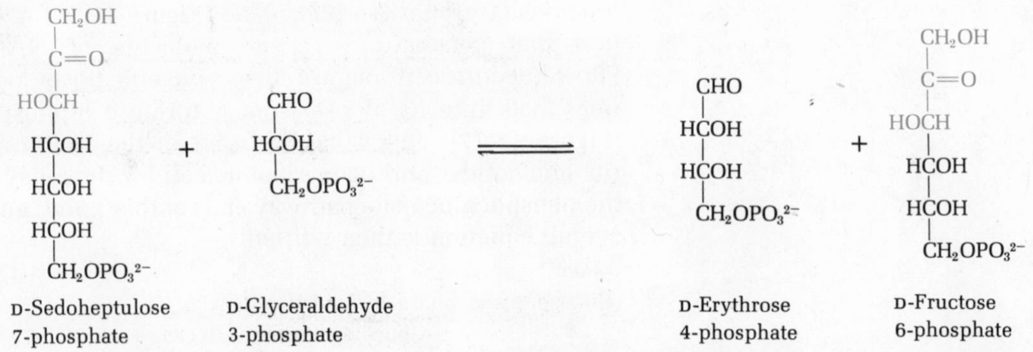

CH₂OH			CHO	CH₂OH	
C=O	CHO		HCOH	C=O	
HOCH	HCOH		HCOH	HOCH	
HCOH	CH₂OPO₃²⁻		CH₂OPO₃²⁻	HCOH	
HCOH				HCOH	
HCOH				CH₂OPO₃²⁻	
CH₂OPO₃²⁻					
D-Sedoheptulose 7-phosphate	D-Glyceraldehyde 3-phosphate		D-Erythrose 4-phosphate	D-Fructose 6-phosphate	

of either two-carbon (glycolaldehyde) or three-carbon (dihydroxyacetone) groups. For example, a prominent reaction catalyzed by transketolase is

D-Xylulose 5-phosphate + D-erythrose 4-phosphate $\longrightarrow$
D-fructose 6-phosphate + D-glyceraldehyde 3-phosphate

in which two intermediates of the glycolytic pathway can be generated (Figure 12-18). The phosphogluconate pathway can also carry out the complete oxidation of glucose 6-phosphate to CO_2 by a complex sequence of reactions in which six molecules of glucose 6-phosphate are oxidized to six molecules each of ribulose 5-phosphate and CO_2; five molecules of glucose 6-phosphate are then regenerated from the six molecules of ribulose 5-phosphate. The overall equation is

6 Glucose 6-phosphate + 12NADP$^+$ $\longrightarrow$
5 glucose 6-phosphate + 6CO$_2$ + 12NADPH + 12H$^+$ + P$_i$

The net equation is then

Glucose 6-phosphate + 12NADP$^+$ $\longrightarrow$
6CO$_2$ + 12NADPH + 12H$^+$ + P$_i$

Summary

Respiration occurs in three stages: (1) the oxidative formation of acetyl CoA from pyruvate, fatty acids, and amino acids, (2) the degradation of acetyl residues by the Krebs tricarboxylic acid cycle to yield CO_2 and H atoms, and (3) the transport of electrons equivalent to these hydrogen atoms to molecular oxygen, a process which is accompanied by coupled phosphorylation of ADP. The tricarboxylic acid cycle, which takes place in the mitochondria, consists of a cyclic series of reactions in which acetyl residues are condensed with the four-carbon oxaloacetic acid to form the six-carbon citric acid. Two carbon atoms of the latter are lost as CO_2 during a sequence of reactions which regenerate a molecule of oxaloacetate. Pyruvic acid, the end product of glycolysis under aerobic conditions and thus the major fuel of the cycle, first undergoes oxidation by the pyruvate dehydrogenase system, which requires five different coenzymes, to yield acetyl CoA and CO_2. The cycle begins when citrate synthase catalyzes the condensation of acetyl CoA with oxaloacetic acid to form citric acid. Aconitase then catalyzes the reversible formation of isocitrate from citrate.

Isocitrate is then oxidized to α-ketoglutarate plus CO_2 by NAD-linked isocitrate dehydrogenase. The succinyl CoA formed is deacylated by GDP and phosphate to form free succinate and GTP, which transfers its terminal phosphate group to ADP. The succinate is then oxidized to fumarate by succinate dehydrogenase, a flavin enzyme. Fumarate is hydrated by fumarase to L-malate, which is oxidized by NAD-linked L-malate dehydrogenase to regenerate a molecule of oxaloacetate. The latter can then combine with another molecule of acetyl CoA and start another revolution of the cycle. Isotopic tracer tests with carbon-labeled fuel molecules or intermediates have established that the tricarboxylic acid cycle is the major mechanism of oxidation in intact cells. The overall rate of the cycle in the liver is controlled by citrate synthase, an allosteric enzyme inhibited by the negative modulators ATP and NADH.

The tricarboxylic acid cycle also can furnish intermediates for biosynthesis. The cycle intermediates are then replenished by anaplerotic reactions, the most important of which is the carboxylation of pyruvate to oxaloacetate. In organisms living on acetate as sole carbon source, a variation of the tricarboxylic acid cycle, the glyoxylate cycle, comes into play and makes possible the net formation of succinate (and other cycle intermediates) from acetate.

Glucose 6-phosphate can be oxidized via the 6-phosphogluconate pathway to pentose phosphates, particularly D-ribose 5-phosphate, by a sequence of enzymes in the soluble cytoplasm. The electron acceptor for these reactions is $NADP^+$; the NADPH so formed is a major reductant in the synthesis of hydrogen-rich biomolecules such as fatty acids and cholesterol.

References

LEHNINGER, A. L.: *Biochemistry*, Worth Publishers, New York, 1970. Chapter 16 gives a more advanced textbook treatment.

LOWENSTEIN, J. M.: "The Tricarboxylic Acid Cycle," in D. M. Greenberg (ed), *Metabolic Pathways*, 3d ed., vol. 1, pp. 146–270, Academic Press, New York, 1967. An excellent review of the present knowledge of the cycle, with special reference to stereochemical relationships; the glyoxylate cycle is also reviewed.

REED, L. J.: "Pyruvate Dehydrogenase Complex," in *Current Topics in Cellular Regulation*, Academic Press, New York, 1969, Vol 1, pages 233–251. A short review of the structure, function, and regulation of this important multienzyme complex.

Problems

1. Show the position of the carbon isotope in the citric acid formed when the following isotopically labeled compounds are incubated with a respiring muscle suspension (a) 3-^{14}C-pyruvate, (b) 2-^{14}C-pyruvate, (c) 1-^{14}C-pyruvate, (d) 2-^{14}C-acetyl portion of acetyl CoA, (e) 3-^{14}C-glyceraldehyde 3-phosphate, (f) 5-^{14}C-fructose 6-phosphate.

2. The substrates indicated below were added to suspensions of minced pigeon flight muscle amply supplied with oxygen but in which succinate dehydrogenase was completely inhibited by malonate. The reactions were allowed to take place until no further oxygen uptake occurred. Write balanced equations for the oxidation of (a) citrate, (b) pyruvate + fumarate under the conditions described.

3. Write an equation for the complete oxidation of fructose 6-phosphate via the glycolytic and tricarboxylic acid cycle pathways.

4. Write the overall equation for the conversion of acetate into succinate by *E. coli* cells when acetate is the sole source of carbon. If the acetate is labeled with ^{14}C in the carboxyl group, in what atoms of succinic acid will the isotope be found?

5. Write a balanced equation for the conversion of glucose into D-ribose 5-phosphate in the liver.

CHAPTER **13** **ELECTRON TRANSPORT AND**
OXIDATIVE PHOSPHORYLATION

Electron transport and oxidative phosphorylation represent the last phase of biological oxidation. Electrons derived from the intermediates of the tricarboxylic acid cycle flow down a multimembered chain of electron-carrier enzymes of successively lower energy levels to molecular oxygen, the ultimate electron acceptor in respiration. During this process much of the free energy of the electrons is conserved, by oxidative phosphorylation, in the form of the phosphate-bond energy of ATP. In animal and plant tissues the enzymes catalyzing electron transport and oxidative phosphorylation are located in the mitochondria.

The Respiratory Chain and its Collecting Function

Figure 13-1 shows the electron-transport chain, the final common pathway taken by all electrons removed from substrate molecules during respiration. These electrons arise not only from the four dehydrogenation steps in the tricarboxylic acid cycle, but also from the dehydrogenation of pyruvate and of fatty acids to yield acetyl CoA, and from the dehydrogenation of glyceraldehyde 3-phosphate in the glycolytic sequence, when it is functioning under aerobic conditions. Most of the electrons arrive via the coenzyme NAD, which serves to collect electrons and funnel them into the chain. Electrons brought in by the flavin dehydrogenases are collected by coenzyme Q, which will be discussed later.

Figure 13-1
The respiratory chain and the points of entry of electrons from various substrates. Also shown are the sites of inhibition of electron transport and the probable sites of energy conservation leading to ATP formation. FP indicates flavoprotein dehydrogenase.

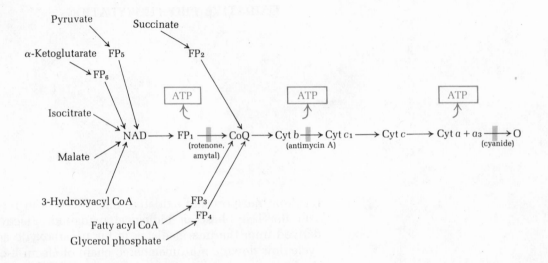

The various members of the respiratory chain are capable of accepting electrons from the preceding component and transferring them to the following one. They are thus called electron carriers. All have prosthetic groups capable of reversibly accepting and releasing electrons.

Oxidation–Reduction Reactions

Electron-transferring reactions, such as those taking place in the respiratory chain, are more generally called oxidation-reduction reactions, or redox reactions. They characteristically proceed with a transfer of electrons from an electron donor, also called the reducing agent or reductant, to an electron acceptor, also called an oxidizing agent or oxidant. In some oxidation-reduction reactions the transfer of electrons is made via the transfer of hydrogen atoms, each of which carries an electron. In this case dehydrogenation is equivalent to oxidation and hydrogenation is equivalent to reduction. In other oxidation–reduction reactions both an electron and a hydrogen atom may be transferred. Electrons and/or hydrogen atoms participating in oxidoreductions are often referred

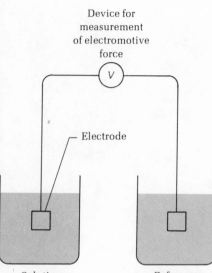

Figure 13-2
Measurement of the standard reduction potential of biological systems (pH 7.0).

Device for measurement of electromotive force

V

Electrode

Solution containing 1.0 M oxidant and 1.0 M reductant at 25° and pH 7.0

Reference half-cell of known potential

to by the neutral term _reducing equivalents_. Oxidizing and reducing agents function as conjugate reductant-oxidant pairs, just as Bronsted acids and bases function as conjugate acid-base pairs:

Acid-base reactions:

$$\text{Proton donor} \rightleftharpoons H^+ + \text{proton acceptor}$$

Oxidation-reduction reactions:

$$\text{Electron donor} \rightleftharpoons e^- + \text{electron acceptor}$$

We have seen that the dissociation constant (or the pK) may be used to express quantitatively the tendency of an acid to dissociate a proton (Chapter 2). In a similar way we may also express the tendency of a reducing agent to lose electrons. For this purpose the _standard reduction potential_ is used. It is defined as the electromotive force (emf) in volts given by an electrode in a solution of a reductant and its conjugate oxidant, both present at 1.0 M concentration, at 25°C, and at pH 7.0. The electrode can accept electrons from the reductant species (Figure 13-2) according to the equation

$$\text{Reductant} \rightleftharpoons \text{oxidant} + ne^-$$

where n is the number of electrons transferred. The standard reduction potential is a measure of the electron pressure a given reductant-oxidant (redox) pair generates at equilibrium under these specified conditions. The standard reduction potentials of a number of biologically important redox couples are given in Table 13-1. Systems having a relatively negative standard reduction potential have a greater tendency to lose electrons (that is, they have a greater electron pressure) than those with a more positive potential. For example, acetaldehyde ($E_0' = -0.60$ volts) has a relatively great tendency to lose electrons and become oxidized to acetic acid. Conversely, the strongly positive standard reduction potential of water, $+0.815$ volt, indicates that the water molecule has very little tendency to lose electrons and form molecular oxygen. Put in another way, molecular oxygen has a very high affinity for electrons and thus for hydrogen atoms.

The standard reduction potentials of various biological oxidation-reduction systems allow us to predict the direction of flow of electrons from one redox couple to another under standard conditions, just as the phos-

Table 13-1 Standard reduction potentials of some conjugate redox couples†

Reductant	Oxidant	E_0', volts	Direction of electron flow
Acetaldehyde	Acetate	−0.60	
H_2	$2H^+$	−0.42	
Isocitrate	α-Ketoglu-tarate + CO_2	−0.38	
$NADH + H^+$	NAD^+	−0.32	
$NADPH + H^+$	$NADP^+$	−0.32	
Lactate	Pyruvate	−0.19	
NADH dehydrogenase (reduced)	(oxidized)	−0.11	
Cytochrome b [Fe(II)]	[Fe(III)]	0.00	
Cytochrome c [Fe(II)]	[Fe(III)]	+0.26	
H_2O	$\frac{1}{2}O_2$	+0.82	

† The data are calculated on the basis of two-electron transfers at pH ≅ 7.0 and T = 25° to 37°C.

phate-group transfer potential allows us to predict the direction in which phosphate groups will be enzymatically transferred (Chapter 10). Electrons will tend to flow from relatively electronegative donors, such as NADH, to more electropositive acceptors, such as the cytochromes, and finally to oxygen. Figure 13-3 is a diagram showing the standard reduction potentials of the electron carriers of the respiratory chain and the direction of electron flow.

NAD (NADP)-Linked Dehydrogenases

We shall now consider the properties of the important electron-transferring enzymes associated with electron transport. Most of the electron pairs entering the respiratory chain arise from the action of a class of dehydrogenases which employ either NAD^+ or $NADP^+$ as electron acceptors; they are also called *pyridine-linked dehydrogenases* (Chapter 8, 11, 12). The structures of NAD and NADP were given in Chapters 8 and 11. Some of the important dehydrogenases of this class are listed in Table 13-2. They catalyze the following general reactions:

Reduced substrate + NAD^+ $\rightleftharpoons$

oxidized substrate + $NADH + H^+$

Table 13-2 Some pyridine-linked dehydrogenases

	E_0' of substrate couple
NAD-linked:	
Isocitrate	−0.38
D-β-Hydroxybutyrate	−0.32
Glyceraldehyde 3-phosphate	−0.29
Dihydrolipoyl	−0.29
L-β-Hydroxyacyl CoA	−0.24
Ethanol	−0.20
Lactate	−0.19
Glycerol 3-phosphate	−0.19
L-Malate	−0.17
NADP-linked:	
Isocitrate	−0.38
Glucose 6-phosphate	−0.32
NAD or NADP:	
L-Glutamate	−0.14

Figure 13-3

The direction and energetics of electron flow in the respiratory chain. Q stands for coenzyme Q, B for cytochrome b, C for cytochrome c, and A for cytochrome a. Shown in color are the segments of the chain in which there are particularly large decreases in free energy.

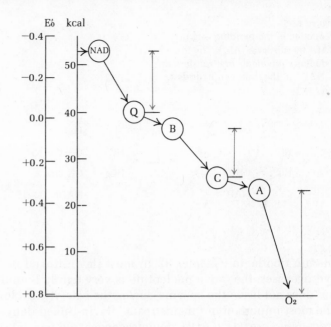

Figure 13-4

Absorption spectra of NAD$^+$ and NADH. Measurements of the reduction of NAD$^+$ to NADH are carried out at 340 nm.

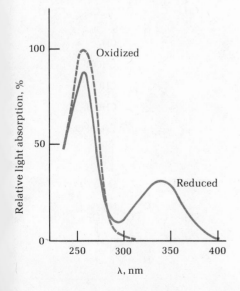

Reduced substrate + NADP$^+$ $\rightleftharpoons$
oxidized substrate + NADPH + H$^+$

Most of the pyridine-linked dehydrogenases are specific for either NAD or NADP, but a few, such as glutamate dehydrogenase, can react with either coenzyme (Table 13-2). NAD and NADP are only loosely bound to the enzyme protein during the catalytic cycle. Thus the oxidized and reduced forms of NAD and NADP should be regarded as substrates rather than as prosthetic groups. The reduced forms have a characteristic spectrum with an absorption band at 340 nm (Figure 13-4). The pyridine-linked dehydrogenases reversibly transfer two reducing equivalents as a hydride ion (Chapter 12) from the substrate to the oxidized form of the pyridine nucleotide; one of these appears in the reduced pyridine nucleotide as a hydrogen atom, the other as an electron (Figure 13-5). The second hydrogen atom from the substrate molecule appears as a free H$^+$ ion in the medium.

Flavin-Linked Dehydrogenases

The flavin-linked dehydrogenases contain either *flavin adenine dinucleotide* (FAD) or *flavin mononucleotide* (FMN) as prosthetic groups. Both coenzymes contain the

241

Figure 13-5
Reduction of the pyridine ring in
NAD$^+$ by substrate (AH$_2$). The two
reducing equivalents are transferred
to NAD$^+$ in the form of a hydride
ion (:H$^-$).

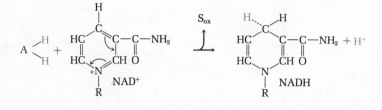

vitamin riboflavin (Chapter 8). In most flavin-linked dehydrogenases, the flavin nucleotide is very tightly bound
and does not leave the enzyme during the catalytic cycle.
The most important or "mainstream" flavin-linked dehydrogenases are (1) NADH dehydrogenase, which catalyzes transfer of electrons from NADH to coenzyme Q in
the respiratory chain, (2) succinic dehydrogenase of the
tricarboxylic acid cycle (Chapter 12), (3) dihydrolipoyl
dehydrogenase of the pyruvate and α-ketoglutarate dehydrogenase systems (Chapter 12), and (4) the flavoprotein
that catalyzes the first dehydrogenation step during fatty
acid oxidation (Chapter 14).

The active portion of FAD or FMN that participates in
the oxidoreduction is the isoalloxazine ring of the riboflavin moiety, which may be reduced (Figure 13-6) by
transfer of a pair of hydrogen atoms from the substrate to
yield the reduced forms, designated FADH$_2$ and FMNH$_2$,
respectively. The reduction of the flavin dehydrogenases
is accompanied by loss of light absorption at 450 nm.
Many flavoenzymes contain metal ions as well as the
flavin prosthetic group.

Both NADH dehydrogenase and succinic dehydrogenase either contain or are bound tightly to nonheme-
iron proteins (Chapter 12), which undergo reversible
reduction and oxidation and which are therefore thought
to participate in electron transport. Figure 13-7 shows the
electron-transferring reactions of the respiratory chain.

Coenzyme Q

Coenzyme Q, which has the function of "collecting"
reducing equivalents from flavin-linked dehydrogenases,
is also called ubiquinone because it is a quinone and is

Figure 13-6
Flavin mononucleotide. Reduction
of the isoalloxazine ring takes place
at the points shown in color.

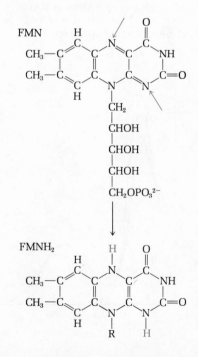

Figure 13-7
The sequential reactions of electron transport. The following abbreviations are used: E_1–FAD, NADH dehydrogenase; E_2–Fe(III), ferric or oxidized form of nonheme-iron protein; cyt b (II), ferrous or reduced form of cytochrome b, etc.

$$NADH + H^+ + E_1\!-\!FAD \rightleftharpoons NAD^+ + E\!-\!FADH_2$$

$$E\!-\!FADH_2 + 2E_2\!-\!Fe(III) \rightleftharpoons E_1\!-\!FAD + 2E_2\!-\!Fe(II) + 2H^+$$

$$2E_2\!-\!Fe(II) + 2H^+ + CoQ \rightleftharpoons 2E_2\!-\!Fe(III) + CoQH_2$$

$$CoQH_2 + 2\ cyt\ b(III) \rightleftharpoons CoQ + 2H^+ + 2\ cyt\ b(II)$$

$$2\ cyt\ b(II) + 2\ cyt\ c(III) \rightleftharpoons 2\ cyt\ b(III) + 2\ cyt\ c(II)$$

$$2\ cyt\ c(II) + 2\ cyt\ a(III) \rightleftharpoons 2\ cyt\ c(III) + 2\ cyt\ a(II)$$

$$2\ cyt\ a(II) + 2\ cyt\ a_3(III) \rightleftharpoons 2\ cyt\ a(III) + 2\ cyt\ a_3(II)$$

$$2\ cyt\ a_3(II) + \tfrac{1}{2}O_2 + 2H^+ \rightleftharpoons 2\ cyt\ a_3(III) + H_2O$$

ubiquitous in all cells. Coenzyme Q is a fat-soluble quinone with a long isoprenoid side chain (Figure 13-8), which in most mammalian tissues has 10 five-carbon isoprenoid units and is thus designated CoQ_{10}. In other organisms it has only 6 or 8 isoprene units (CoQ_6 and CoQ_8).

The Cytochromes

The cytochromes are a group of iron-containing, electron-transferring proteins of aerobic cells that act sequentially in the transport of electrons from coenzyme Q to molecular oxygen. They contain iron-porphyrin prosthetic groups and thus resemble hemoglobin and myoglobin; all are members of the class of *heme proteins*.

Porphyrins (Figure 13-9) are named and classified on the basis of their side-chain substituents. Protoporphyrin IX is the most abundant and contains four methyl groups, two vinyl groups, and two propionic acid groups. It is the porphyrin present in hemoglobin, myoglobin, and most of the cytochromes. Protoporphyrin forms very stable complexes with di- and trivalent metal ions. Such a complex of protoporphyrin with Fe(II) is called *protoheme*, or more simply, *heme;* a similar complex with Fe(III) is called *hemin*, or *hematin*. The iron-porphyrin prosthetic group endows heme proteins with a characteristic reddish or brown color, which is useful in their spectrophotometric measurement.

The cytochromes were first studied systematically by D. Keilin beginning in 1925. With a spectroscope he observed that insect muscles contain pigments resembling hemoglobin which undergo oxidation and reduction during respiration. He named them cytochromes, postulated that they act in a chain to carry electrons from fuel

Figure 13-8
Coenzyme Q. The subscript n indicates the number of isoprenoid units in the side-chain.

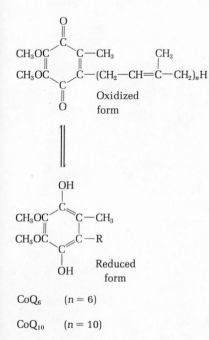

CoQ_6 $(n = 6)$

CoQ_{10} $(n = 10)$

Figure 13-9
Structures of porphin and protoporphyrin,
and iron-binding in cytochrome c.

Porphin

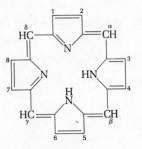

Protoporphyrin IX

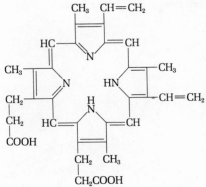

Binding of iron atom in cytochrome c. The
four nitrogen atoms of the porphyrin ring
bind to the iron in a planar arrangement.
X and Y represent binding groups contributed by the protein.

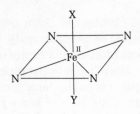

molecules to oxygen, and classified them into three
major classes, a, b, and c, depending on the characteristic
maxima in their light absorption spectra. Each type of cytochrome in its reduced state has three distinctive absorption bands in the visible range (Figure 13-10). The
cytochromes undergo Fe(II)–Fe(III) valence changes
during their function as electron carriers.

With one exception cytochromes are very tightly
bound to the mitochondrial membrane and are difficult
to obtain in soluble and homogeneous forms. The exception is cytochrome c, which is easily extracted from mitochondria and has been obtained in pure crystalline
form (Chapter 3). The single iron-protoporphyrin group
of cytochrome c is covalently linked to the protein. It is
the only common heme protein in which there is such a
covalent linkage.

Figure 13-10
The absorption spectrum of
cytochrome c.

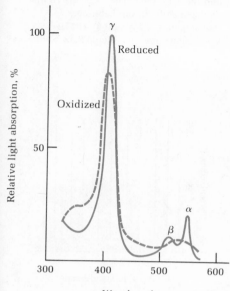

Cytochromes a and a_3, together called *cytochrome oxidase*, deserve special attention. Instead of protoporphyrin they contain porphyrin A, which differs from protoporphyrin in having different substituent groups, particularly a 15-carbon hydrocarbon side chain. The oxidized or Fe(III) form of cytochrome $a + a_3$ can accept electrons from reduced cytochrome c to become the Fe(II) form; these electrons are then transferred to molecular oxygen, thus regenerating the Fe(III) form. Cytochrome $a + a_3$ also contains copper. Of all the cytochromes in the chain only cytochrome a_3 can react directly with oxygen.

The Sequence of Electron Carriers in the Respiratory Chain

The sequence of electron-transfer reactions in the respiratory chain shown in Figures 13-1 and 13-7 is supported by several lines of evidence. First, it is consistent with the standard reduction potentials of the different electron carriers, in that the potentials become more positive as electrons pass from substrate to oxygen. Second, experiments on the specificity of the isolated electron carriers support the sequence; thus NADH can reduce NADH dehydrogenase but cannot directly reduce cytochromes b, c, or $a + a_3$. Third, complexes containing functionally linked carriers have been isolated, among them a complex of cytochromes b and c_1 and a complex of cytochromes a and a_3.

Sensitive spectrophotometric measurements have also shown that in normal respiring mitochondria in the steady state the electron carrier nearest the substrate end of the chain, namely NAD, is the most reduced member of the chain, whereas the carriers at the oxygen end (cytochromes $a + a_3$) are almost entirely in the oxidized form. The intermediate carriers are present in successively more oxidized steady states going from substrate to oxygen. When the respiratory chain is blocked with a specific inhibitor, the electron carriers at the substrate side of the block in the chain become more reduced and those at the oxygen side become more oxidized (Figure 13-11).

The discovery of inhibitors specific for certain points in the chain has also greatly helped in the study of electron transport. The most important are *rotenone*, an extremely toxic insecticide and fish poison, and *sodium amytal*, a barbiturate drug, both of which block electron transfer from NAD to coenzyme Q; the antibiotic *antimycin A*, isolated from a strain of *Streptomyces*, which blocks transfer of electrons from cytochrome b to c; and cyanide, which blocks the terminal cytochrome $a + a_3$ step (Figure 13-1).

Oxidative Phosphorylation

The coupling of phosphorylation to respiration was first discovered in the late 1930s and was called *aerobic* or *oxidative phosphorylation*, to distinguish it from glycolytic phosphorylation. It was found that tissue suspensions oxidizing pyruvate via the tricarboxylic acid cycle caused the formation of three molecules of ATP from ADP and phosphate for each atom of oxygen consumed. Later it was discovered that oxidative phosphorylation of ADP occurs at the expense of the large amount of energy delivered as electrons pass down the respiratory chain from NADH to oxygen. Figures 13-1 and 13-4 show that there are three points along the respiratory chain where the oxidation–reduction energy of electron transport is converted into the phosphate bond energy of ATP. For this reason oxidative phosphorylation is also called *respiratory chain phosphorylation*. The overall complete equation for electron transport and the coupled respiratory phosphorylations is

$$NADH + H^+ + 3ADP + 3P_i + \tfrac{1}{2}O_2 \longrightarrow NAD^+ + 4H_2O + 3ATP$$

Let us now see how efficiently the three phosphorylations of ADP recover the energy released in electron transport. The amount of energy delivered by a reaction in which there is a transfer of electrons can be calculated from the expression

$$\Delta G^{\circ\prime} = -nF\Delta E_0'$$

in which $\Delta G^{\circ\prime}$ is the standard free energy change in calories, n is the number of electrons transferred, F is the faraday (23,062 cal), and $\Delta E_0'$ is the difference between the standard reduction potential of the donor system and that of the acceptor system. All components are assumed to be at $1.0\,M$ concentration at 25°C and pH 7.0. The standard free energy change as a pair of electrons pass from NADH $(E_0' = -0.32$ volt$)$ to oxygen $(E_0' = +0.82$ volt$)$ is

$$\Delta G^{\circ\prime} = -2 \times 23{,}062 \times [0.82 - (-0.32)] = -52{,}700 \text{ cal}$$
$$= -52.7 \text{ kcal}$$

Since 7.3 kcal of energy is required to make 1 mole of ATP from ADP and phosphate, the coupled phosphorylation

Figure 13-11
Hydraulic analogy of the respiratory chain. When electron transport is inhibited, all the carriers to the substrate side are reduced and those on the oxygen side are in the oxidized state. (From Lehninger, Bioenergetics, Copyright © 1965, W. A. Benjamin, Inc., Menlo Park, California.)

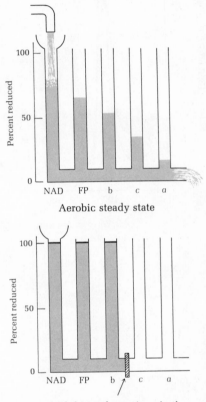

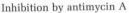

of three molecules of ATP during passage of a pair of electrons from NADH to oxygen thus conserves $21.9/52.7 \times 100$, or about 40 percent of the total free energy decline during electron transport from NADH to oxygen, under standard conditions.

The multimembered respiratory chain thus is a device for breaking up the rather large decline in free energy occurring as a pair of electrons moves from NADH to molecular oxygen into a series of smaller energy drops, of which three are approximately the size of the energy currency of the cell, that is, the 7.3 kcal "quantum" required to generate a molecule of ATP from ADP and phosphate (Figure 13-4). The respiratory chain is therefore a kind of cascade delivering free energy in useful packets.

All the enzymes required for electron transport and oxidative phosphorylation occur in the mitochondria, whose membrane must be intact in order for the energy-conserving mechanisms to function (see below).

Uncoupling Agents and Inhibitors of Oxidative Phosphorylation

Oxidative phosphorylation can be uncoupled from electron transport by specific chemical agents, particularly certain phenols such as 2,4-dinitrophenol (Figure 13-12). In the presence of these agents, respiration of tissue suspensions may continue at a normal rate, or may even be stimulated, but coupled phosphorylation of ADP to ATP does not take place. Such compounds are known as uncoupling agents. They are believed to function by promoting the passage of protons (H^+) across the membrane of the mitochondria, thus using up the energy of electron transport. These agents do not uncouple glycolytic phosphorylation, nor do they affect cellular reactions other than oxidative phosphorylation.

Another class of agents, represented by the antibiotic oligomycin, blocks both oxidative phosphorylation and electron transport. Still another class of inhibitors is represented by the antibiotics valinomycin and gramicidin. Like the uncoupling agents, they permit respiration to occur but not phosphorylation. They allow K^+ to be "pumped" through the mitochondrial membrane, thus diverting respiratory energy from ATP production.

The Energy Balance Sheet for Glycolysis and Respiration

We can now construct the overall equation for the complete oxidation of glucose to CO_2 and water by molecular

Figure 13-12
2.4-Dinitrophenol, an uncoupling agent.

oxygen via the glycolytic sequence and the tricarboxylic acid cycle, including all the phosphorylations. Glycolysis of one molecule of glucose under aerobic conditions yields two molecules of pyruvate, two of NADH and two of ATP, formed in the soluble cytoplasm:

$$\text{Glucose} + 2P_i + 2ADP + 2NAD^+ \longrightarrow$$
$$2 \text{ pyruvate} + 2NADH + 2H^+ + 2ATP + 2H_2O$$

Then we have the equation for the oxidation of two molecules of pyruvate to CO_2 and H_2O, together with the phosphorylations, which take place in the mitochondria

$$2 \text{ Pyruvate} + 5O_2 + 30ADP + 30P_i \longrightarrow 6CO_2 + 30ATP + 34H_2O$$

To these we must now add the equation for electron transport from the two molecules of extramitochondrial NADH formed in the glycolytic conversion of glucose to pyruvate, a process which generates two molecules of ATP per pair of electrons in most tissues

$$2NADH + 2H^+ + O_2 + 4P_i + 4ADP \longrightarrow 2NAD^+ + 4ATP + 6H_2O$$

The sum of these three equations is the overall equation of glycolysis plus respiration

$$\text{Glucose} + 6O_2 + 36P_i + 36ADP \longrightarrow 6CO_2 + 36ATP + 42H_2O$$

Breaking up the overall equation into its exergonic and endergonic components gives

Exergonic component:

$$\text{Glucose} + 6O_2 \longrightarrow 6CO_2 + 6H_2O \qquad \Delta G^{\circ\prime} = -680 \text{ kcal}$$

Endergonic component:

$$36P_i + 36ADP \longrightarrow 36ATP + 36H_2O \qquad \Delta G^{\circ\prime} = +263 \text{ kcal}$$

The overall efficiency of energy recovery in the complete oxidation of glucose is thus $263/680 \times 100 = 39$ percent.

The Mechanism of Oxidative Phosphorylation

Despite many years of research the enzymatic mechanism by which oxidoreduction energy is converted into the phosphate-bond energy of ATP during electron transport is still unsolved; it remains one of the most challenging problems in modern biology. Two major hypothesis have been postulated for this fundamental process.

The *chemical-coupling hypothesis* postulates that ATP is formed from ADP and phosphate as the end result of a sequence of consecutive enzyme-catalyzed reactions with common intermediates, in such a way that transfer of electrons from one electron carrier to the next causes the generation of a high-energy covalent bond which becomes the precursor of the high-energy phosphate bond of ATP. According to this hypothesis oxidative phosphorylation occurs by the action of a multienzyme sequence functioning in principle like all other metabolic sequences.

However, after many years of research, no one has yet succeeded in detecting or isolating the hypothetical high-energy intermediates that are essential to the chemical coupling hypothesis. Moreover, this hypothesis does not account for the fact that oxidative phosphorylation occurs only in preparations of mitochondria retaining reasonably intact membrane structure, a fact suggesting that the membrane is an important part of the phosphorylation mechanism.

The *chemiosmotic hypothesis,* proposed by P. Mitchell, a British biochemist, involves a radically different principle. Mitchell postulates that the mitochondrial membrane is an essential element in the mechanism of oxidative phosphorylation and that it is normally impermeable to H^+ ions. He proposes that the transport of electrons along the carriers of the respiratory chain, which are located in the membrane, causes outward transport of H^+ ions across the membrane. The electron-transport chain thus is viewed as a "pump" for the H^+ ions. Such an energy-rich gradient of H^+ ions, rather than a chemical intermediate, is postulated to be the immediate driving force for the formation of ATP from ADP and phosphate.

Many experimental observations are consistent with the chemiosmotic hypothesis, but it is not yet possible to conclude which of these or other hypotheses is correct.

Mitochondria and Their Enzymatic Organization

We have seen that the tricarboxylic acid cycle, electron transport, and oxidative phosphorylation take place in the mitochondria, organelles in the cytoplasm of all aerobic eukaryotic cells. A single rat liver cell contains about 1000 mitochondria, which make up about 20 percent of the total cytoplasmic volume.

Mitochondria are often strategically located near struc-

Figure 13-13
Electron micrograph of mitochondrion in thin section of a bat
pancreas cell. The inner membrane folds inward to form the cristae.

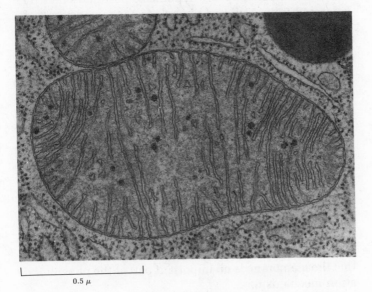

0.5 μ

tures that require ATP, the product of their activity. In highly active muscles, the ATP-producing mitochondria are regularly arranged in rows along the ATP-requiring contractile filaments. Mitochondria are also frequently located near a fuel source; for example, they are often found adjacent to cytoplasmic fat droplets, which serve as a source of fuel for oxidation.

Mitochondria vary considerably in size and shape. Those from liver cells are spherical or football-shaped; they are about 2 μ long and 1 μ in width. All mitochondria have two membranes (Figure 13-13), an outer surrounding membrane that is smooth and an inner membrane with many inward folds called *cristae*. Within the inner membrane is the <u>*matrix*</u>, a gel-like phase containing about 50 percent protein. The enzymes for the tricarboxylic acid cycle are located in the matrix; the electron carriers are located in the inner membrane. Electron microscopy shows that the inner surface of the inner membrane is covered with regularly spaced spherical particles (diameter 80 to 90 Å) connected to the membrane by narrow stalks. These knoblike structures, called <u>*inner-membrane spheres*</u>, are molecules of the enzyme which makes ATP, called <u>*ATP synthetase*</u>, or F_1. It is a relatively large protein (mol wt 280,000) with a diameter of about 80 Å. Oxidative phosphorylation thus takes place within the inner compartment.

The outer mitochondrial membrane is freely permeable to most low molecular weight solutes. However, the inner membrane is relatively impermeable and allows only certain specific substances to pass. Among these are ADP and ATP, which pass through the membrane via a membrane transport system. This system is inhibited by <u>atractyloside,</u> a toxic glycoside from the Mediterranean thistle. The ATP–ADP transport system allows one molecule of ADP to enter the mitochondria in exchange for one molecule of ATP coming out after its phosphorylation has taken place.

The Regulation of Respiration and Glycolysis

We have seen that both the glycolytic sequence (Chapter 11) and the tricarboxylic acid cycle (Chapter 12) contain regulatory enzymes which respond to ADP and ATP as positive and negative modulators, respectively. ADP and ATP also serve to regulate the rate of electron transport and thus the rate of ATP production. When the supply of respiratory substrate and phosphate is ample, the maximal rate of oxygen consumption by mitochondria occurs when the ADP concentration in the medium is high and that of ATP low. On the other hand, when the concentration of ADP is low and ATP is high, mitochondria show only a very low respiratory rate, as little as 5 to 10 percent of the maximal rate. Figure 13-14 shows that addition of even small amounts of ADP evokes a maximal rate of respiration, which continues until all the ADP is phosphorylated to ATP, thus yielding a high ATP to ADP ratio, at which point oxygen consumption abruptly returns to a low resting rate. This dependence of the rate of electron transport on ADP concentration is called <u>acceptor control</u> of respiration.

Figure 13-15 summarizes the important regulatory mechanisms for the integration of the rate of glycolysis with the rate of the tricarboxylic acid cycle. Phosphofructokinase, the major regulatory enzyme of the glycolytic sequence, catalyzes the rate-limiting step in glycolysis, the phosphorylation of fructose 6-phosphate to fructose 1,6-diphosphate. The activity of phosphofructokinase is stimulated by ADP and inhibited by ATP. When the ATP/ADP ratio in the cell is high, the enzyme is severely inhibited; when this ratio is low, the enzyme increases in activity. Thus, whenever there is ample ATP in the cell, phosphofructokinase "turns off" and both glycolysis and the tricarboxylic acid cycle must slow down. Moreover, whenever citrate is overproduced in

Figure 13-14
Control of rate of respiration by ADP concentration.

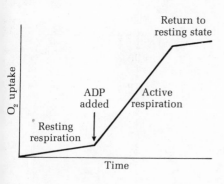

the tricarboxylic acid cycle, it acts as negative modulator of phosphofructokinase and also slows down the rate of glycolysis and thus the rate of formation of acetyl CoA. Phosphofructokinase is thus the "pacemaker" enzyme of glycolysis. Figure 13-15 also shows that when the ATP/ADP ratio is high, the rate of the tricarboxylic acid must also slow down, since ATP is an inhibitor of citrate synthesis as well as isocitrate oxidation. Through these allosteric enzymes a set of tight controls regulates this major energy-producing metabolic pathway of the cell.

Summary

In oxidation-reduction reactions electrons are transferred from the reductant, or electron donor, to the oxidant, or electron acceptor. The tendency of any biological reductant to donate electrons is given by its standard reduction potential (E_0'), the electromotive force given by 1.0 M reductant in the presence of 1.0 M oxidant species at an inert electrode at 25° and pH 7.0. The E_0' values of any given conjugate redox pair allows prediction of the direction of net electron flow in its reaction with another redox pair.

There are three major classes of oxidation-reduction enzymes. (1) The pyridine-linked dehydrogenases catalyze reversible transfer of electrons from substrates to the loosely bound coenzymes NAD^+ or $NADP^+$, to form NADH and NADPH, respectively. (2) The flavin-linked dehydrogenases contain tightly bound FMN or FAD as prosthetic groups and usually a metal ion. In oxidized form they are intensely colored; on reduction they are colorless. The most important are succinate dehydrogenase and NADH dehydrogenase. (3) The cytochromes, acting in series, transfer electrons from flavoproteins to oxygen. They contain iron protoporphyrin prosthetic groups and undergo Fe(II)-Fe(III) transitions.

The respiratory chain of mitochondria consists of the sequence NADH, flavoproteins, nonheme iron proteins, coenzyme Q, and cytochromes b, c_1, c, a, and a_3, which donates electrons to oxygen. Electron transport along this chain is characteristically inhibited at specific sites by amytal, rotenone, antimycin A, and cyanide. The large free-energy change on passage of a pair of electron equivalents from NADH to oxygen is broken up into three packets by the respiratory chain, each capable of causing the phosphorylation of ADP to ATP. Oxidative phosphorylation, which recovers some 40 percent of the energy released during electron transport, is uncoupled by 2,4-dinitrophenol and inhibited by oligomycin.

The tricarboxylic acid cycle, electron transport, and oxidative phosphorylation occur in the mitochondria. The electron car-

Figure 13-15
Summary of the important control mechanisms in glycolysis and respiration. The dotted colored arrows show the origin of feedback inhibitors (ATP, NADH, citrate, glucose 6-phosphate) and the colored bars indicate the sites of feedback inhibition. The points of action of positive or stimulatory modulators (AMP, ADP, acetyl CoA, NAD^+) are shown by colored arrows parallel to the black reaction arrows.

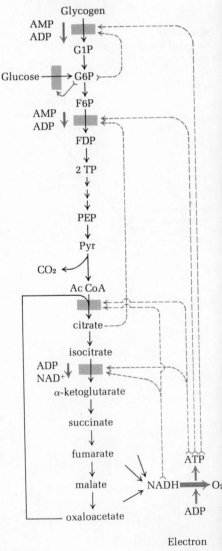

Electron transport and oxidative phosphorylation

riers of the respiratory chain are located in the inner mitochondrial membrane, as are the enzymes required to generate ATP.

The rate of electron transport is regulated by the concentration of ADP. The concentration of ADP (and ATP) is also important in regulating the rate of glycolysis and the tricarboxylic acid cycle.

References

KEILIN, D.: *The History of Cell Respiration and Cytochromes,* Cambridge University Press, New York, 1966.

LEHNINGER, A. L.: *Biochemistry,* Worth Publishers, New York, 1970. Chapters 17 and 18 provide more advanced text material.

LEHNINGER, A. L.: *The Mitochondrion: Molecular Basis of Structure and Function,* W. A. Benjamin, Menlo Park, Cal., 1964.

Problems

1. Assuming that all reactants and products are maintained at 1.0 M concentration, pH 7.0, and $T = 25°$, calculate the standard free energy change as a pair of electron equivalents passes from (a) NADH to cytochrome c, (b) succinate to cytochrome c, (c) cytochrome c to oxygen, (d) cytochrome b to cytochrome c.

2. If the following reactions start from 1 mM concentrations of all reactants and all products and are allowed to proceed to equilibrium, in which direction would they go, left or right, as written?
   (a) Isocitrate $+ NAD^+ \rightleftharpoons \alpha$-ketoglutarate $+ CO_2 + NADH + H^+$
   (b) Malate $+ NAD^+ \rightleftharpoons$ oxaloacetate $+ NADH + H^+$

3. Calculate the theoretical efficiency of energy conservation in the form of ATP as succinate is oxidized to fumarate by molecular oxygen in intact mitochondria.

4. How many ATP molecules are generated during the complete oxidative degradation of one molecule of each of the following to CO_2 and H_2O?
   (a) fructose 6-phosphate; (b) acetyl CoA; (c) glyceraldehyde 3-phosphate; (d) sucrose.

CHAPTER 14 OXIDATION OF FATTY ACIDS

The oxidation of fatty acids is an important source of energy in higher animals and plants, which can store large amounts of neutral fat in certain tissues as fuel reserve. Neutral fat has a high caloric value (9 kcal/gram) and can be stored in nearly anhydrous form in intracellular fat droplets, whereas glycogen and starch (caloric value, 4 kcal/gram) are highly hydrated and cannot be stored in such a concentrated form nor in as large amounts. In vertebrates, it is estimated that fatty acid oxidation normally provides at least half of the oxidative energy in the liver, kidneys, heart muscle, and resting skeletal muscle. Moreover, in fasting or hibernating animals and in migrating birds, fat is virtually the sole source of energy.

Triacylglycerols and phospholipids must first undergo hydrolytic cleavage to yield their fatty acid components by the action of lipases and phospholipases (Chapter 6), before they undergo oxidation to carbon dioxide and water.

The Fatty Acid Oxidation Cycle

Because most natural fatty acids have an even number of carbon atoms, it was long suspected that fatty acids are synthesized and degraded in the cell by addition or subtraction of two-carbon fragments. Moreover, early experiments showed that various nonbiological derivatives of fatty acids fed to animals yield end products in the urine which suggest that fatty acids are degraded by removal of successive two-carbon fragments starting from the car-boxyl end. From such considerations it was postulated that fatty acids are oxidized by successive β oxidations, that is, oxidation at the β carbon to yield a β-keto acid, which then undergoes cleavage to form acetic acid and a fatty acid shorter by two carbon atoms (Figure 14-1).

For many years no success attended efforts to demon-strate fatty acid oxidation in cell-free extracts or ho-mogenates of animal tissues. However, in the 1940s the missing link was found. Addition of ATP restored the ability of liver homogenates to oxidize fatty acids. It was therefore postulated that ATP was required to prime or activate the fatty acid by an enzymatic reaction. It was also found that the oxidation of fatty acids in liver ho-mogenates yielded two-carbon units that can enter the tricarboxylic acid cycle. The next important clue, which led to recognition of the nature of the enzymatic steps of fatty acid oxidation, came from the work of F. Lynen and his colleagues in the early 1950s. They found that the ATP-dependent activation of fatty acids involves their esterification with the thiol group of coenzyme A and that all the subsequent intermediates in fatty acid oxida-tion are esters of this coenzyme.

Figure 14-1
The β-oxidation of myristic acid to yield lauric acid and acetic acid.

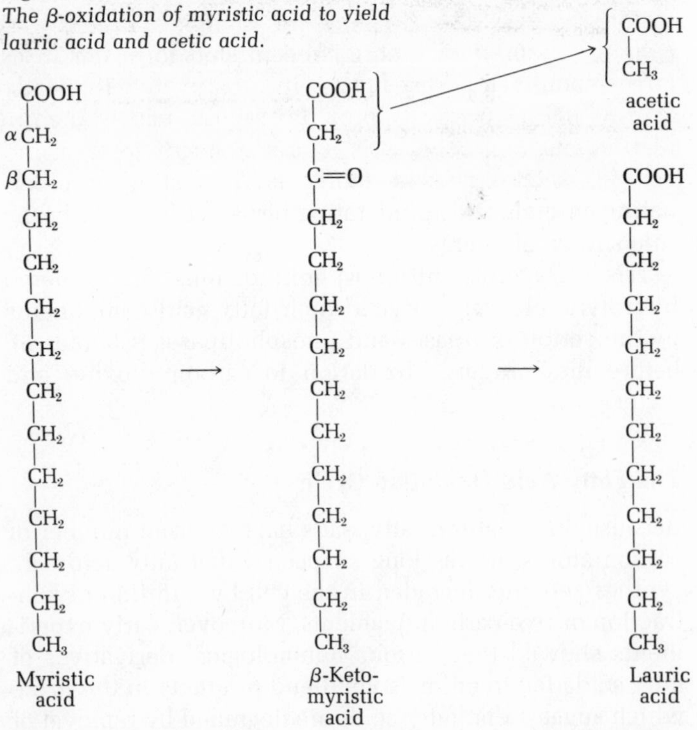

The various enzymes catalyzing the successive steps in fatty acid oxidation have been isolated in highly purified form and the complete pathway of oxidation has been elucidated (Figure 14-2). The oxidation of fatty acids proceeds by the successive removal of two-carbon fragments in the form of acetyl CoA. Oxidation of the 16-carbon palmitic acid, for example, yields altogether 8 molecules of acetyl CoA and 14 pairs of hydrogen atoms. The acetyl CoAs so formed then may enter the tricarboxylic acid cycle, whereas the hydrogen atoms, or their equivalent electrons, enter the respiratory chain. Their passage to oxygen leads to oxidative phosphorylation of ADP. Fatty acid oxidation occurs exclusively in the mitochondria, which contain all the requisite enzymes and coenzymes.

The separate enzymatic steps in fatty acid oxidation will now be examined in detail.

Activation and Entry of Fatty Acids into Mitochondria

There are three stages in the entry of fatty acids into mitochondria from the extramitochondrial cytoplasm: (1) the enzymatic esterification of the free fatty acid with extramitochondrial CoA at the expense of ATP, which occurs in the outer membrane; (2) the transfer of the fatty acyl group from CoA to the carrier molecule *carnitine*, which carries it across the inner membrane; and (3) the transfer of the fatty acyl group from carnitine to intramitochondrial CoA.

The first step is catalyzed by *fatty acid activating enzyme*, also called *fatty acid: CoA ligase* or *fatty acid thiokinase*, which promotes the reversible reaction

$$RCOOH + ATP + CoASH \rightleftharpoons R-\underset{\underset{O}{\|}}{C}-SCoA + AMP + PP_i$$

As the thioester linkage is formed between the fatty acid carboxyl group and the thiol group of CoA (Figure 14-3), the ATP undergoes cleavage to yield AMP and inorganic pyrophosphate. The pyrophosphate formed in the activation reaction is then hydrolyzed by inorganic pyrophosphatase

$$Pyrophosphate + H_2O \longrightarrow 2\ phosphate$$

The net effect of this hydrolysis is the utilization of two

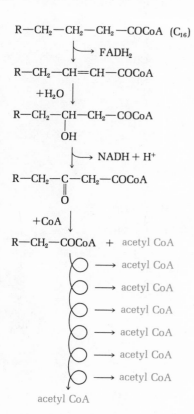

Figure 14-2
The fatty acid oxidation cycle. One acetyl CoA is removed during each pass through the sequence. One molecule of palmitic acid (C_{16}), after activation to palmitoyl CoA, yields eight molecules of acetyl CoA.

high-energy phosphate bonds of ATP to pull the overall equilibrium of the preceding activation reaction far in the direction of the formation of fatty acyl CoA.

An enzyme-bound intermediate has been identified in the action of fatty acid activating enzymes. It is a mixed anhydride of the fatty acid and the phosphate group of AMP, a *fatty acyl adenylate* (Figure 14-4). It is formed on the active site and then reacts with free CoA—SH to yield fatty acyl CoA and free AMP as products. The fatty acid thiokinases are found in the outer mitochondrial membrane.

Neither higher fatty acids nor their CoA esters can cross the inner mitochondrial membrane. However, the enzyme *acyl CoA: carnitine fatty acid transferase* now catalyzes transfer of the fatty acyl group from its thioester linkage with CoA to the hydroxyl group of *carnitine* (Figure 14-5)

$$\text{Fatty acyl CoA} + \text{carnitine} \rightleftharpoons \text{CoA} + \text{fatty acyl carnitine}$$

The fatty acyl carnitine ester so formed readily passes through the inner membrane into the internal or matrix compartment of the mitochondria. In the latter the fatty acyl group is enzymatically transferred from carnitine to intramitochondrial CoA by a fatty acyl CoA: carnitine fatty acid transferase present in the matrix, to complete the process of fatty acid activation and entry

$$\text{Fatty acyl carnitine} + \text{CoA} \rightleftharpoons \text{fatty acyl CoA} + \text{carnitine}$$

This entry mechanism has the effect of keeping extra-mitochondrial and intramitochondrial CoA in separate pools or compartments. The internal fatty acyl CoA now becomes the substrate for the enzymes of the fatty acid oxidation cycle, which occurs exclusively in the inner mitochondrial compartment.

The First Dehydrogenation Step

The fatty acyl CoA ester first undergoes enzymatic dehydrogenation at the α and β carbon atoms (carbon atoms 2 and 3) to form $\Delta^{2,3}$-*trans*-unsaturated fatty acyl CoA as product (Figure 14-6)

$$\text{Fatty acyl CoA} + \text{E—FAD} \longrightarrow \Delta^{2,3}\text{-}trans\text{-enoyl CoA} + \text{E—FADH}_2$$

The symbol $\Delta^{2,3}$- designates the position of the double bond. The hydrogen atoms from the substrate are trans-

Figure 14-3
General structure of fatty acyl CoA esters.

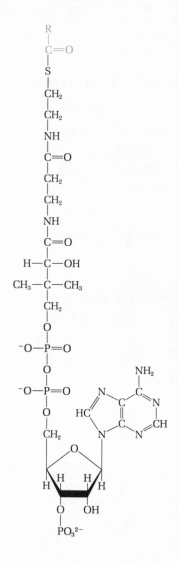

Figure 14-4
Structure of a fatty acyl adenylate.

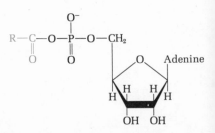

Figure 14-5
Acyl CoA: carnitine fatty acid transferase reaction.

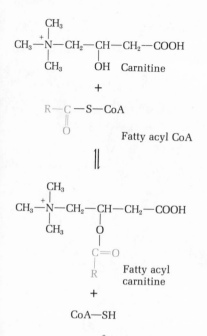

Figure 14-6
The fatty acyl CoA dehydrogenase reaction.

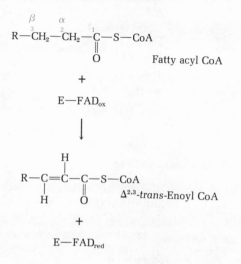

Figure 14-7
Electron transport from fatty acyl CoA to cytochrome b.

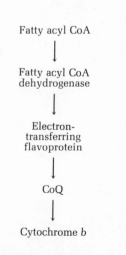

ferred to FAD, the tightly bound prosthetic group of fatty acyl CoA dehydrogenase. The reduced form of this enzyme then donates its electrons to the respiratory chain via a connecting electron carrier, called the *electron-transferring flavoprotein*, which in turn is believed to transfer its electrons to coenzyme Q of the respiratory chain (Figure 14-7).

The Hydration Step

In the next step water adds to the double bond of $\Delta^{2,3}$-unsaturated CoA esters to form β-hydroxy (3-hydroxy) acyl CoA esters, catalyzed by the enzyme *enoylhydratase*, also called *3-hydroxyacyl CoA hydrolyase*, which has been isolated in crystalline form

$$\Delta^{2,3}\text{-trans-enoyl CoA} + H_2O \rightleftharpoons \text{L-3-hydroxyacyl CoA}$$

The reaction catalyzed is shown in Figure 14-8. The addition of water across the $\Delta^{2,3}$-*trans* double bond results in the formation of the L stereoisomer of the 3-hydroxyacyl CoA ester.

The Second Dehydrogenation Step

In the next step of the fatty acid cycle, the L-3-hydroxyacyl CoA is dehydrogenated to form 3-ketoacyl CoA (Figure

14-9) by L-3-*hydroxyacyl CoA dehydrogenase*; NAD$^+$ is its specific electron acceptor. The reaction is

L-3-Hydroxyacyl CoA + NAD$^+$ $\rightleftharpoons$
$$\text{3-ketoacyl CoA + NADH + H}^+$$

This enzyme is absolutely specific for the L-stereoisomer. The NADH formed in the reaction donates its reducing equivalents to the NADH dehydrogenase of the respiratory chain.

The Thiolytic Cleavage Step

In the last step of the fatty acid oxidation sequence, which is catalyzed by *β-ketothiolase* (more simply, *thiolase*), the 3-ketoacyl CoA undergoes cleavage by interaction with a molecule of free CoA, to yield the carboxyl-terminal two-carbon fragment as free acetyl CoA and the CoA ester of a fatty acid shortened by two carbon atoms (Figure 14-10)

3-Ketoacyl CoA + CoA $\rightleftharpoons$
$$\text{shortened fatty acyl CoA + acetyl CoA}$$

This reaction is highly exergonic; thus, the equilibrium favors cleavage.

The Balance Sheet for Fatty Acid Oxidation

We have now completed one round of the fatty acid oxidation sequence, in which one molecule of acetyl CoA and two pairs of hydrogen atoms have been removed from a long-chain fatty acyl CoA. The overall equation for one turn of the fatty acid oxidation sequence acting on palmitoyl CoA is

Palmitoyl CoA + CoA + FAD + NAD$^+$ + H$_2$O $\longrightarrow$
$$\text{myristoyl CoA + acetyl CoA + FADH}_2\text{ + NADH + H}^+$$

Altogether, seven turns of the fatty acid oxidation cycle are required to convert one molecule of palmitoyl CoA to eight molecules of acetyl CoA

Palmitoyl CoA + 7CoA + 7FAD + 7NAD$^+$ + 7H$_2$O $\longrightarrow$
$$\text{8 acetyl CoA + 7FADH}_2\text{ + 7NADH + 7H}^+$$

Each molecule of FADH$_2$ formed during oxidation of the fatty acid donates a pair of electron equivalents to the respiratory chain at the level of CoQ; two molecules of ATP

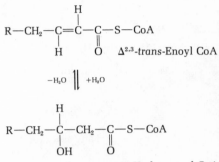

Figure 14-8
The enoylhydratase reaction.

$\Delta^{2,3}$-*trans*-Enoyl CoA

L-3-Hydroxyacyl CoA

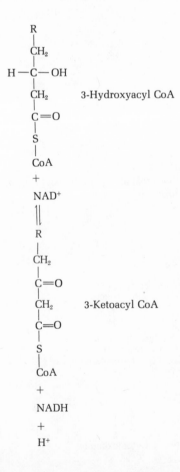

Figure 14-9
The second dehydrogenation step.

3-Hydroxyacyl CoA

3-Ketoacyl CoA

Figure 14-10
The thiolase reaction.

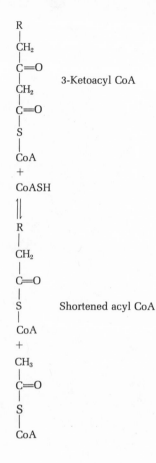

are generated during the ensuing electron transport (Chapter 13). Similarly, each molecule of NADH formed delivers its electrons to NADH dehydrogenase and subsequently results in formation of three molecules of ATP as the electrons pass to oxygen. Thus five molecules of ATP are formed per molecule of acetyl CoA cleaved. We can therefore write an equation for the oxidation of palmitoyl CoA to eight molecules of acetyl CoA which includes the phosphorylations, as follows:

$$\text{Palmitoyl CoA} + 7\text{CoA} + 7\text{O}_2 + 35\text{P}_i + 35\text{ADP} \longrightarrow$$
$$8 \text{ acetyl CoA} + 35\text{ATP} + 42\text{H}_2\text{O} \quad (1)$$

The eight molecules of acetyl CoA formed in the fatty acid cycle may now enter the tricarboxylic acid cycle. The following equation represents the balance sheet for oxidation of the acetyl CoA molecules and the coupled phosphorylations (Chapter 12):

$$8 \text{ Acetyl CoA} + 16\text{O}_2 + 96\text{P}_i + 96\text{ADP} \longrightarrow$$
$$8\text{CoA} + 96\text{ATP} + 104\text{H}_2\text{O} + 16\text{CO}_2 \quad (2)$$

Combining Equations (1) and (2), we get the overall equation for the complete oxidation of palmitoyl CoA to carbon dioxide and water

$$\text{Palmitoyl CoA} + 23\text{O}_2 + 131\text{P}_i + 131\text{ADP} \longrightarrow$$
$$\text{CoA} + 16\text{CO}_2 + 146\text{H}_2\text{O} + 131\text{ATP}$$

Over 40 percent of the standard free energy of oxidation of palmitic acid is recovered as the phosphate-bond energy of ATP.

Ketone Bodies and Their Oxidation

In many vertebrates the liver has the enzymatic capacity to divert some of the acetyl CoA derived from fatty acid or pyruvate oxidation, presumably during periods of excess formation, into free acetoacetate and D-β-hydroxybutyrate, which are transported via the blood to the peripheral tissues, where they may be oxidized via the tricarboxylic acid cycle. These compounds are called the *ketone bodies* (Figure 14-11). Acetoacetate is formed by the condensation of two molecules of acetyl CoA, catalyzed by thiolase

$$\text{Acetyl CoA} + \text{acetyl CoA} \rightleftharpoons \text{acetoacetyl CoA} + \text{CoA}$$

Figure 14-11
The ketone bodies.

Acetoacetic acid

$$\underset{\quad\;\; \text{O}}{\text{CH}_3\overset{\|}{\text{C}}\text{CH}_2\text{COOH}}$$

D-β-Hydroxybutyric acid

$$\underset{\quad\; \text{OH}}{\text{CH}_3\overset{|}{\text{C}}\text{HCH}_2\text{COOH}}$$

The acetoacetyl CoA formed in this reaction then undergoes loss of CoA to become acetoacetate in a sequence of reactions whose end result is the following reaction:

$$\text{Acetoacetyl CoA} + H_2O \longrightarrow \text{acetoacetate} + \text{CoA}$$

The free acetoacetic acid so produced is enzymatically reduced by D-*β-hydroxybutyrate dehydrogenase* to D-β-hydroxybutyric acid. This enzyme is specific for the D-stereoisomer and it does not act on the CoA esters of β-hydroxy acids

$$\text{Acetoacetate} + \text{NADH} + H^+ \rightleftharpoons \text{D-}\beta\text{-hydroxybutyrate} + \text{NAD}^+$$

The mixture of free acetoacetic and D-β-hydroxybutyric acids resulting from this reaction then may diffuse out of the liver cells into the bloodstream to the peripheral tissues. Normally the concentration of ketone bodies in the blood is very low, but in fasting or in the disease diabetes mellitus, it may reach extremely high levels. This condition, known as ketosis, arises when the rate of formation of the ketone bodies exceeds the capacity of the peripheral tissues to utilize them.

In the peripheral tissues the D-β-hydroxybutyric acid is oxidized to acetoacetic acid, which is then activated to form its CoA ester. This occurs by transfer of CoA from succinyl CoA

$$\text{Succinyl CoA} + \text{acetoacetate} \rightleftharpoons \text{succinate} + \text{acetoacetyl CoA}$$

The acetoacetyl CoA so formed is then cleaved by thiolase action

$$\text{Acetoacetyl CoA} + \text{CoA} \rightleftharpoons 2 \text{ acetyl CoA}$$

and the resulting acetyl CoA enters the tricarboxylic acid cycle.

Summary

Fatty acids are ultimately oxidized via the tricarboxylic acid cycle. Free fatty acids are first activated by esterification with CoA—SH to form acyl CoA esters at the outer mitochondrial membrane and are then converted into O-fatty acyl carnitine esters, which can cross the inner mitochondrial membrane into the matrix, where fatty acyl CoA esters are reformed. All subsequent steps in the oxidation of fatty acids take place upon their CoA esters within the mitochondrial matrix. Four reaction steps are required to remove each acetyl CoA residue: (1) the dehydrogenation of carbon atoms 2 and 3 by FAD-linked fatty acyl CoA dehydrogenases, (2) hydration of the resulting 2,3-*trans* double bond by enoylhydratase, (3) dehydrogenation of the resulting L-β-hydroxy fatty acyl CoA by an NAD$^+$-linked dehydrogenase, and (4) CoA-requiring cleavage by thiolase of the resulting β-keto fatty acyl CoA to form acetyl CoA and the CoA ester of a fatty acid shortened by two carbons. The shortened fatty acid CoA ester can then reenter the sequence. The 16-carbon palmitic acid yields ultimately 8 molecules of acetyl CoA, which are then oxidized to CO_2 via the tricarboxylic acid cycle. About 40 percent of the standard free energy of oxidation of palmitic acid is recovered by oxidative phosphorylation as ATP energy. The ketone bodies acetoacetate and β-hydroxybutyrate are formed in the liver and are carried to other tissues, where they are oxidized via acetyl CoA and the tricarboxylic acid cycle.

References

DAWSON, R. M. C., and R. N. RHODES (eds.): *Metabolism and Physiological Significance of Lipids,* John Wiley & Sons, New York, 1964.

GREVILLE, G. D., and P. K. TUBBS: "The Catabolism of Long-Chain Fatty Acids in Mammalian Tissues," in *Essays in Biochemistry,* vol. IV, Academic Press, New York, 1968. Excellent and readable review.

LEHNINGER, A. L.: *Biochemistry,* Worth Publishers, New York, 1970. Chapter 19 gives further details regarding fatty acid oxidation.

Problems

1. Write overall equations for the formation of acetyl CoA from the following substances in the liver. Include all necessary activation steps but neglect water terms.
   (a) Stearic acid
   (b) Myristoyl CoA
   (c) D-β-Hydroxybutyric acid

2. Write overall equations for the complete oxidation to CO_2 of the following substrates in the kidney. Include any activation steps required and all oxidative phosphorylations, but neglect the water terms.
   (a) Lauric acid
   (b) *n*-Butyric acid
   (c) Acetoacetic acid

3. If palmitic acid labeled with ^{14}C in carbon atom 9 is oxidized under conditions in which the tricarboxylic acid cycle is operating, which carbon atoms of the following intermediates will become labeled (a) acetyl CoA, (b) citric acid?

   Which carbon atoms of (a) and (b) would be labeled if the palmitic acid were labeled at carbon atom 16?

4. Write the overall equation for the formation of free acetoacetate from glucose in the liver.

CHAPTER **15** OXIDATIVE DEGRADATION OF AMINO
ACIDS; THE UREA CYCLE

Although amino acids function primarily as precursors
of proteins, they are often oxidized by animals as a
source of energy, particularly when they are ingested in
excess or when body proteins are utilized as fuel, as in
fasting and in the disease diabetes mellitus. In the
process of amino acid catabolism the carbon skeletons of
the amino acids are oxidized to carbon dioxide and
water, whereas their amino groups are converted into
urea or other nitrogenous excretory products, which
differ depending on the species of organism. The tricarboxylic acid cycle is the ultimate pathway for oxidation
of the carbon skeletons of the amino acids to CO_2 and
water. The oxidative degradation of amino acids occurs
largely in the liver of vertebrates, and to some extent in
the kidneys.

Removal of α-Amino Groups: Transamination

The α-amino groups of the 20 amino acids are ultimately
removed at some stage in their oxidative degradation
and, if not reused for synthesis of new amino acids, are
collected and converted into a form that may be excreted.
The removal of the α-amino groups of most of the amino
acids occurs by enzymatic reactions called generically
transaminations. In these reactions, which are catalyzed
by *transaminases* or *aminotransferases,* the α-amino
group is transferred to the α-carbon atom of an α-keto
acid, usually α-ketoglutaric acid, leaving behind the corresponding α-keto acid analog of the amino acid and

causing the amination of the α-ketoglutaric acid to glutamic acid (Figure 15-1). _Glutamic transaminase_ catalyzes the general reaction

α-Amino acid + α-ketoglutaric acid $\rightleftharpoons$

α-keto acid + glutamic acid

The reactions catalyzed by transaminases are freely reversible; they have an equilibrium constant of about 1.0. It will be noted that there is no _net_ deamination in such reactions, since the α-keto acid acceptor becomes aminated as the α-amino acid is deaminated. However, the whole point of transamination reactions is to _collect_ the amino groups of various amino acids in the form of only one α-amino acid, usually glutamic acid. Glutamic transaminase is specific for α-ketoglutarate and glutamate as one substrate pair, but is nonspecific for the other substrate pair. It therefore can bring about transfer of amino groups from many amino acids to α-ketoglutarate. There are still other transaminases, for example, _alanine transaminase_ and _aspartic transaminase_, which transfer amino groups from alanine and aspartate, respectively, to α-ketoglutarate

Alanine + α-ketoglutarate $\rightleftharpoons$ pyruvate + glutamate
Aspartate + α-ketoglutarate $\rightleftharpoons$ oxaloacetate + glutamate

Thus α-ketoglutarate is the acceptor of amino groups from most of the other amino acids and serves to channel them into a final sequence of reactions by which excretory nitrogenous end products are formed.

All transaminases have the same tightly bound coenzyme and share a common reaction mechanism. The coenzyme is _pyridoxal phosphate_, a derivative of vitamin B_6 (Chapter 8). Pyridoxal phosphate is the prosthetic group not only of the transaminases, but also of a number of other enzymes catalyzing reactions involving α-amino acids. In the transaminases pyridoxal phosphate functions as a carrier of amino groups. During the catalytic cycle this coenzyme undergoes reversible transitions between its free aldehyde form, _pyridoxal phosphate_, and its aminated form, _pyridoxamine phosphate_ (Figure 15-2), and in this manner acts as a carrier of amino groups from an α-amino acid to a α-keto acid (Figure 15-3).

Another pathway for removal of α-amino acids is by oxidative deamination; it is probably of minor impor-

Figure 15-1
The glutamate transaminase reaction. The transferred amino group is in color.

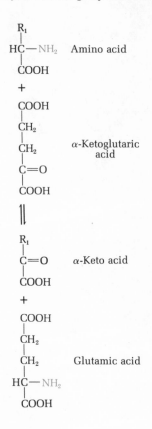

Figure 15-2
The prosthetic group of transaminases. See also Chapter 8. The transferred amino group is in color.

Pyridoxal phosphate

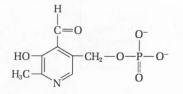

Pyridoxamine phosphate

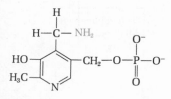

Figure 15-3
Role of pyridoxal phosphate in the action of transaminases. Ⓔ *symbolizes the enzyme-pyridoxal phosphate complex.*

$$HC-NH_2 + O=C-Ⓔ \rightleftharpoons C=O + NH_2-CH_2-Ⓔ$$

(with R_1 above and COOH below the respective carbons, and H below the central carbon)

$$C=O + NH_2-CH_2-Ⓔ \rightleftharpoons HC-NH_2 + O=C-Ⓔ$$

(with R_2 above and COOH below the respective carbons, and H below)

tance. L-*Amino acid oxidase*, a flavoprotein found in the kidney, catalyzes the reaction

$$\text{L-Amino acid} + O_2 \longrightarrow \alpha\text{-keto acid} + NH_3 + H_2O_2$$

A major role of this enzyme, which contains FMN (flavin mononucleotide) as prosthetic group, is to remove an amino group from lysine. D-*Amino acid oxidase* catalyzes the reaction

$$\text{D-Amino acid} + O_2 \longrightarrow \alpha\text{-keto acid} + NH_3 + H_2O_2$$

Presumably this enzyme functions to deaminate D-amino acids, which are found in combined form in the cell walls of bacteria and which are released in and absorbed from the digestive tract of vertebrates. The hydrogen peroxide generated in these reactions is destroyed by the enzyme *catalase*, which promotes the reaction

$$H_2O_2 \longrightarrow H_2O + \tfrac{1}{2}O_2$$

Degradation of the Carbon Skeletons of Amino Acids

There are 20 different multienzyme sequences for the oxidative degradation of the 20 different amino acids. However, they ultimately converge into five terminal pathways leading to the tricarboxylic acid cycle (Figure 15-4). The carbon skeletons of 10 of the amino acids are ultimately converted into acetyl CoA via either pyruvate

Figure 15-4
Pathways of entry of the carbon skeletons of amino acids into Krebs tricarboxylic acid cycle.

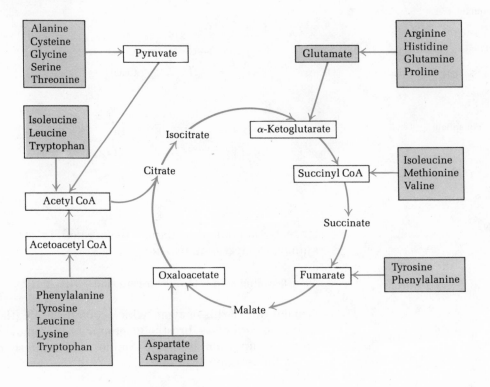

or acetoacetyl CoA. Five amino acids are converted into α-ketoglutarate, 3 to succinyl CoA, and 2 to oxaloacetate. Two amino acids, phenylalanine and tyrosine, are so degraded that one portion of the carbon skeleton enters the cycle as acetyl CoA and the other as fumarate. However, not all the carbon atoms of each of the 20 amino acids enter the tricarboxylic acid cycle, since some are lost en route by decarboxylation reactions.

We shall not describe the individual pathways for all the amino acids in detail. Rather, we shall summarize these pathways by means of "flow sheets" which show the pathways leading to each point of entry in the tricarboxylic acid cycle. A few of the enzymatic reactions that are particularly noteworthy for their mechanisms or biological significance will be emphasized.

Figure 15-5
Pathways to acetyl CoA via pyruvic acid.

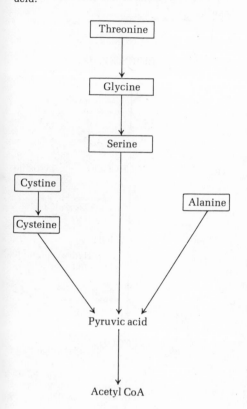

Pathways Leading to Acetyl CoA

Ten amino acids enter the tricarboxylic acid cycle by way of acetyl CoA; five are degraded to acetyl CoA via pyruvic acid and the remainder via acetoacetyl CoA. The five amino acids entering via pyruvic acid are alanine, cysteine, glycine, serine, and threonine (Figure 15-5). Alanine yields pyruvate directly on transamination with α-ketoglutarate. The four-carbon amino acid threonine is degraded to the two-carbon glycine. However, glycine is then converted to serine, a three-carbon compound, by enzymatic addition of a hydroxymethyl group carried by the coenzyme tetrahydrofolic acid (Chapter 8). Tetrahydrofolic acid serves as a coenzyme in many reactions involving the transfer of one-carbon groups, such as methyl, formyl, formimino, and hydroxymethyl groups (Figure 15-6).

The amino acids phenylalanine, tyrosine, lysine, tryptophan, and leucine yield acetyl CoA via acetoacetyl CoA (Figure 15-7). Four carbon atoms of phenylalanine and of tyrosine yield free acetoacetate (Figure 15-8), which then is activated at the expense of ATP and CoA to form acetoacetyl CoA (Chapter 14). Another four carbon atoms of tyrosine and phenylalanine are recovered as fumarate, an intermediate of the tricarboxylic acid cycle. Eight of the nine carbon atoms of these two amino acids thus may enter the tricarboxylic acid cycle; the remaining carbon atom is lost as CO_2.

Many genetic defects in amino acid metabolism have been found in human beings. Two enzymes in the phenylalanine-tyrosine pathway deserve special mention in

Figure 15-6
Tetrahydrofolic acid (right) and its role in the formation of serine from glycine (below).

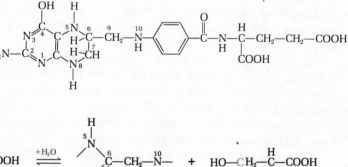

Formation of serine from glycine

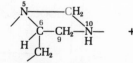   +

N^5,N^{10}-Methylenetetrahydrofolic acid
(methylene carbon in color)   Glycine   Tetrahydrofolic acid   Serine
                                                                                                  (Added carbon atom in color)

Figure 15-7
Pathways to acetyl CoA via acetoacetyl CoA.

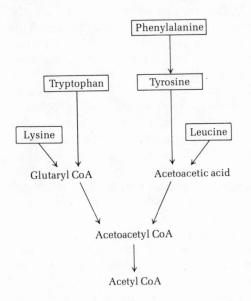

Figure 15-8
Conversion of phenylalanine and tyrosine to acetoacetic and fumaric acids.

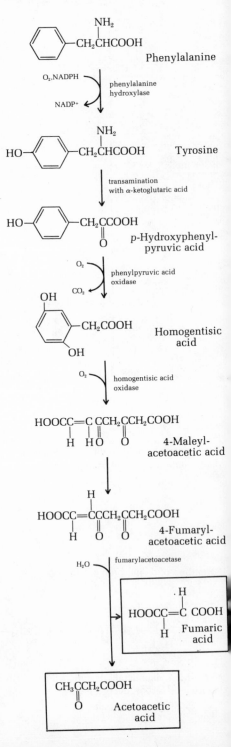

this respect. The first, *phenylalanine hydroxylase*, is absent in 1 out of every 10,000 human beings, owing to a recessive gene. In such individuals, a secondary pathway of phenylalanine metabolism that is normally little used comes into play. In this minor pathway, phenylalanine undergoes transamination to yield *phenylpyruvic acid*, which accumulates in the blood and is excreted in the urine. The excess phenylpyruvic acid in the blood during childhood impairs normal development of the brain and causes severe mental retardation. This condition is called *phenylketonuria*, often abbreviated PKU; it was among the first genetic defects of metabolism recognized in man.

The second enzyme in the phenylalanine-tyrosine pathway, *homogentisic acid oxidase*, also is defective in some humans due to a heritable genetic mutation. Persons with this defect excrete homogentisic acid in the urine, which when made alkaline and exposed to oxygen turns dark because the homogentisic acid is oxidized to a black pigment. This condition is known as *alkaptonuria*.

The pathway from tryptophan to acetyl CoA is the most complex of the pathways of amino acid catabolism; it has 13 steps. Certain intermediates in tryptophan catabolism are important precursors for biosynthesis of many other important substances, including the vitamin

Figure 15-9
Derivatives of tryptophan.

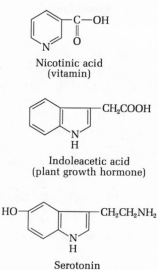

Nicotinic acid
(vitamin)

Indoleacetic acid
(plant growth hormone)

Serotonin
(contricts blood
vessels)

nicotinic acid (Chapter 8) and the plant hormone indole-acetic acid (Figure 15-9).

The α-Ketoglutaric Acid Pathway

The carbon skeletons of five amino acids (arginine, histidine, glutamic acid, glutamine, and proline) enter the tricarboxylic acid cycle via α-ketoglutarate, whose immediate precursor is glutamic acid. Figure 15-10 shows a schematic diagram of these pathways.

The Succinic Acid Pathway

The carbon skeletons of methionine, isoleucine, and valine are ultimately degraded by pathways which yield succinic acid, an intermediate of the tricarboxylic acid cycle (Figure 15-11). Isoleucine and valine have rather similar patterns of degradation. Both undergo transamination followed by oxidative decarboxylation of the resulting α-keto acids. Four of the five carbon atoms of valine are converted to succinic acid, as are three of the six carbon atoms of isoleucine.

Figure 15-10
Pathways to α-ketoglutarate.

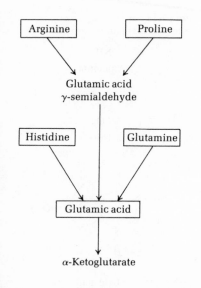

Figure 15-11
Pathways from methionine, isoleucine,
and valine to succinate.

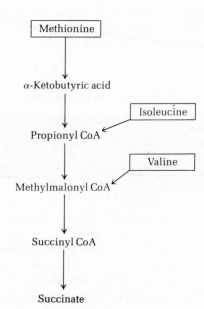

The oxidative decarboxylation of the α-keto acids derived from valine, isoleucine, and leucine by deamination is catalyzed by the same enzyme. This enzyme is genetically defective in some humans, leading to the excretion of these α-keto acids in the urine. This rare condition, which causes mental retardation in infants, is called maple syrup urine disease, because of the characteristic odor imparted to the urine by these keto acids. Several of the heritable genetic defects involving amino acid metabolism in man cause failure of certain nerve bundles to develop properly and thus lead to mental retardation.

The Fumarate Pathway

As was pointed out above, phenylalanine and tyrosine each yield two 4-carbon products, acetoacetate and fumarate (Figure 15-12). The fumarate so formed is then oxidized further in the tricarboxylic acid cycle.

The Oxaloacetic Acid Pathway

The carbon skeletons of asparagine and aspartic acid ultimately enter the tricarboxylic acid cycle via oxaloacetate (Figure 15-4). The enzyme asparaginase catalyzes the hydrolysis of asparagine to yield aspartate

$$\text{Asparagine} + H_2O \longrightarrow \text{aspartate} + NH_3$$

The aspartate then donates its amino group to α-ketoglutarate to yield glutamate; the carbon skeleton of aspartate, oxaloacetate, then enters the cycle.

Formation of Nitrogenous Excretion Products

Most organisms salvage some of the amino groups derived from the catabolism of amino acids and use them over again in the synthesis of new amino acids. The remainder are ultimately excreted by most invertebrates and vertebrates in one of three forms: urea, ammonia, or uric acid (Figure 15-13). Most terrestrial vertebrates excrete nitrogen from amino acids as urea; such organisms are termed ureotelic. Many aquatic animals, such as the teleost fishes, excrete amino groups as ammonia; they are termed ammonotelic. Birds and land-dwelling reptiles, whose water intake is limited, excrete amino nitrogen in a semisolid form as suspensions of solid uric acid; these

Figure 15-12
Formation of fumaric acid from phenylalanine and tyrosine

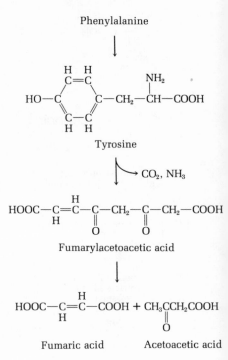

Figure 15-13
Excretory forms of amino groups.

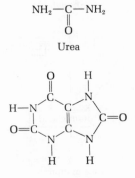

organisms are termed uricotelic. The amphibia occupy a midposition. The tadpole, which is aquatic in habits, excretes ammonia. After metamorphosis, during which the liver acquires the necessary enzymes, the adult frog forms and excretes urea.

The Urea Cycle

Urea formation, which takes place almost entirely in the liver of terrestrial vertebrates, is catalyzed by a cyclic mechanism, the *urea cycle,* first postulated by Krebs and Henseleit. In this cycle one molecule of carbon dioxide and two amino groups originally derived from α-amino acids by deamination enter the cycle to form arginine from the non-protein amino acid *ornithine* (Figure 15-14). This part of the urea cycle occurs in nearly all organisms capable of arginine biosynthesis. However, only ureotelic animals possess large amounts of the enzyme *arginase,* which catalyzes the irreversible hydrolysis of arginine to form urea and ornithine. Urea, a neutral water-soluble molecule, is excreted in the urine.

Figure 15-14
The flow sheet of the urea cycle.

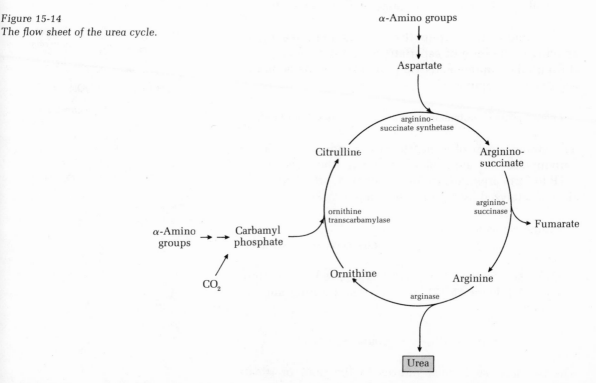

The ornithine formed then is ready for the next turn of the urea cycle.

The first amino group to enter the urea cycle arises as free ammonia by the oxidative deamination of glutamate in the mitochondria by glutamate dehydrogenase

$$\text{Glutamate} + NAD^+ \rightleftharpoons \alpha\text{-ketoglutarate} + NH_3 + NADH + H^+$$

The free ammonia is then utilized, together with carbon dioxide, to form *carbamyl phosphate* in a reaction catalyzed by *carbamyl phosphate synthetase*

$$CO_2 + NH_3 + 2ATP + H_2O \longrightarrow NH_2{-}\overset{\displaystyle\|}{\underset{\displaystyle O}{C}}{-}O{-}\overset{\displaystyle O^-}{\underset{\displaystyle \|}{\overset{\displaystyle |}{P}}}{-}O^- + 2ADP + P_i$$

Carbamyl phosphate

Carbamyl phosphate is a high-energy compound. It donates its carbamyl group to ornithine to form *citrulline* and phosphate (Figure 15-15) in a reaction catalyzed by *ornithine transcarbamylase*

Figure 15-15
Formation of citrulline.

$$\text{Carbamyl phosphate} + \text{ornithine} \longrightarrow \text{citrulline} + P_i$$

The second amino group then enters the urea cycle, arriving in the form of aspartate, which in turn acquired it from other amino acids via glutamate by the action of aspartate transaminase

$$\text{Oxaloacetate} + \text{glutamate} \rightleftharpoons \text{aspartate} + \alpha\text{-ketoglutarate}$$

The amino group of aspartate next condenses with the carbonyl carbon atom of citrulline in the presence of ATP to form argininosuccinate (Figure 15-16). This reaction is catalyzed by *argininosuccinate synthetase*

$$\text{Citrulline} + \text{aspartate} + ATP \longrightarrow$$
$$\text{argininosuccinate} + AMP + PP_i$$

Argininosuccinate is then reversibly cleaved by *argininosuccinase* (Figure 15-17) to form free arginine and fumarate

$$\text{Argininosuccinate} \longrightarrow \text{arginine} + \text{fumarate}$$

The fumarate so formed returns to the pool of tricar-

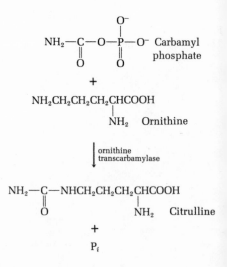

Figure 15-16
Formation of argininosuccinic acid.

$$NH_2-\underset{\underset{O}{\|}}{C}-NHCH_2CH_2CH_2\underset{\underset{NH_2}{|}}{C}HCOOH \quad Citrulline$$

+

$$HOOCCH_2\underset{\underset{NH_2}{|}}{C}HCOOH \quad Aspartic\ acid$$

ATP ⟶
AMP, PP$_i$ ⟵
Argininosuccinate synthetase

$$HN=\underset{\underset{NH}{|}}{C}-NHCH_2CH_2CH_2\underset{\underset{NH_2}{|}}{C}HCOOH$$

$$H-\underset{\underset{CH_2COOH}{|}}{C}-COOH \quad \begin{matrix}Argininosuccinic\\ acid\end{matrix}$$

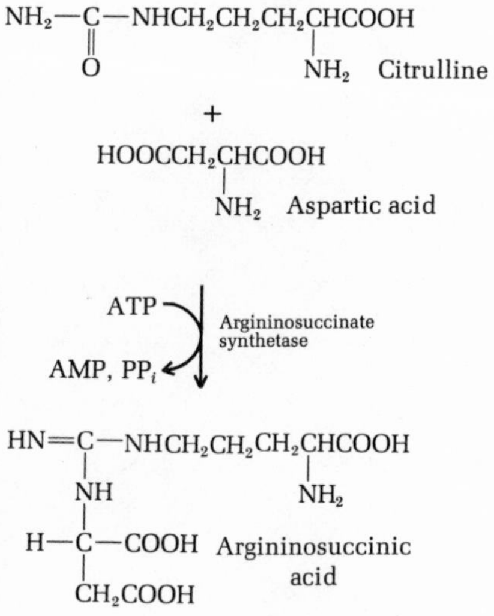

Figure 15-17
The argininosuccinase reaction.

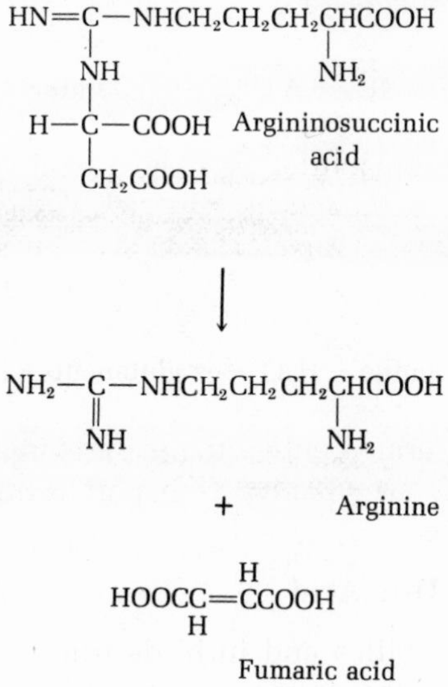

$$HN=\underset{\underset{NH}{|}}{C}-NHCH_2CH_2CH_2\underset{\underset{NH_2}{|}}{C}HCOOH$$

$$H-\underset{\underset{CH_2COOH}{|}}{C}-COOH \quad \begin{matrix}Argininosuccinic\\ acid\end{matrix}$$

↓

$$NH_2-\underset{\underset{NH}{\|}}{C}-NHCH_2CH_2\ CH_2\underset{\underset{NH_2}{|}}{C}HCOOH$$

+   Arginine

$$HOOCC\overset{H}{=}\underset{H}{C}COOH$$

Fumaric acid

Figure 15-18
Formation of urea in the arginase reaction.

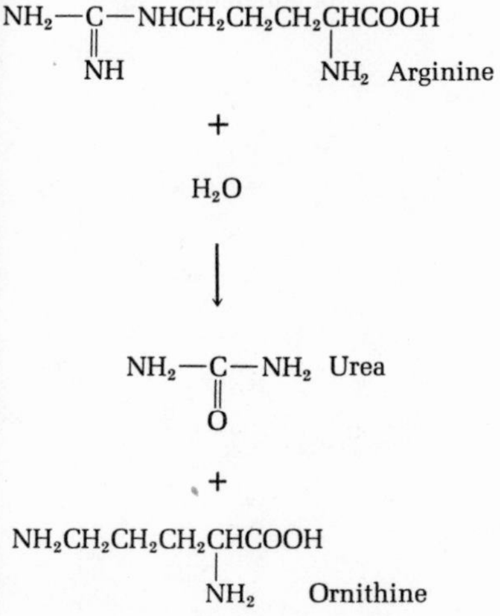

$$NH_2-\underset{\underset{NH}{\|}}{C}-NHCH_2CH_2CH_2\underset{\underset{NH_2}{|}}{C}HCOOH \quad Arginine$$

+

$$H_2O$$

↓

$$NH_2-\underset{\underset{O}{\|}}{C}-NH_2 \quad Urea$$

+

$$NH_2CH_2CH_2CH_2\underset{\underset{NH_2}{|}}{C}HCOOH \quad Ornithine$$

boxylic acid cycle intermediates. Arginase then cleaves urea from arginine and regenerates ornithine (Figure 15-18)

$$Arginine + H_2O \longrightarrow ornithine + urea$$

The overall equation of the urea cycle is

$$2NH_3 + CO_2 + 3ATP + 2H_2O \longrightarrow$$
$$urea + 2ADP + 2P_i + AMP + PP_i$$

Since the pyrophosphate formed is hydrolyzed to phosphate, the formation of one molecule of urea ultimately requires four high-energy phosphate bonds.

Ammonia Excretion

In aquatic ammonotelic animals, the amino groups derived from various α-amino acids are transaminated to α-ketoglutarate to form glutamate, which then undergoes oxidative deamination via glutamate dehydrogenase to yield free ammonia in the liver. However, because free ammonia is very toxic and is not transported to the kidneys as such, it is first converted into glutamine by *glutamine synthetase*

$$\text{Glutamate} + NH_3 + ATP \longrightarrow \text{glutamine} + ADP + P_i$$

Glutamine is transported via the blood to the kidney, where it loses its amide nitrogen as ammonium ion (NH_4^+) into the kidney tubules by the action of *glutaminase*

$$\text{Glutamine} + H_2O \longrightarrow \text{glutamate} + NH_4^+$$

The NH_4^+ so formed enters the urine. Since glutamine is nontoxic, it is an effective transport form of ammonia.

Formation of Uric Acid

In terrestrial reptiles and in birds uric acid is the chief form in which the amino groups of α-amino acids are excreted. Uric acid also happens to be the chief end product of purine metabolism in primates, birds, and terrestrial reptiles. The pathway of formation of uric acid is complex, since the purine ring must first be built from simple precursors. The origin of the nitrogen atoms of uric acid is shown schematically in Figure 15-19.

Figure 15-19
Origin of nitrogen atoms of excretory uric acid.

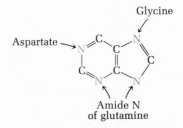

Summary

The carbon skeletons of the amino acids undergo oxidative degradation to compounds that may enter the tricarboxylic acid cycle for oxidation. The amino groups of most amino acids are removed by transamination to α-ketoglutarate, yielding glutamate. The amino groups of certain other amino acids may be removed by oxidative deamination. There are five pathways by which carbon skeletons of amino acids enter the tricarboxylic acid cycle, namely, via: (1) acetyl CoA, (2) α-ketoglutarate, (3) succinate, (4) fumarate, and (5) oxaloacetate. The amino acids entering via acetyl CoA are divided into two groups. The first, which includes alanine, cysteine, glycine, serine, and threonine, yields acetyl CoA via pyruvate, and the second (leucine, lysine, phenylalanine, tyrosine, and tryptophan) yields acetyl CoA via acetoacetyl CoA. The amino acids proline, histidine, arginine, glutamine and glutamic acid enter via α-ketoglutarate; methionine, isoleucine and valine enter via succinate; four carbon atoms of phenylalanine and tyrosine enter via fumarate; and asparagine and aspartic acid enter via oxaloacetate. In man, a number of genetic defects in amino acid catabolic pathways occur.

In ureotelic animals (terrestrial mammals and amphibia), urea is the final excretion product of amino nitrogen. It is formed in the urea cycle. Urea is formed in the liver by the action of arginase on arginine, the other cleavage product being ornithine. Arginine is resynthesized from ornithine by carbamylation of the latter to citrulline at the expense of carbamyl phosphate, followed by addition of an amino group to citrulline at the expense of aspartic acid. Ammonotelic animals (most fishes) excrete amino nitrogen as ammonia, which derives from the hydrolysis of glutamine in the kidney. Uricotelic animals (birds, terrestrial reptiles) excrete amino nitrogen as uric acid, a derivative of purine.

References

LEHNINGER, A. L.: *Biochemistry*, Worth Publishers, New York, 1970. Chapter 20 gives more detail on pathways of amino acid oxidation.

MEISTER, A.: *Biochemistry of the Amino Acids*, 2nd ed., vols. I and II, Academic Press, New York, 1965. Comprehensive and detailed treatise.

NYHAN, W. L. (ed.): *Amino Acid Metabolism and Genetic Variation*, McGraw-Hill, New York, 1967. Collection of articles on genetic alterations in amino acid metabolism.

Problems

1. Write a series of equations for the conversion of one molecule of phenylalanine into acetyl CoA and fumaric acid. How many ATP molecules will be synthesized when these products are subsequently oxidized via the tricarboxylic acid cycle?

2. Write the sequential equations for the conversion of the α-amino group of valine into urea in the liver.

3. When glutamic acid labeled with ^{14}C in the α-carbon atom and ^{15}N in the amino group undergoes oxidative degradation in the liver of a rat, in which atoms of the following metabolites will each isotope be found?

   (a) urea (b) succinic acid (c) arginine (d) citrulline (e) ornithine (f) aspartic acid.

16 PHOTOSYNTHETIC ELECTRON TRANSPORT
AND PHOTOPHOSPHORYLATION

We now turn to the molecular mechanisms by which
solar energy is captured and converted into chemical
energy by photosynthetic cells. Because solar energy is
the ultimate source of all biological energy, it might appear more logical to have considered photosynthesis
first, before examining biological oxidations in heterotrophs. However, photosynthesis is best approached after
the study of respiration, since it involves electron transport and coupled phosphorylations that are very similar
to those occurring in respiration, but not as well known.

The capacity to carry out photosynthesis is found in
both prokaryotic and eukaryotic organisms. The photosynthetic eukaryotes include not only the higher green
plants, but also the multicellular green, brown, and red
algae, as well as unicellular organisms such as eu-
glenoids and dinoflagellates. The photosynthetic prokaryotes include the blue-green algae, the green bacteria,
and the purple bacteria. More than half of all the photosynthesis on the surface of the earth is carried out by
unicellular organisms, particularly the algae.

The Equation of Photosynthesis

All photosynthetic organisms except bacteria use water
as electron or hydrogen donor to reduce carbon dioxide
or other electron acceptors; as a consequence they evolve
molecular oxygen. The overall equation for this group of
photosynthetic organisms, which includes higher plants,
may be given as

$$nH_2O + nCO_2 \xrightarrow{\text{light}} (CH_2O)_n + nO_2 \qquad (1)$$

in which n is often assigned the value 6 to correspond with formation of hexose $(C_6H_{12}O_6)$ as end product of photosynthesis.

The photosynthetic bacteria neither produce nor use molecular oxygen; in fact, most of them are strict anaerobes and are poisoned by oxygen. Instead of water, they use as electron donors either inorganic compounds, such as hydrogen sulfide, or organic compounds, such as lactic acid or isopropanol. For example, the green sulfur bacteria use hydrogen sulfide according to the equation

$$2H_2S + CO_2 \xrightarrow{\text{light}} (CH_2O) + H_2O + 2S$$

C. Van Niel, a pioneer in the study of the comparative aspects of metabolism, postulated that plant and bacterial photosynthesis are fundamentally similar processes despite the difference in electron donors. This similarity is evident if the equation of photosynthesis is written in a more general form

$$2H_2D + CO_2 \xrightarrow{\text{light}} (CH_2O) + H_2O + 2D$$

in which H_2D symbolizes a hydrogen donor and D its oxidized form. H_2D may thus be water, hydrogen sulfide, isopropanol, or any one of a number of different hydrogen donors; the nature of the hydrogen donor is characteristic for each species of photosynthetic cells. Van Niel also predicted that the molecular oxygen formed during plant photosynthesis must be derived exclusively from the oxygen atoms of water and not from the carbon dioxide. Isotopic experiments with the use of ^{18}O-labeled water and carbon dioxide are consistent with this prediction. For this reason Equation (1) for plant photosynthesis may be rewritten to show that the oxygen evolved in photosynthesis derives from water

$$2H_2\overset{\rule{0.9cm}{0.4pt}}{O} + CO_2 \xrightarrow{\text{light}} (CH_2O) + H_2O + \overset{\downarrow}{O}_2$$

The Light and Dark Reactions

Photosynthesis in green plants consists of two phases or stages, the *light reactions*, which are directly dependent on light energy, and the subsequent *dark reactions*, which can occur in the absence of light. In the light reactions, light energy is absorbed by chlorophyll and is used to generate oxygen and two energy-rich end products, ATP and the reducing agent NADPH. In the ensuing dark reactions, ATP and NADPH are then used to reduce carbon dioxide to form glucose and other organic products. The formation of oxygen and the reduction of carbon dioxide thus are distinct and separate processes.

In this chapter we shall deal largely with the light reactions in the photosynthetic cells of higher green plants. The subsequent formation of glucose in the dark reactions will be described in Chapter 17.

Intracellular Organization of Photosynthetic Systems

Before we turn to the molecular details of photosynthesis, we must sketch the intracellular organization of photosynthetic processes, since this aspect has become of central importance in the experimental approaches to the biochemical mechanisms.

In cells of higher plants the photosynthetic apparatus is localized in the chloroplasts. Chloroplasts are usually much larger than mitochondria (1–10 μ in diameter); they are often globular or diskoid in shape (Figure 16-1). They are surrounded by a single continuous outer membrane, which is rather fragile. The inner-membrane system, which is folded in a highly complex manner, encloses an internal compartment. The inner membrane is arranged in the form of many flattened vesicles called *thylakoid disks*, which are usually stacked transversely across the chloroplast. These stacks are called grana (Figure 16-1). The grana contain all the photosynthetic pigments of the chloroplast and also the enzymes required for the primary light-dependent reactions. Chloroplasts may be isolated by differential centrifugation of extracts of plant cells.

Excitation of Molecules by Light

Visible light is a form of electromagnetic radiation of wavelength 4000 to 7000 Å. It arises from the nuclear fusion of hydrogen atoms to form helium atoms and electrons in the sun. The overall reaction is

$$4{}_1^1H \longrightarrow {}^4He + 2e^- + h\nu$$

The symbol $h\nu$ represents a *quantum* of light energy. The

Figure 16-1
Electron micrograph of a chloroplast of Nitella and schematic drawing showing arrangement of the membranes in the thylakoid disks. Studies of developing chloroplasts indicate that the thylakoid vesicles are derived from the inner membrane.

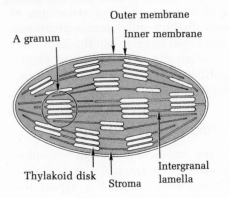

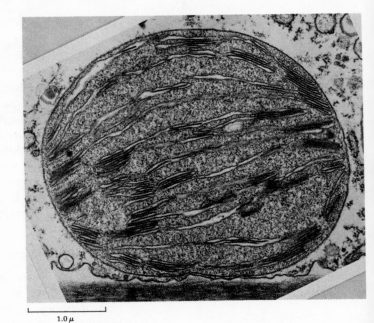

1.0 μ

energy of light quanta, which are also called *photons,* is inversely proportional to their wavelength (Table 16-1). Thus, photons of short wavelength, at the violet end of the spectrum, have the greatest energy.

The ability of a substance to absorb light depends on its atomic structure, particularly on the arrangement of electrons surrounding the nucleus. When light strikes an atom or molecule capable of absorbing light at a given wavelength, light energy is absorbed by some of the electrons, which are boosted to a higher energy level; the atom or molecule thus goes into an energy-rich *excited state.* Only photons of certain wavelengths can excite a given atom or molecule because the excitation of molecules is quantized; that is, light energy is absorbed only in discrete packets on an all-or-none basis, thus the term *quantum.* Excited molecules are usually very unstable; when the incident light is turned off, the high-energy electrons return to their normal orbitals and the excited molecule reverts to its original state, called the *ground state.* This return is accompanied by loss of the energy originally absorbed during excitation, either as heat or light; the light so emitted is called *fluorescence.* Now let us examine the characteristic light-absorbing pigments of green-plant cells.

Table 16-1 Energy content of photons

Wavelength nm	Color	kcal per einstein*	
400	Violet	71.8	(71.8)
500	Blue	57.7	(57.7)
600	Yellow	47.8	
700	Red	40.6	

* One einstein is 6.023×10^{23} photons.

The Photosynthetic Pigments

All photosynthetic cells contain one or more types of the class of green pigments known as *chlorophylls*, but not all photosynthetic cells are green; photosynthetic algae and bacteria may be brown, red, or purple. This variety of colors results because, besides chlorophyll, most photosynthetic cells also contain members of two other classes of light-trapping pigments, the yellow *carotenoids* and the blue or red *phycobilins*, often called *accessory pigments*.

Chlorophylls

All oxygen-producing photosynthetic cells of higher plants contain two types of chlorophyll, one of which is always chlorophyll *a*. Chlorophyll *a* (Figure 16-2) contains four substituted pyrrole rings, one of which (ring IV) is reduced. The pyrrole rings are arranged in a macrocyclic structure in which the four central nitrogen atoms are coordinated with Mg^{2+} to form an extremely stable, essentially planar complex. Chlorophyll has a long, hydrophobic terpenoid side chain, consisting of the alcohol *phytol* esterified to a propionic acid substituent in ring IV (Figure 16-2).

The second type of chlorophyll in oxygen-producing cells is either chlorophyll *b* (green plants), chlorophyll *c* (brown algae, diatoms, and dinoflagellates), or chlorophyll *d* (red algae). Photosynthetic cells that produce no oxygen do not contain a second chlorophyll.

That chlorophyll must be the primary light-trapping molecule in green cells has been established by means of a *photochemical action spectrum,* a plot of the efficiency of different wavelengths of visible light in supporting photosynthetic oxygen evolution. Figure 16-3 shows the action spectrum of photosynthesis in green algae and the light absorption spectrum of the cells, which is largely a function of their chlorophyll content. Since the action spectrum corresponds closely with the absorption spectrum, it has been concluded that chlorophyll is quantitatively the predominant light-trapping molecule in photosynthesis.

The Accessory Pigments

The carotenoids and phycobilins, photosynthetic pigments which absorb light at wavelengths other than those absorbed by the chlorophylls, serve as supplemen-

Figure 16-2
Chlorophyll a.

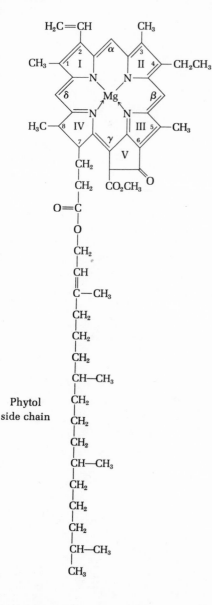

Phytol
side chain

Figure 16-3
Correspondence between the light absorption
spectrum and the photochemical action
spectrum of the green alga Ulva taeniata.

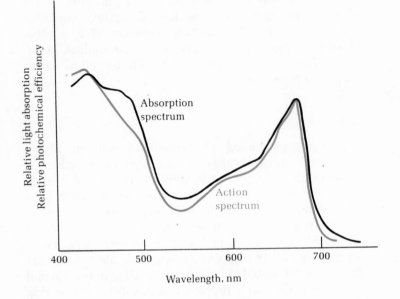

tary light receptors for portions of the visible spectrum not completely covered by chlorophyll.

Among the photosynthetic carotenoids are β-*carotene* and *lutein* (Figure 16-4). The phycobilin pigments (Figure 16-5) occur only in red and blue-green algae. Phycobilins are open-chain tetrapyrroles (Figure 16-5), in contrast to chlorophyll, which is a cyclic tetrapyrrole. Phycobilins are conjugated to specific proteins.

The photosynthetic pigments in the chloroplasts of plants are connected with characteristic electron-transport chains (see below).

The Hill Reaction and Photoinduced Electron Transport

The first experimental evidence that absorption of light energy causes electron flow came from the discovery of R. Hill in 1937 that light-dependent oxygen evolution can take place in cell-free preparations obtained from photosynthetic organisms. Illumination of these preparations, which contained chloroplasts, caused evolution of oxygen and simultaneous reduction of an electron acceptor, according to the general equation

Figure 16-4
Important carotenoid pigments in photosynthetic cells.

β-Carotene   Lutein

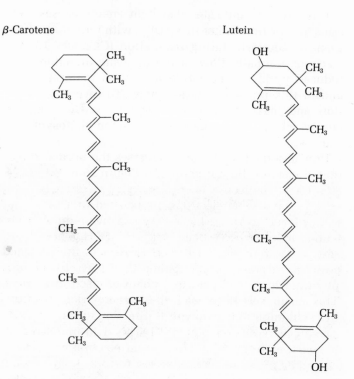

Figure 16-5
Phycoerythrobilin, a red photosynthetic pigment.

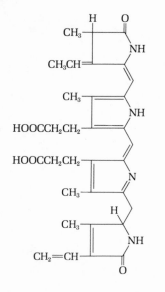

$$2H_2O + 2A \xrightarrow{\text{light}} 2AH_2 + O_2 \qquad (2)$$

in which A is the hydrogen or electron acceptor and AH_2 its reduced form. No electron donor other than water was required. Various artificial electron acceptors, such as reducible dyes, were found to accept electrons (and hydrogen) from water. Most important, Hill found that carbon dioxide was not required for this reaction nor was it reduced to a stable form that accumulated, demonstrating that oxygen evolution and carbon dioxide reduction can be dissociated from each other. The reaction summarized in Equation (2) is universally known as the *Hill reaction* and the artificial acceptor A as a *Hill reagent*. Subsequently it was found that NADP is the natural biological electron acceptor in green plants, according to the equation

$$2H_2O + 2NADP^+ \xrightarrow{\text{light}} 2NADPH + 2H^+ + O_2$$

This equation indicates that light energy causes electrons to flow from water to $NADP^+$, with formation of oxygen. In contrast, during respiration (Chapters 12, 13) electrons normally flow in the opposite direction from reduced pyridine nucleotide to oxygen to form water, and in this process release energy. To reverse electron flow and make it go from oxygen to $NADP^+$ requires energy, which is furnished by light in the photosynthetic process.

How absorbed light energy reverses the normal direction of electron flow is shown in Figure 16-6. When the chlorophyll molecule is excited by absorption of light energy, some of its electrons are boosted to a very high energy level, so that it has a very negative reduction potential, more negative than that of NADPH. These high-energy electrons do not directly return to their normal positions or ground state. Rather, they leave the chlorophyll molecule and pass to a chain of electron carriers. This chain serves to lead these energy-rich electrons from chlorophyll in a downhill direction toward $NADP^+$, causing its reduction to NADPH. This light-induced flow of electrons from chlorophyll to an electron acceptor is known as photosynthetic electron transport. The reduction of an electron acceptor by such a chain is called photoreduction. The electron "holes" left in chlorophyll after its excitation and loss of electrons to $NADP^+$ must be refilled in order for the process to continue. The requisite electrons ultimately come from water, by a pathway we shall consider below. As water loses electrons to refill these electron holes, free molecular oxygen is formed.

Noncyclic Electron Flow and Noncyclic Photophosphorylation

The light-induced electron flow taking place in the Hill reaction is called noncyclic electron flow, since electrons pass from H_2O via chlorophyll to $NADP^+$ or some other electron acceptor, causing its reduced form to accumulate. This process continues as long as electron acceptor is still available. We shall see below that there is another type of photoinduced electron flow that is cyclic in nature.

In 1957, D. Arnon and his colleagues discovered that noncyclic electron flow in isolated chloroplast preparations is accompanied by the coupled formation of ATP from ADP and phosphate. They postulated that the energy required for this process, called photophosphorylation, ultimately came from the light energy absorbed by

Figure 16-6
Noncyclic photoinduced electron transport from H_2O to $NADP^+$ with evolution of oxygen. Movement of electrons upward requires input of light energy.

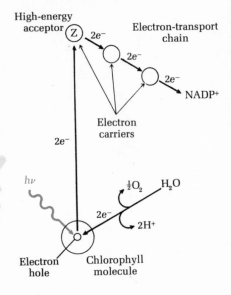

the chlorophyll. It is now believed that two molecules of ATP are formed during the passage of each pair of electrons from H_2O to $NADP^+$. This type of light-induced photophosphorylation is called *noncyclic photophosphorylation*. The overall equation of noncyclic electron transport and photophosphorylation is

$$H_2O + A + 2P_i + 2ADP \xrightarrow{\text{light}} AH_2 + \tfrac{1}{2}O_2 + 2ATP + 2H_2O$$

where A designates the electron acceptor and AH_2 its reduced form. Two molecules of water are formed during the dehydrating condensation of P_i and ADP.

The Two Light Reactions in Plant Photosynthesis

Several lines of evidence indicate that there are two light reactions in oxygen-evolving plant photosynthesis. Plant chloroplasts contain two distinct types of chlorophyll, whereas the photosynthetic bacteria, which do not evolve oxygen, usually contain only one. Moreover, it has been found that plants must be illuminated simultaneously at two different portions of the spectrum for maximal efficiency, one region at 680 nm and the other between 500 and 600 nm. From these and other observations it has been postulated that green plants have two photosystems. Photosystem I, which is activated by light at 680 nm, is associated with chlorophyll *a* and is not involved in oxygen evolution. Photosystem II, which is activated by shorter wavelengths of light, appears to be involved in oxygen evolution; it utilizes a second type of chlorophyll (*b*, *c*, or *d*), as well as accessory pigments. All oxygen-evolving photosynthetic cells contain both photosystems I and II, whereas the photosynthetic bacteria, which do not evolve oxygen, contain only photosystem I. Each photosystem has its own characteristic assembly of light-absorbing pigments clustered on the thylakoid membranes of the chloroplast.

Interrelationships between Photosystem I and Photosystem II

Now some important questions arise. How are photosystems I and II related in green plants? Do they function independently, or are they connected?

The available experimental evidence is best explained by the concept that the two photosystems operate in a connected, sequential fashion as shown in Figure 16-7. This formulation, which is accepted by most inves-

Figure 16-7
Series arrangement of photosystems I and II. The two systems are connected by an electron-transport chain between acceptor Q of photosystem II and P700 of photosystem I. Non-cyclic electron flow employs both systems, starting from water and ending in NADPH. Cyclic electron flow requires only photosystem I; electrons boosted to Z in photosystem I can return to ground state P700 via a shunt provided by cytochrome b₆. The entire diagram is drawn to conform to the scale of reduction potentials at the left. Movement of electrons upward requires input of energy from light; movement downward proceeds spontaneously. End-products are in colored boxes.

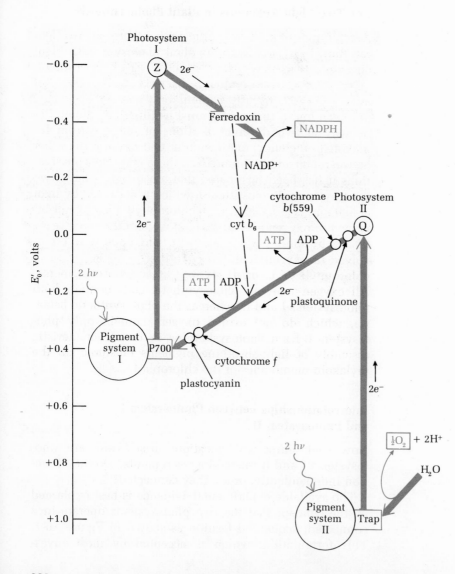

tigators, postulates that when photosystem I is excited, it loses electrons which are passed along a chain of electron carriers to NADP$^+$, causing its reduction. The electron hole left in photosystem I must then be refilled with electrons. It is postulated that the electrons required ultimately come from water via another chain of electron carriers extending from photosystem II to photosystem I. When photosystem II is illuminated, some of its electrons are boosted to a high-energy level and then flow downhill via the connecting electron-transferring chain to the electron hole in photosystem I, restoring the latter to its reduced state. During this process of electron flow from photosystem II to photosystem I, two molecules of ATP are generated per pair of electrons, in energy-coupling reactions resembling those of oxidative phosphorylation in mitichondria (Chapter 13). The restoration of the active center of photosystem II to its reduced state results from downhill transfer of electrons from water, with evolution of oxygen.

The overall equation for these reactions is

$$H_2O + NADP^+ + 2P_i + 2ADP \xrightarrow{\text{light}}$$
$$\tfrac{1}{2}O_2 + NADPH + H^+ + 2ATP + 2H_2O$$

The molecule of H_2O on the left hand side is the electron (hydrogen) donor; the two molecules of H_2O on the right result from the formation of ATP from ADP and phosphate.

Cyclic Electron Flow and Cyclic Photophosphorylation

There is another type of light-induced electron flow that takes place in chloroplasts of green plants, namely, *cyclic electron flow*, which can be explained by the formulations in Figures 16-7 and 16-8. Cyclic electron flow can be recognized only by an effect produced by the flow, namely, the phosphorylation of ADP to ATP. When isolated chloroplasts are illuminated in the absence of any added electron donor or electron acceptor no accumulation of a reduced substance such as NADPH occurs. Yet under these conditions a light-dependent phosphorylation of ADP takes place with formation of ATP. It has therefore been concluded that the energy required for the phosphorylation of ADP comes from light and that the phosphorylation must be coupled to the flow of electrons from excited chlorophyll along a chain of electron carriers in such a way that the electrons return to the electron holes left in the deexcited chlorophyll. The return is made possible by a shunt or bypass,

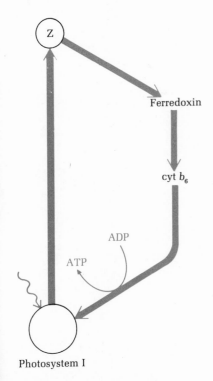

Figure 16-8
Cyclic photoinduced electron transport and phosphorylation. This process is made possible by a shunt. Compare with Figure 16-7.

289

so that electrons from photosystem I return to chlorophyll a directly and do not pass to NADP. This process is called cyclic photoinduced electron flow. Although it is not possible to measure the amount of cyclic electron flow directly, since the electrons do not accumulate in any product, it is possible to measure the ATP produced in this process, called *cyclic photophosphorylation*. It is inhibited by poisons acting primarily on the electron carriers. The equation for cyclic electron flow is

$$P_i + ADP \xrightarrow{\text{light}} ATP + H_2O$$

Cyclic electron flow and cyclic photophosphorylation are believed to occur when the plant cell is amply supplied with reducing power in the form of NADPH, but requires additional ATP for its metabolic needs.

Electron Carriers in Photosynthetic Electron Transport

At least three electron carriers are involved in the transfer of electrons from photosystem I to $NADP^+$. The first is a pigment called *P700*, because it has a light absorption band at 700 nm which is bleached on illumination. P700 is believed to be a special derivative of chlorophyll a. The second carrier is *ferredoxin*, a nonheme-iron protein, which resembles the nonheme iron protein of mitochondrial electron transport (Chapter 13). Spinach ferredoxin has a molecular weight of about 11,600. It contains two iron atoms, which are bound to two specific sulfur atoms; the latter are released from ferredoxin as H_2S on acidification. Ferredoxin can be reduced and reoxidized via one-electron steps.

Ferredoxin in its reduced state cannot pass its electrons directly to $NADP^+$, but an enzyme capable of bridging this gap has been isolated from spinach chloroplasts. It is a flavoprotein called *ferredoxin-NADP oxidoreductase*. It participates as follows:

$$2Fd_{reduced} + 2H^+ + NADP_{ox} \xrightarrow{\text{oxidoreductase}} 2Fd_{ox} + NADPH + H^+$$

Fd is the symbol for ferredoxin. It is possible that there is another unidentified electron carrier, designated Z in Figure 16-7, between P700 and ferredoxin.

Several electron carriers different from those above participate in the transfer of electrons from photosystem II to photosystem I. Three distinctive types of cytochromes occur in this chain. They are not the same as those which participate in mitochondrial electron transport. The first is *cytochrome f* (Latin, *frons,* leaf); it is bound to chloroplast structure and can be released only by use of alkaline nonpolar solvents. It has a molecular weight of about 100,000 and contains two hemes per molecule. Cytochrome b_6, or b_{563}, has not been obtained in water-soluble form, being very tightly bound to the chloroplast membrane structure. Another b cytochrome, cytochrome b_3, or b_{559}, has been identified in the chain between photosystems I and II. This chain also contains a blue copper protein called *plastocyanin* (Figure 16-7) and two quinones, vitamin K_1 and *plastoquinone* (Figure 16-9), the latter an analog of the coenzyme Q of mitochondria. Very likely other carriers occur which remain to be identified.

The carrier sequence in the photosynthetic electron-transport chain between photosystems II and I has been studied by many of the same methods used in analysis of mitochondrial electron transport; however, there are some large gaps in our knowledge of this chain.

Figure 16-9

Plastoquinone. In different plant species n varies from 6 to 10.

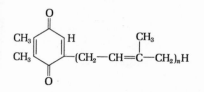

The Energetics of Photosynthesis

The maximum thermodynamic efficiency of photosynthesis has been one of the most hotly debated problems in biochemistry. The standard free energy change for the synthesis of hexose from CO_2 and H_2O is

$$6CO_2 + 6H_2O \longrightarrow C_6H_{12}O_6 + 6O_2 \qquad \Delta G^{\circ\prime} = +686 \text{ kcal}$$

Since the caloric value of a light quantum depends on its wavelength and may range from 70 kcal at 400 nm to about 42 kcal at 700 nm (Table 16-1), the lower of these values is used for calculations of efficiency.

The quantum requirement of photosynthesis in the intact cell is determined by experimental measurement of the amount of light absorbed by a suspension of photosynthesizing cells in relation to the amount of carbon dioxide reduced. In order to determine the efficiency of photosynthesis it is necessary to know how many light quanta are required by the photosynthetic cell to yield one molecule of hexose. Although there are many experimental difficulties in such measurements, most investigators agree that 48 quanta are required per molecule of hexose or 8 quanta for each molecule of oxygen evolved. Thus $48 \times 42 = 2016$ kcal are required, which represents a thermodynamic efficiency of $686/2016 = 38$ percent, or nearly the same as that of the overall process of glucose combustion and oxidative phosphorylation (Chapter 13).

We may now anticipate a point developed more fully in Chapter 17 and give the overall equation for the reduction of CO_2 to glucose during the second or dark phase of photosynthesis

$$6CO_2 + 12NADPH + 12H^+ + 18ATP + 12H_2O \longrightarrow$$
$$C_6H_{12}O_6 + 12NADP^+ + 18ADP + 18P_i$$

This equation requires 18 molecules of ATP and 12 of NADPH to reduce six molecules of CO_2. If two phosphorylations take place per pair of electrons during light-induced electron transport, as most investigators postulate, then 24 ATPs and 12 NADPHs will be formed, as in the equation

$$48h\nu + 12H_2O + 12NADP^+ + 24P_i + 24ADP \longrightarrow$$
$$6O_2 + 12NADPH + 12H^+ + 24ATP + 12H_2O$$

Obviously, more ATP is formed by this reaction than is needed for the reduction of 6 molecules of carbon dioxide in the dark reactions, but the plant cell has many other uses for ATP.

The efficiency of photosynthesis in nature is much lower than the figures calculated for the basic molecular process. It has been calculated from the output of fixed carbon by a field of corn that only about 1 to 2 percent of the solar energy falling on the field is recovered. Sugar cane is much more efficient; it can recover up to 8 percent of the captured light in the form of organic products.

Summary

Water is the electron donor and the source of the oxygen evolved by photosynthesis in green plants. However, the photosynthetic bacteria, which do not evolve oxygen, employ H_2S, or organic compounds as electron donors. Green plant photosynthesis can be divided into two stages, light reactions and dark reactions. In the former, absorbed light energy is used to cause electrons to flow from water to $NADP^+$ and simultaneously generate ATP. In the dark reactions, the NADPH and ATP are used to generate glucose from carbon dioxide. In green plant cells photosynthesis takes place in the thylakoids, flattened vesicles within chloroplasts, which are stacked to form the grana. Photosynthetic cells contain three types of light-capturing pigments, chlorophylls, carotenoids, and phycobilins. Light energy is absorbed by the pigment assembly, which goes into an excited state and yields high energy electrons. There are two photosystems in oxygen-evolving green plant cells. Photosystem I contains chlorophyll a and is activated by longer wavelengths of light. Photosystem II contains a second type of chlorophyll and is activated by shorter wavelengths; it is responsible for oxygen evolution. Organisms that do not evolve O_2 lack photosystem II. Photosystems I and II are linked in series. Boosting an electron to a highly reducing potential by excitation of photosystem I leads to reduction of $NADP^+$ via a chain of carriers, including ferredoxin and ferredoxin-NADP oxidoreductase. The electrons required to fill the electron holes left in photosystem I come from excited photosystem II via a central electron-transport chain. The electrons required to fill the electron holes in photosystem II comes from H_2O.

Phosphorylation of ADP is coupled to the central electron-transport chain. Apparently eight light quanta at 700 nm are required to evolve each molecule of oxygen and to yield 2NADPHs and 4ATPs, which are then utilized to reduce each molecule of CO_2 used to form hexose.

References

ARNON, D. I.: "Photosynthetic Activity of Isolated Chloroplasts," *Physiol. Rev.*, **47**:317–358 (1967). The role of ferredoxin and the "parallel" hypothesis for photosystems I and II.

BOARDMAN, N. K.: "The Photochemical System of Photosynthesis," *Advan. Enzymol.*, **30**:1–80 (1968). A more comprehensive review; up to date and impartial.

HILL, R.: "The Biochemists' Green Mansions: The Photosynthetic Electron Transport Chain in Plants," in P. N. Campbell and G. D. Greville (eds.), *Essays in Biochemistry*, vol. 1, pp. 121–152, Academic Press, New York, 1965. An interesting account written by a pioneer.

PART III BIOSYNTHESIS AND THE
UTILIZATION OF PHOSPHATE-BOND
ENERGY

PART III BIOSYNTHESIS AND THE
UTILIZATION OF PHOSPHATE-BOND
ENERGY

The cellular processes that require free energy are (1) bio-synthesis, in which chemical work is carried out, (2) contraction and motility, which are forms of mechanical work, and (3) active transport of nutrients or inorganic ions against concentration gradients, which is osmotic or concentration work. The free energy required for these processes is furnished primarily by the phosphate-bond energy of ATP.

Fundamental to all living organisms is the biosynthesis of cell constituents from simple precursors, the principal process involved in the creation and maintenance of the intricate orderliness of living cells. We shall now, in Part III of this book, consider the biosynthesis of carbohydrates, lipids, amino acids, and nucleotides. In Part IV we shall examine the biosynthesis of the informational macromolecules—the nucleic acids and proteins—into which genetic information is incorporated.

As we now turn from the biochemistry of catabolism and ATP synthesis to the biochemistry of anabolism and ATP utilization, some of the organizing principles of biosynthetic pathways require re-emphasis. The first is that the chemical pathway taken in the biosynthesis of a biomolecule is not usually identical to the pathway taken in its degradation. The two pathways may contain one or even several identical steps, but there is always at least one enzymatic step that is dissimilar in the anabolic and catabolic pathways leading to and from a given biomolecule. If the reactions of catabolism and anabolism were catalyzed by the same set of enzymes acting reversibly, no stable biological structure of any complexity could exist, because cell macromolecules necessarily would change in amount whenever the concentrations of their precursors fluctuated.

The second organizing principle underlying biosynthetic reactions is that energy-requiring biosynthetic processes are obligatorily coupled to the energy-yielding breakdown of ATP, in such a way that the overall coupled reaction is exergonic and is thus essentially irreversible in the direction of biosynthesis. The total amount of phosphate-bond energy utilized in a given biosynthetic pathway usually exceeds the minimum amount of free energy required to bring about the biosynthesis.

The third important principle is that biosynthetic reactions are regulated independently of the mechanisms regulating the corresponding catabolic reactions. Such independent control is made possible by the fact that the regulatory or allosteric enzyme(s) controlling the rate of the catabolic pathway do not participate in the anabolic pathway. Biosynthetic pathways are primarily regulated by the concentration of their end products. The regulatory enzyme that is under allosteric control by the end product is almost always the first enzyme in the sequence, starting from some key precursor or from a branch point in a metabolic chain. This arrangement avoids wasting precursors to make unneeded intermediates. These relationships emphasize again the principle of maximum economy in the molecular logic of living cells.

The biosynthesis of glucose and other carbohydrates from simpler precursors is one of the most prominent biosynthetic processes carried out by living organisms. Heterotrophic animals bring about the conversion of pyruvate, lactate, amino acids, and other simple precursors into glucose and also glycogen. Photosynthetic organisms generate enormous amounts of hexoses from carbon dioxide and water; these hexoses in turn are converted into starch, cellulose, and other plant polysaccharides.

The Synthesis of Glucose 6-Phosphate from Pyruvic acid

We have seen that the conversion of glucose 6-phosphate to pyruvate is the central pathway of carbohydrate breakdown in most organisms, under either aerobic or anaerobic conditions. In a comparable manner, the reverse process, the conversion of pyruvate to glucose 6-phosphate (Figure 17-1) is a central pathway in the biosynthesis of monosaccharides and polysaccharides, in both plants and animals. The formation of carbohydrate from non-carbohydrate precursors, such as amino acids and lactic acid, is called *gluconeogenesis*, literally, the formation of new sugar. An example is the synthesis of glucose in the liver from lactic acid of the blood, which occurs during recovery from intense muscular activity, in which much glucose is broken down to lactic acid by skeletal muscles. Plants also use most of the pathway between pyruvate and glucose 6-phosphate in the photosynthetic formation of hexoses from carbon dioxide (Figure 17-1), as we shall see.

Most of the reaction steps in the biosynthetic pathway from pyruvic acid to glucose are catalyzed by enzymes of the glycolytic cycle and thus proceed by reversal of the steps employed in glycolysis (Figure 17-2). However, there are three irreversible steps in the normal "downhill" glycolytic pathway (Chapter 11) which cannot be utilized in the "uphill" conversion of pyruvate to glucose. In the biosynthetic direction these steps are bypassed by alternative reactions (Figure 17-2) which are more favorable for synthesis. The first of these is the conversion of pyruvate to phosphoenolpyruvate, which obviously cannot occur by direct reversal of the pyruvate kinase reaction because of its large positive standard free energy change

$$\text{Pyruvate} + \text{ATP} \rightarrow$$
$$\text{phosphoenolpyruvate} + \text{ADP} \qquad \Delta G^{\circ\prime} = +7.5 \text{ kcal}$$

Instead, the phosphorylation of pyruvate is achieved by a roundabout sequence of reactions which requires the cooperation of enzymes in both the cytoplasm and the mitochondria (Figure 17-2). The first step is catalyzed by *pyruvate carboxylase* of mitochondria

$$\text{Pyruvate} + CO_2 + \text{ATP} \xrightarrow{\text{acetyl CoA}} \text{oxaloacetate} + \text{ADP} + P_i$$

Pyruvate carboxylase is a regulatory enzyme which is completely inactive in the absence of its positive modulator acetyl CoA.

The oxaloacetate formed from pyruvate is next reduced to malate in the mitochondria at the expense of NADH

$$\text{NADH} + H^+ + \text{oxaloacetate} \rightleftharpoons \text{NAD}^+ + \text{malate}$$

The malate then diffuses out of the mitochondria into the surrounding cytoplasm, where it is reoxidized by the cytoplasmic form of NAD-linked malate dehydrogenase to yield extramitochondrial oxaloacetate

$$\text{Malate} + \text{NAD}^+ \rightleftharpoons \text{oxaloacetate} + \text{NADH} + H^+$$

The oxaloacetate so formed is acted upon by *phosphoenolpyruvate carboxykinase* to yield phosphoenolpyruvate, a reaction in which guanosine triphosphate (GTP) serves as the phosphate donor

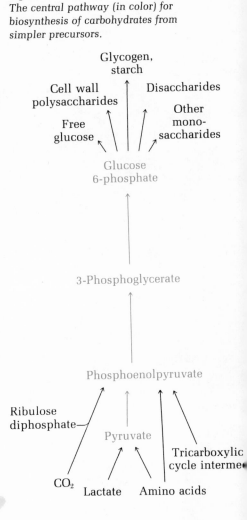

Figure 17-1
The central pathway (in color) for biosynthesis of carbohydrates from simpler precursors.

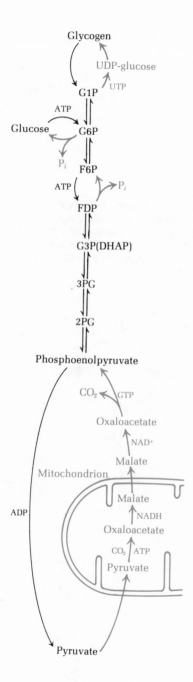

Figure 17-2
Bypass reactions (in color) in the
synthesis of glucose and glycogen
from pyruvate.

$$Oxaloacetate + GTP \xrightarrow{Mg^{2+}}$$
$$phosphoenolpyruvate + CO_2 + GDP$$

This enzyme is found in the cytoplasm of the liver in the rat and mouse. We may now write the overall equation of this bypass pathway for the formation of phosphoenolpyruvate from pyruvate

$$Pyruvate + ATP + GTP \rightleftharpoons$$
$$phosphoenolpyruvate + ADP + GDP + P_i$$

Two high-energy phosphate bonds, one from ATP and one from GTP, each equivalent to -7.3 kcal, must ultimately be expended to phosphorylate one molecule of pyruvate to phosphoenolpyruvate. We shall see other biosynthetic reactions in which two or more high-energy phosphate bonds of ATP are ultimately consumed to bring about the formation of but a single covalent bond.

The second reaction of the downhill glycolytic sequence which is bypassed during gluconeogenesis is the phosphorylation of fructose 6-phosphate by phosphofructokinase. During glucose synthesis this irreversible reaction is bypassed (Figure 17-2) by the enzyme *diphosphofructose phosphatase*, also known as *fructose diphosphatase*, which carries out the essentially irreversible hydrolysis of the 1-phosphate group to yield fructose 6-phosphate

$$Fructose\ 1,6\text{-}diphosphate + H_2O \rightarrow$$
$$fructose\ 6\text{-}phosphate + P_i \qquad \Delta G^{\circ\prime} = -3.9 \text{ kcal}$$

This is a regulatory enzyme and it is strongly inhibited by the negative modulators AMP or ADP. In the next step in the pathway to glucose, fructose 6-phosphate is reversibly converted into glucose 6-phosphate by phosphohexoisomerase

$$Fructose\ 6\text{-}phosphate \rightleftharpoons glucose\ 6\text{-}phosphate$$

The glucose 6-phosphate so formed has several possible fates. In the liver of vertebrates it yields free blood glucose. This does not occur by reversal of the hexokinase reaction, but rather by the action of *glucose 6-phosphatase*, which catalyzes the irreversible hydrolytic reaction

$$Glucose\ 6\text{-}phosphate + H_2O \rightarrow$$
$$glucose + P_i \qquad \Delta G^{\circ\prime} = -3.3 \text{ kcal}$$

This enzyme is characteristically found in the endoplasmic reticulum of the liver of vertebrates. Glucose 6-phosphatase is not present in muscles or in the brain, which have no capacity to furnish free glucose to the blood.

We may now sum up the biosynthetic reactions leading from pyruvate to glucose in the following overall equation:

$$2CH_3COCOOH + 4ATP + 2GTP + 2NADH + 2H^+ + 6H_2O \rightarrow$$
$$glucose + 2NAD^+ + 4ADP + 2GDP + 6P_i$$

For each molecule of glucose formed, six high-energy phosphate bonds are consumed and two molecules of NADH are required as reductant. This equation is clearly not the simple reverse of the equation for the conversion of glucose into pyruvate

$$Glucose + 2ADP + 2P_i + 2NAD^+ \rightarrow$$
$$2CH_3COCOOH + 2ATP + 2NADH + 2H^+ + 2H_2O$$

In order to drive the "uphill" formation of glucose from pyruvate, six high-energy phosphate bonds must be invested in order to provide sufficient "push."

Regulation of the Pathway from Pyruvate to Glucose 6-Phosphate

Figure 17-3 summarizes the control points in the pathway leading from pyruvate to glucose 6-phosphate. As in most biosynthetic pathways, the first step is catalyzed by a regulatory enzyme, pyruvate carboxylase, which is stimulated by its positive modulator acetyl CoA. As a consequence, glucose synthesis is promoted whenever excess mitochondrial acetyl CoA builds up beyond the immediate need of the cell for fuel. The secondary control point of this pathway, the reaction catalyzed by diphosphofructose phosphatase, is inhibited by AMP but stimulated by ATP. Thus control is exerted over the pathway leading from pyruvate to glucose both by the level of respiratory fuel (acetyl CoA) and by ATP.

Gluconeogenesis from Tricarboxylic Acid Cycle Intermediates

The just-described pathway from pyruvate to glucose also allows the net synthesis of glucose from various precursors of pyruvate or phosphopyruvate. Chief among them are the tricarboxylic acid cycle intermediates citrate, isocitrate, cis-aconitate, α-ketoglutarate, succinate, and

Figure 17-3
Control points in the pathway between pyruvate and glucose 6-phosphate. Positive modulation of regulatory enzymes is designated by parallel arrows in color; negative modulation by ▬▬

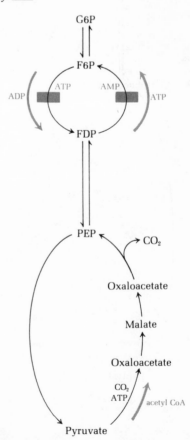

fumarate, which may undergo conversion to malate via the tricarboxylic acid cycle. Malate then may leave the mitochondria and undergo oxidation to oxaloacetate in the extramitochondrial cytoplasm, where the formation of phosphoenolpyruvate takes place by the action of the cytoplasmic phosphoenolpyruvate carboxykinase described above.

The net conversion of tricarboxylic cycle intermediates into new glucose is readily observed in animals treated with the toxic glycoside *phloridzin*. This poison blocks reabsorption of glucose from the kidney tubule and thus causes the blood glucose to be excreted nearly quantitatively into the urine. Feeding of succinic acid or other cycle intermediates to phloridzin-poisoned animals causes excretion of an amount of glucose equivalent to three carbon atoms of the intermediate fed. However, acetyl CoA cannot yield "new" glucose in animal tissues owing to the irreversibility of formation of acetyl CoA in the pyruvate dehydrogenase reaction (Chapter 12).

Gluconeogenesis from Amino Acids

As has been shown in Chapter 15, some or all of the carbon atoms of the amino acids derived from proteins are ultimately convertible by animals into either acetyl CoA or into certain intermediates of the tricarboxylic acid cycle. Amino acids which can serve as precursors of phosphoenolpyruvate and thus of glucose are called *glycogenic* amino acids. Examples are alanine, glutamic acid, and aspartic acid, which on deamination yield pyruvic, α-ketoglutaric, and oxaloacetic acids, respectively, and thus gain entry into the tricarboxylic acid cycle to become precursors of malate and ultimately phosphoenolpyruvate.

Gluconeogenesis from Acetyl CoA via the Glyoxylate Cycle

Plants and many microorganisms are, however, able to carry out the net synthesis of carbohydrate from fatty acids by way of acetyl CoA, a process made possible by the reactions of the glyoxylate cycle (Chapter 12). This cycle permits the net conversion of acetyl CoA to succinic acid according to the overall reaction

2 Acetyl CoA + NAD$^+$ + 2H$_2$O $\rightarrow$

$$\text{succinate} + 2\text{CoA} + \text{NADH} + \text{H}^+$$

Two specific enzymes are required for this pathway,

isocitrate lyase and malate synthetase (Chapter 12); they do not occur in higher animals. The succinate formed in the glyoxylate pathway yields oxaloacetate, which in turn is the precursor of phosphoenolpyruvate. By this pathway stored fat is converted into glucose by germinating seeds.

Photosynthetic Formation of Hexoses by Reduction of Carbon Dioxide

The photosynthetic formation of hexoses from carbon dioxide is a massive biosynthetic process in the plant world. We have seen that absorbed light energy is conserved by photosynthetic cells as the phosphate-bond energy of ATP and as reducing power in the form of NADPH (Chapter 16). The ATP and NADPH so generated are then utilized to bring about the reduction of carbon dioxide to form glucose and other reduced products in the dark phase of photosynthesis. Although the photosynthetic formation of hexoses utilizes part of the biosynthetic pathway from pyruvate to glucose we have discussed above, special mechanisms are first required to convert or "fix" carbon dioxide in an organic form.

An important clue to the nature of the CO_2 fixation mechanism in photosynthesis first came from the work of M. Calvin and his associates. They illuminated green algae in the presence of radioactive carbon dioxide ($^{14}CO_2$) for very short intervals (only a few seconds) and then quickly killed the cells, extracted them, and with the aid of chromatographic methods searched for the earliest radioactive metabolites in which the labeled carbon could be found. One of the compounds which became labeled very early was 3-phosphoglyceric acid, a known intermediate of glycolysis; the isotope was found predominantly in its carboxyl carbon atom. (This carbon atom does not become labeled rapidly in animal tissues in the presence of radioactive CO_2.) These findings strongly suggested that the labeled 3-phosphoglyceric acid is an early intermediate in photosynthesis in algae, a view supported by the fact that 3-phosphoglyceric acid is readily converted into glucose, as we have seen above. Later, an enzyme that catalyzes incorporation of $^{14}CO_2$ into the carboxyl carbon of 3-phosphoglycerate was found in large amounts in green leaves. This enzyme, called diphosphoribulose carboxylase or ribulose diphosphate carboxylase, catalyzes the carboxylation and cleavage of the five-carbon sugar ribulose 1,5-diphosphate to form two molecules of 3-phosphoglyceric acid, one of which

Figure 17-4
Fixation of CO_2 by diphosphoribulose carboxylase. The carbon atom of the entering CO_2 is in color.

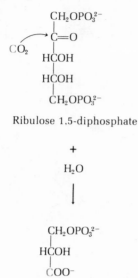

Ribulose 1,5-diphosphate

+

H_2O

CH$_2$OPO$_3^{2-}$

HCOH

COO$^-$

+

COO$^-$

HCOH

CH$_2$OPO$_3^{2-}$

3-Phosphoglycerate
(two molecules)

bears the isotopic carbon introduced as CO_2 (Figure 17-4). This enzyme has a molecular weight of 550,000 and is extremely abundant in the cell, making up about 15 percent of the total protein of the chloroplast.

The 3-phosphoglycerate formed by this enzyme can then be converted into glucose 6-phosphate by reversal of the glycolytic reactions and the diphosphofructose phosphatase bypass reaction described above. It is noteworthy that the glyceraldehyde 3-phosphate dehydrogenase of green plants, which is required to reduce 3-phosphoglycerate to glyceraldehyde 3-phosphate, requires NADPH as reductant rather than NADH.

This set of reactions does not by itself account for the fact that all six carbon atoms of hexose are ultimately formed from CO_2 during photosynthesis. To provide such a pathway, Calvin proposed a cyclic mechanism for free hexose synthesis (Figure 17-5) in which one molecule of ribulose 1,5-diphosphate is regenerated for each molecule of CO_2 reduced. Figure 17-6 shows the reactions of this pathway. It employs seven of the enzymes of the synthetic glycolytic pathway, three enzymes of the phosphogluconate pathway (Chapter 12), and phosphoribulokinase (see below). One way of writing the overall equation of this complex cycle is

6 Ribulose 1,5-diphosphate + 6CO$_2$

+ 18ATP + 12NADPH + 12H$^+$ →

6 ribulose 1,5-diphosphate + hexose

+ 18P$_i$ + 18ADP + 12NADP$^+$

Ribulose 1,5-diphosphate is written on both sides of the equation only to show that it is a necessary component which is regenerated at the end of the cycle. The net reaction, after canceling out the ribulose 1,5-diphosphate on both sides, is

6CO$_2$ + 18ATP + 12NADPH + 12H$^+$ →

hexose + 18P$_i$ + 18ADP + 12NADP$^+$

The sequential reactions contributing to this overall equations may now be examined (Figure 17-6). Reactions (1) to (8) are reactions already given for the formation of glucose from CO_2 and ribulose 1,5-diphosphate; reactions (9) to (15) are concerned with the regeneration of ribulose 1,5-diphosphate. The glucose formed as end product [reaction (8)] is shown in a box. The seven-carbon sugar phosphate sedoheptulose 1,7-diphosphate is a charac-

Figure 17-5
The photosynthetic formation of glucose from CO₂ via the Calvin
cycle in spinach leaves. The inputs are shaded in gray and the
products in color. Abbreviations are 3PG = 3-phosphoglyceric
acid; G3P = glyceraldehyde 3-phosphate; DHAP = dihydroxyace-
tone phosphate, FDP = fructose 1,6-diphosphate, F6P = fructose
6-phosphate, G6P = glucose 6-phosphate, E4P = erythrose 4-phos-
phate, X5P = xylulose 5-phosphate; SDP = sedoheptulose 1,7-
diphosphate; S7P = sedoheptulose 7-phosphate; R5P = ribose
5-phosphate; Ru5P = ribulose 5-phosphate; RuDP = ribulose 1,5-
diphosphate.

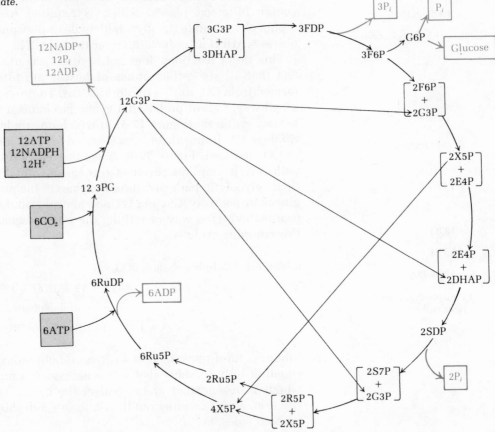

teristic intermediate in this pathway. This complex
sequence of reactions is now believed to represent the
major pathway for photosynthetic formation of hexoses
in many plants.

In certain crop plants, such as corn, sugar cane, and
sorghum, as well as many tropical plants, the initial fixation of carbon dioxide into organic form does not occur
by the ribulose diphosphate carboxylase reaction described above, but rather by the carboxylation of phos-

Figure 17-6
Sequence of reactions during photo-
synthetic formation of glucose
(in box) *from* CO_2.

$6CO_2$ + 6 ribulose 1,5-diphosphate →

12 3-phosphoglycerate   (1)

12 3-Phosphoglycerate + 12ATP →

12 1,3-diphosphoglycerate   (2)

12 1,3-Diphosphoglycerate
+ 12NADPH + $12H^+$ →
12 glyceraldehyde 3-phosphate + $12NADP^+$ (3)

5 Glyceraldehyde 3-phosphate →

5-dihydroxyacetone phosphate   (4)

3 Glyceraldehyde 3-phosphate
+ 3 dihydroxyacetone phosphate →
3 fructose 1,6-diphosphate   (5)

3 Fructose 1,6-diphosphate →

3 fructose 6-phosphate + $3P_i$ (6)

Fructose 6-phosphate → glucose 6-phosphate   (7)

Glucose 6-phosphate → glucose + P_i (8)

2 Fructose 6-phosphate
+ 2 glyceraldehyde 3-phosphate $\xrightarrow{\text{transketolase}}$
xylulose 5-phosphate + 2 erythrose 4-phosphate   (9)

2 Erythrose 4-phosphate
+ 2 dihydroxyacetone phosphate $\xrightarrow{\text{aldolase}}$
2 sedoheptulose 1,7-diphosphate   (10)

2 Sedoheptulose 1,7-diphosphate $\xrightarrow{\text{phosphatase}}$
2 sedoheptulose 7-phosphate + $2P_i$ (11)

2 Sedoheptulose 7-phosphate
+ 2 glyceraldehyde 3-phosphate $\xrightarrow{\text{transketolase}}$
2 ribose 5-phosphate + 2 xylulose 5-phosphate   (12)

2 Ribose 5-phosphate $\xrightarrow{\text{isomerase}}$ 2 ribulose 5-phosphate   (13)

4 Xylulose 5-phosphate $\xrightarrow{\text{epimerase}}$ 4 ribulose 5-phosphate   (14)

6 Ribulose 5-phosphate + 6ATP $\xrightarrow{\text{phosphoribulokinase}}$
6 ribulose 1,5-diphosphate + 6ADP   (15)

phoenolpyruvate, catalyzed by *phosphoenolpyruvate carboxykinase*

Phosphoenolpyruvate + CO_2 + H_2O → oxaloacetate + P_i

The oxaloacetate so formed is ultimately converted into glucose via 3-phosphoglycerate.

Synthesis of Glycogen and Starch

Glucose 6-phosphate is not only the precursor of free blood glucose, but also of the storage polymers glycogen and starch. It is first converted into glucose 1-phosphate by phosphoglucomutase (Chapter 11)

Glucose 6-phosphate $\rightleftharpoons$ glucose 1-phosphate

$$\Delta G^{\circ\prime} = +1.74 \text{ kcal}$$

Although glycogen phosphorylase (Chapter 11) can convert glucose 1-phosphate into glycogen in the test tube, under intracellular conditions it catalyzes only the breakdown of glycogen to glucose 1-phosphate. In the cell there is a different pathway for the conversion of glucose 1-phosphate to glycogen, one which involves a reaction principle widely utilized in the biosynthesis of disaccharides, oligosaccharides, and polysaccharides. Through the work of L. Leloir and his colleagues it is now recognized that the donor of the monosaccharide residue in most such biosynthetic reactions is a nucleoside diphosphate sugar derivative, which is formed from a nucleoside 5'-triphosphate (NTP) and the sugar 1-phosphate by the action of enzymes known generically as *pyrophosphorylases*

NTP + sugar 1-phosphate $\rightleftharpoons$ NDP-sugar + PP_i

The nucleoside diphosphate so formed now serves as a sugar carrier. In glycogen synthesis in animals glucose 1-phosphate reacts with uridine triphosphate (UTP) to yield uridine diphosphate glucose (UDP-glucose)

α-D-Glucose 1-phosphate + UTP $\rightleftharpoons$ UDP-glucose + PP_i

by the action of UDP-glucose pyrophosphorylase. The structure of UDP-glucose is shown in Figure 17-7. The resulting pyrophosphate undergoes hydrolysis to two molecules of phosphate.

Figure 17-7
Uridine diphosphate glucose.

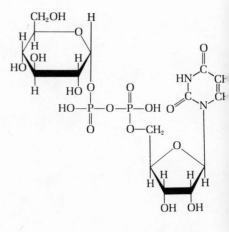

Figure 17-8
Action of amylo (1,4 → 1,6) trans-glycosylase (branching enzyme) on a branch of glycogen.

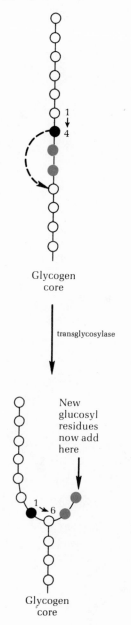

In the second step leading to glycogen formation, the glycosyl group of UDP-glucose is transferred to the terminal glucose residue at the nonreducing end of an amylose chain, to form an $\alpha(1 \to 4)$ glycosidic linkage between carbon atom 1 of the added glycosyl residue and the 4-hydroxyl of the terminal glucose residue of the chain. This reaction is catalyzed by *glycogen synthetase*

$$\text{UDP-glucose} + (\text{glucose})_n \to \text{UDP} + (\text{glucose})_{n+1}$$

The overall equilibrium of this set of three reactions greatly favors synthesis of glycogen. Glycogen synthetase requires as a primer an $\alpha(1 \to 4)$ polyglucose chain having at least four glucose residues, to which it adds successive glucosyl groups. However, it is more active with long-chain glucose polymers as primers.

Glycogen synthetase exists in two forms, one of which is inactive in the absence of glucose 6-phosphate. When the glucose 6-phosphate concentration in the cell becomes elevated, this metabolite functions as a positive allosteric modulator which can increase the activity of this form of glycogen synthetase over 40-fold, thus causing glycogen to be made from the excess glucose 6-phosphate. Glycogen storage and breakdown are also regulated by certain hormones (Chapter 20).

Glycogen synthetase cannot make the $\alpha(1 \to 6)$ bonds found in the branch points of glycogen (Chapters 5, 11). However, a glycogen-branching enzyme, *amylo(1,4 → 1,6) transglycosylase*, which is present in many animal tissues, catalyzes transfer of a terminal oligosaccharide fragment of six or seven glucosyl residues from the end of the main glycogen chain to the 6-hydroxyl group of a glucose residue of the same or another glycogen chain, in such a manner as to form an $\alpha(1 \to 6)$ linkage and thus create a branch point (Figure 17-8), to which further glucosyl residues may be added by glycogen synthetase.

In plant tissues starch synthesis occurs by an analogous pathway catalyzed by *amylose synthetase*. However, ADP-glucose (ADPG) rather than UDP-glucose (UDPG) is the active glucose donor in most plants

$$\text{ATP} + \alpha\text{-D-glucose 1-phosphate} \rightleftharpoons \text{ADP-glucose} + \text{PP}_i$$

$$\text{ADP-glucose} + (\text{glucose})_n \to \text{ADP} + (\text{glucose})_{n+1}$$

Conversion of Glucose 6-Phosphate into Other Hexoses

Pathways for the conversion of glucose 6-phosphate into D-fructose 6-phosphate and then into D-mannose 6-phosphate have been described in Chapter 11. Another important pathway is the conversion of D-glucose into D-galactose. For this purpose, UDP-glucose is first formed from glucose 1-phosphate. Enzymatic epimerization of UDP-glucose then occurs at carbon atom 4 of the glucose residue to form uridine diphosphate galactose

$$\text{UDP-glucose} \rightleftharpoons \text{UDP-galactose}$$

UDP-galactose formed in this reaction is a precursor in the synthesis of the disaccharide lactose in the mammary gland.

Synthesis of the Disaccharides Lactose and Sucrose

Glucose 6-phosphate formed by gluconeogenesis is also the ultimate precursor of various disaccharides. In animals both the galactose and glucose portions of lactose or milk sugar are derived from glucose 6-phosphate. UDP-galactose, formed by the reactions outlined above, reacts with D-glucose to yield lactose

$$\text{UDP-galactose} + \text{D-glucose} \rightarrow \text{UDP} + \text{lactose}$$

In many plants the abundant disaccharide sucrose (Chapters 5, 11) is formed by the following reactions:

$$\text{ATP} + \text{glucose} \rightarrow \text{glucose 6-phosphate} + \text{ADP}$$

$$\text{Glucose 6-phosphate} \rightleftharpoons \text{glucose 1-phosphate}$$

$$\text{UTP} + \text{glucose 1-phosphate} \rightarrow \text{UDP-glucose} + \text{PP}_i$$

$$\text{ATP} + \text{fructose} \rightarrow \text{fructose 6-phosphate} + \text{ADP}$$

$$\text{UDP-glucose} + \text{fructose 6-phosphate} \rightarrow$$
$$\text{UDP} + \text{sucrose 6}'\text{-phosphate}$$

$$\text{Sucrose 6}'\text{-phosphate} + \text{H}_2\text{O} \rightarrow \text{sucrose} + \text{P}_i$$

Sum: $2\text{ATP} + \text{UTP} + \text{glucose} + \text{fructose} \rightarrow$
$$\text{sucrose} + 2\text{ADP} + \text{UDP} + \text{PP}_i + \text{P}_i$$

Summary

The central common pathway in the biosynthesis of all carbohydrates from noncarbohydrate precursors is the route from pyruvate to glucose. However, the three irreversible reactions of the glycolysis sequence are replaced by "bypass" reactions that are energetically favorable for synthesis. In the first, pyruvate is converted to phosphoenolpyruvate by the mitochondrial sequence pyruvate $\xrightarrow{CO_2}$ oxaloacetate → malate followed by the cytoplasmic sequence malate → oxaloacetate $\xrightarrow{GTP}$ phosphoenolpyruvate. The second bypass is the hydrolysis of fructose 1,6-diphosphate to fructose 6-phosphate catalyzed by fructose diphosphatase, and the third is the hydrolysis of glucose 6-phosphate by glucose 6-phosphatase to yield free glucose. The overall pathway, which requires input of six high-energy phosphate bonds, is largely regulated by the first reaction of the sequence, catalyzed by pyruvate carboxylase.

Tricarboxylic acid cycle intermediates can undergo net conversion to form new glucose, a process called gluconeogenesis, since each can form oxaloacetate, as can lactate. Similarly, those amino acids capable of yielding oxaloacetate also may be converted into glucose. Neither acetyl CoA nor CO_2 can undergo net conversion into glucose in animal tissues. However, in plants and microorganisms, acetyl CoA can be converted into glucose by operation of the glyoxylate cycle.

In plant photosynthesis, CO_2 gains entry into the carbon backbone of glucose by a dark reaction with ribulose 1,5-diphosphate to yield two molecules of 3-phosphoglycerate, which can be converted into glucose. At the expense of ATP and NADPH generated in the light reactions, six molecules of CO_2 can ultimately be converted into glucose by cyclic operation of the Calvin cycle, which consists of reactions of the phosphogluconate and glycolytic pathways.

The pathway from glucose 6-phosphate to starch and glycogen involves the regulatory enzyme glycogen synthetase, with uridine diphosphate glucose as intermediate glucose carrier. Glycogen synthetase and glycogen phosphorylase activities are independently controlled in muscle and liver.

Nucleoside diphosphate sugars also serve as precursors of some monosaccharides such as D-galactose, of disaccharides such as sucrose and lactose, and of other polysaccharides.

References

GREENBERG, D.M. (ed.): *Metabolic Pathways*, vols. I and II, 3d ed., Academic Press, New York, 1967. A series of articles and reviews; especially good on biosynthetic role of the tricarboxylic acid cycle.

HASSID, W. Z.: "Biosynthesis of Oligosaccharides and Polysaccharides in Plants," *Science*, **165**:137–144 (1969).

LEHNINGER, A. L.: *Biochemistry*, Worth Publishers, New York, 1970. Chapter 22 gives further information on the biosynthesis of carbohydrates.

ZELITCH, I.: *Photosynthesis, Photorespiration, and Plant Productivity*, Academic Press, New York, 1971.

Problems

1. Pyruvic acid labeled with the isotope ^{14}C in the carboxyl carbon atom was injected into a rat. The rat was sacrificed 30 minutes later and some glycogen isolated from the liver. The glycogen was hydrolyzed to D-glucose by boiling in dilute HCl. Which carbon atoms of the resulting D-glucose would be most strongly labeled with ^{14}C?

2. The experiment in Problem 1 is repeated, but instead of labeled pyruvic acid the rat is given some L-alanine labeled with ^{14}C in the α-carbon atom. Which carbon atoms of the D-glucose residues of the liver glycogen will become most strongly labeled?

3. Write a series of equations for the enzymatic conversion of the carbon skeleton of L-alanine into free glucose in the liver.

4. How many high-energy phosphate bonds are required to convert two molecules of alanine into one glucose residue of liver glycogen?

5. Write a balanced equation for the addition of one glucose residue to a glycogen molecule in the liver starting from free glucose.

CHAPTER 18 THE BIOSYNTHESIS OF FATTY ACIDS AND LIPIDS

Because the capacity of higher animals to store polysaccharides such as glycogen is rather limited, glucose ingested in excess of their immediate caloric needs and storage capacity is converted into fatty acids and they in turn into triacyglycerols or neutral fats, which may be deposited in large amounts in the adipose or fat tissues. Triacyglycerols are also made and stored in large amounts in the seeds of higher plants. Phospholipids are synthesized by both plants and animals to serve as components of membranes.

Biosynthesis of the Saturated Fatty Acids

After the discovery of the pathway of fatty acid oxidation in mitochondria and the role of CoA in this process (Chapter 14), it was widely expected that the biosynthesis of fatty acids would be found to occur by reversal of the same enzymatic steps employed for their oxidation. However, it was ultimately discovered that the biosynthesis of fatty acids occurs by the action of a different pathway, catalyzed by different enzymes, and proceeds in a different part of the cell.

Figure 18-1
Malonyl CoA.

COOH
CH₂
C—S—CoA
O

The fatty acid synthetase system catalyzes the following overall reaction, in which one molecule of acetyl CoA and seven molecules of malonyl CoA (Figure 18-1) are combined and reduced to make a molecule of palmitic acid. The reducing power required is furnished by NADPH

Acetyl CoA + 7 malonyl CoA + 14NADPH + 14H⁺ →

$$CH_3(CH_2)_{14}COOH + 7CO_2 + 8CoA + 14NADP^+ + 6H_2O$$

Palmitic acid

A novel feature of this pathway is that the three-carbon acid malonic acid (as its CoA ester) is the precursor of the two-carbon units from which the fatty acid chain is built. The single molecule of acetyl CoA required in the process serves as a *primer* or *starter*; the methyl and carboxyl carbon atoms of the acetyl group become carbon atoms 16 and 15, respectively, of the palmitic acid formed. Chain growth during fatty acid synthesis thus starts at the carboxyl group of the acetyl CoA and proceeds by successive addition of acetyl residues at the carboxyl end of the growing chain. Each successive acetyl residue is ultimately derived from the two carbon atoms of malonyl CoA that are nearest the CoA group; the third carbon atom, that is, that of the unesterified carboxyl group, is lost as CO_2.

A second distinctive feature of the mechanism of fatty acid synthesis is that the acyl intermediates in the process are covalently bound thioesters of a small-molecular-weight protein called *acyl carrier protein* (abbreviated ACP), which has an essential —SH group. Acyl ACP forms a complex with one or more of the six other enzyme proteins required for the complete synthesis of palmitic acid. In yeast cells, all seven proteins of the fatty acid synthetase complex are tightly associated in a cluster (Figure 18-2), but in E. coli, they normally occur as single entities. We shall now examine the individual steps in fatty acid synthesis in more detail.

Formation of Malonyl CoA

In animal tissues the immediate source of all the carbon atoms of palmitic acid synthesized by the fatty acid synthetase complex, which is present in the cytoplasm, is cytoplasmic acetyl CoA, which in turn ultimately derives from the intramitochondrial acetyl CoA. The latter first reacts with oxaloacetate to form citrate, which then passes into the cytoplasm and undergoes cleavage to yield acetyl CoA by the ATP-dependent citrate-cleavage enzyme

Citrate + ATP + CoA → acetyl CoA + ADP + P_i + oxaloacetate

The malonyl CoA required as the immediate precursor of 14 of the 16 carbon atoms of palmitic acid is formed

Figure 18-2
The fatty acid synthetase complex
of yeast. It consists of six enzymes
and acyl carrier protein.

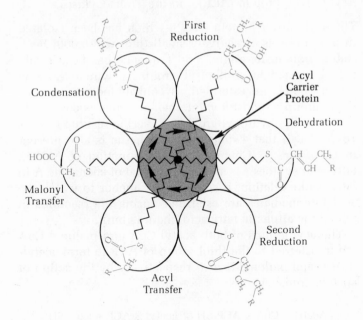

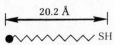

20.2 Å

The 4′-phosphopantetheine
side chain of ACP

Figure 18-3
The acetyl CoA carboxylase reaction.

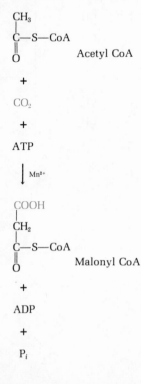

from cytoplasmic acetyl CoA and carbon dioxide (Figure 18-3) by the action of *acetyl CoA carboxylase*

$$ATP + acetyl\ CoA + CO_2 \rightarrow malonyl\ CoA + ADP + P_i$$

The CO_2 becomes the unesterified carboxyl group of malonyl CoA. Acetyl CoA carboxylase contains the vitamin biotin (Chapter 8) as its prosthetic group, covalently bound in amide linkage to the ϵ-amino group of a specific lysine residue of the enzyme protein. The biotin serves as an intermediate carrier of carbon dioxide in a two-step reaction cycle (Figure 8-7)

$$CO_2 + ATP + biotin\text{-}enzyme \rightleftharpoons$$
$$carboxybiotin\text{-}enzyme + ADP + P_i$$

$$Carboxybiotin\text{-}enzyme + acetyl\ CoA \rightleftharpoons$$
$$malonyl\ CoA + biotin\text{-}enzyme$$

Acetyl CoA carboxylase is a regulatory enzyme. It catalyzes the rate-limiting step in the synthesis of fatty acids. The carboxylation of acetyl CoA by the liver enzyme is greatly accelerated by the positive modulator citrate; in fact, the enzyme shows little or no activity in the absence of citrate.

Acyl Carrier Protein (ACP) and the Transacylases

The acyl carrier protein (ACP), which has been isolated in pure form, is a relatively small (mol wt 10,000) heat-stable protein. Its amino acid sequence has been established. Its single sulfhydryl group, to which the acyl intermediates are esterified, is contributed by its prosthetic group *4'-phosphopantetheine*, which is covalently linked to the hydroxyl group of a serine residue (Figure 18-4). Recall that 4'-phosphopantetheine is also present in coenzyme A (Chapter 14). The function of ACP in fatty acid synthesis is analogous to that of coenzyme A in fatty acid oxidation. It serves as an anchor to which the acyl intermediates are esterified during the reactions in which the aliphatic fatty acid chain is built.

The acyl groups of both acetyl CoA and malonyl CoA are transferred to the thiol group of ACP to form *acetyl-S-ACP* and *malonyl-S-ACP* respectively by the action of specific *transacylases*

$$\text{Acetyl-S-CoA} + \text{ACP-SH} \rightleftharpoons \text{acetyl-S-ACP} + \text{CoA-SH}$$

$$\text{Malonyl-S-CoA} + \text{ACP-SH} \rightleftharpoons \text{malonyl-S-ACP} + \text{CoA-SH}$$

The Steps in Fatty Acid Synthesis

Acetyl-S-ACP and malonyl-S-ACP combine with loss of the unesterified carboxyl group of malonyl-S-ACP as carbon dioxide, to yield acetoacetyl-S-ACP (Figure 18-5)

$$\text{Acetyl-S-ACP} + \text{malonyl-S-ACP} \rightleftharpoons$$
$$\text{acetoacetyl-S-ACP} + CO_2 + \text{ACP-SH}$$

The carbon dioxide so formed contains the same carbon atom that was introduced as CO_2 by the acetyl CoA carboxylase reaction. Carbon dioxide thus plays a catalytic role in fatty acid synthesis, since it is regenerated as each two-carbon unit is inserted. The acetoacetyl-S-ACP next undergoes reduction with NADPH to form the D-β-hydroxybutyryl-S-ACP (Figure 18-6), catalyzed by β-*ketoacyl ACP reductase*

$$\text{Acetoacetyl-S-ACP} + \text{NADPH} + H^+ \rightarrow$$
$$\text{D-}\beta\text{-hydroxybutyryl-S-ACP} + \text{NADP}^+$$

It is noteworthy that the product is not the same stereoisomeric form as the β-hydroxyacyl intermediate in fatty acid oxidation, which is the L stereoisomer (Chapter 14).

Figure 18-4
Structure of the prosthetic group of acyl carrier protein (ACP). In the yeast fatty acid synthetase complex, which consists of ACP plus the 6 enzymes of the cycle, the 4'-phosphopantetheine moiety appears to serve as a swinging arm to rotate the fatty acyl group undergoing lengthening from one enzyme to the next.

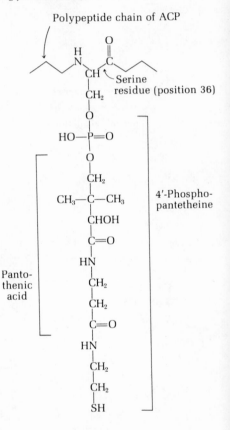

Figure 18-5
The initiation of chain growth in fatty acid synthesis.

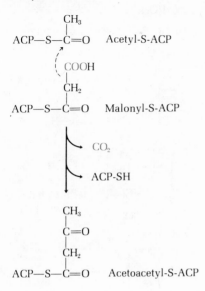

$$CH_3$$
$$|$$
$$ACP—S—C{=}O \quad \text{Acetyl-S-ACP}$$

$$COOH$$
$$|$$
$$CH_2$$
$$|$$
$$ACP—S—C{=}O \quad \text{Malonyl-S-ACP}$$

$$\searrow CO_2$$

$$\searrow ACP\text{-}SH$$

$$CH_3$$
$$|$$
$$C{=}O$$
$$|$$
$$CH_2$$
$$|$$
$$ACP—S—C{=}O \quad \text{Acetoacetyl-S-ACP}$$

Figure 18-6
Intermediates in fatty acid synthesis.

D-β-Hydroxybutyryl-S-ACP

$$OH$$
$$|$$
$$CH_3—C—CH_2—C—S—ACP$$
$$| \qquad\quad \|$$
$$H \qquad\quad O$$

Crotonyl-S-ACP

$$H$$
$$CH_3—C{=}C—C—S—ACP$$
$$\quad\quad H \quad \|$$
$$\qquad\qquad O$$

Butyryl-S-ACP

$$CH_3—CH_2—CH_2—C—S—ACP$$
$$\qquad\qquad\qquad\quad \|$$
$$\qquad\qquad\qquad\quad O$$

D-β-Hydroxybutyryl-S-ACP is then dehydrated to yield a *trans*-$\Delta^{2,3}$-enoyl-S-ACP derivative, namely crotonyl-S-ACP, (Figure 18-6) by enoyl ACP dehydratase

$$\text{D-}\beta\text{-Hydroxybutyryl-S-ACP} \rightleftharpoons \text{crotonyl-S-ACP} + H_2O$$

Crotonyl-S-ACP is then reduced to butyryl-S-ACP (Figure 18-6); the electron donor is NADPH

$$\text{Crotonyl-S-ACP} + \text{NADPH} + H^+ \rightleftharpoons \text{butyryl-S-ACP} + \text{NADP}^+$$

The formation of butyryl-S-ACP completes the first of seven cycles, in each of which a molecule of malonyl-S-ACP enters at the carboxyl end of the growing fatty acid chain, with displacement of the ACP molecule from its carboxyl group and loss of the free carboxyl group of malonyl-S-ACP as carbon dioxide. After seven cycles palmityl-S-ACP is the final end product. Free palmitic acid is then discharged from ACP by the action of a hydrolytic enzyme. The central function of the acyl carrier protein in the yeast fatty acid synthetase complex is shown in Figure 18-2. Its prosthetic group has been postulated to serve as a "swinging arm" which rotates from one enzyme of the complex to the next.

The overall equation for fatty acid synthesis is

$$8 \text{ Acetyl CoA} + 14\text{NADPH} + 14H^+ + 7\text{ATP} \rightarrow$$

$$\text{palmitic acid} + 8\text{CoA} + 14\text{NADP}^+ + 7\text{ADP} + 7P_i + 6H_2O$$

The 14 molecules of NADPH required for the reductive steps in fatty acid synthesis arise largely from reduction of NADP$^+$ by glucose 6-phosphate catalyzed by glucose 6-phosphate dehydrogenase and subsequent reactions of the phosphogluconate pathway (Chapter 12). In liver and adipose tissue of vertebrates, in which the rate of fatty acid synthesis is rather high, the phosphogluconate cycle is very active.

The enzymatic synthesis of palmitic acid differs from the oxidation of palmitic acid in (1) its intracellular location, (2) the nature of the acyl carrier, (3) the form in which the two-carbon units are added or removed, (4) the stereoisomeric configuration of the β-hydroxyacyl intermediate, and (6) the electron donor-acceptor system for the crotonyl-butyryl step. These differences illustrate how opposing metabolic processes may be chemically and physically segregated from each other in the cell.

Elongation of Palmitic Acid and Formation of Unsaturated Fatty Acids

Palmitic acid, the normal end product of the cytoplasmic fatty acid synthetase system, may be lengthened to form stearic acid or still longer saturated fatty acids by two different types of enzyme systems, one in the mitochondria and the other in the endoplasmic reticulum. In the former, lengthening occurs by successive additions of acetyl CoA, whereas in the latter malonyl CoA is the donor of acetyl residues.

In addition, unsaturated acids can be made from palmitic acid. Palmitic and stearic acids serve as precursors of the two most common monounsaturated fatty acids of animal tissues, namely *palmitoleic* and *oleic acids*, both of which possess a *cis* double bond in the $\Delta^{9,10}$ position (Chapter 8). In mammals and most other aerobic organisms the double bonds are introduced by the reaction

Palmitoyl CoA + NADPH + H$^+$ + O$_2$ →
$$\text{palmitoleyl CoA + NADP}^+ + 2\text{H}_2\text{O}$$

Because both palmityl CoA and NADPH undergo oxidation together, this type of reaction is called a *mixed-function oxidation*.

Linoleic and linolenic acids, which have two and three double bonds respectively (Chapter 6), cannot be synthesized by mammals and must be obtained from plant sources; they are therefore called *essential fatty acids*. Lack of these acids in the diet of rats causes a scaly dermatitis. Linoleic and linolenic acids are synthesized in plants from oleic acid. Once ingested by mammals the essential fatty acids become precursors of other polyunsaturated acids such as arachidonic acid (Chapter 6).

Biosynthesis of Triacylglycerols and Phospholipids

Triacylglycerols and the major phospholipids phosphatidylethanolamine and phosphatidylcholine share common steps in their synthesis in animal tissues. Two major precursors are required, L-glycerol 3-phosphate and fatty acyl CoA. L-Glycerol phosphate is derived from two different sources. Its normal precursor is dihydroxyacetone phosphate generated during glycolysis by action of the cytoplasmic NAD-linked *glycerol phosphate dehydrogenase*

Dihydroxyacetone phosphate + NADH + H$^+$ ⇌
$$\text{L-glycerol 3-phosphate + NAD}^+$$

Figure 18-7
The formation of a diacylglycerol.

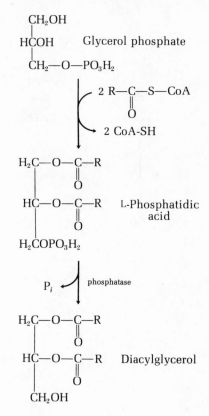

Figure 18-8
Cytidine diphosphate diacylglycerol. The portion of the molecule arising from phosphatidic acid is in color.

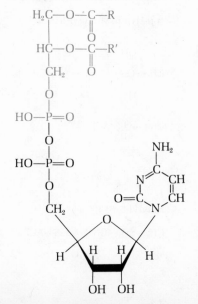

Fatty acyl CoA is formed by fatty acid-activating enzymes (Chapter 14).

The first stage in the synthesis of the major glycerol lipids is the acylation of the free hydroxyl groups of glycerol phosphate by two molecules of fatty acyl CoA to yield an L-*phosphatidic acid* (Figure 18-7). Phosphatidic acid occurs in only trace amounts in cells, but it is an important intermediate common to the biosynthesis of both triacyglycerols and the phosphoglycerides.

To form the triacylglycerols, the phosphatidic acid undergoes hydrolysis by a specific phosphatase to form a diacylglycerol. The latter then reacts with a third molecule of a fatty acyl CoA to yield the triacylglycerol. This sequence is given in the reactions

2 Fatty acyl CoA + glycerol phosphate →

L-phosphatidic acid + 2CoA

L-Phosphatidic acid + H_2O → diacylglycerol + P_i

Fatty acyl CoA + diacylglycerol → triacylglycerol + CoA-SH

The major phosphoglycerides, which serve as components of membranes and of transport lipoproteins, are also built from phosphatidic acid in a branching biosynthetic pathway. The phosphatidic acid is first converted by a reaction with CTP into *cytidine diphosphate diacylglycerol* (Figure 18-8), a common precursor of a number of phosphoglycerides

L-Phosphatidic acid + CTP ⇌

cytidine diphosphate diacylglycerol + PP_i

This reaction provides another example of a nucleotide acting as a carrier of a biochemical group, in this case, a phosphatidic acid. CDP-diacylglycerol then reacts with the amino acid serine to yield the phosphoglyceride *phosphatidylserine*, with loss of the CMP group (Figure 18-9). Phosphatidylserine is then converted into *phosphatidylethanolamine* by loss of carbon dioxide from the carboxyl group of its serine component. *Phosphatidylcholine* is then formed by enzymatic methylation of phosphatidylethanolamine. The methyl groups required are carried by tetrahydrofolate (Chapters 8, 15).

Biosynthesis of Cholesterol

Most of the steps in enzymatic synthesis of cholesterol are now known in some detail. The ultimate precursor of

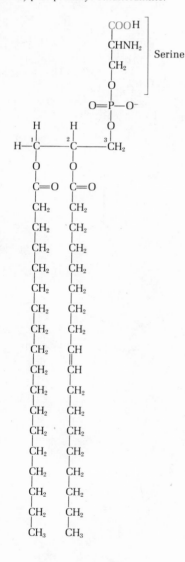

Figure 18-9
Phosphatidylserine. Loss of a molecule of CO₂ (color) leads to formation of phosphatidyl ethanolamine.

Figure 18-10

Steps in the biosynthesis of cholesterol. Three molecules of acetyl CoA combine to yield mevalonic acid, which becomes phosphorylated to 3-phospho-5-pyrophosphomevalonic acid. On loss of CO_2 and phosphate 3-isopentenyl phosphate is formed. Stepwise assembly of six molecules of the latter ultimately yields the linear hydrocarbon squalene, which cyclizes and rearranges to become cholesterol.

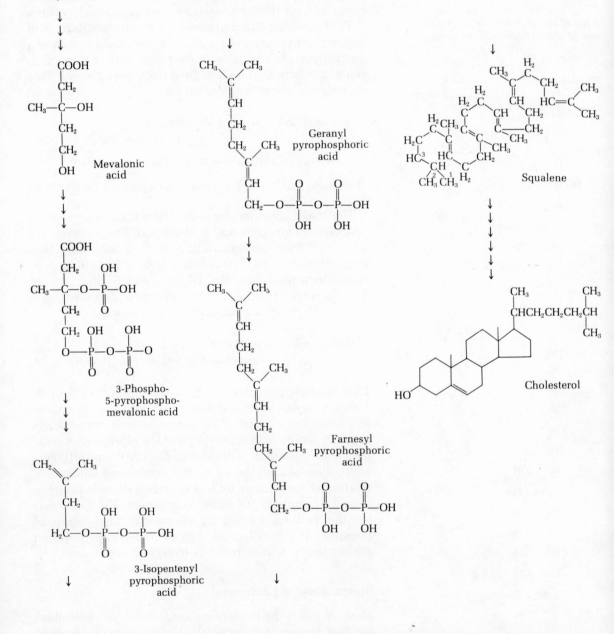

cholesterol is acetyl CoA. Acetyl CoA is first converted to *hydroxymethylglutaryl CoA* and thence to *mevalonic acid,* a direct precursor of a number of isoprenoid compounds (Chapter 6) and also of *squalene,* an open-chain hydrocarbon which is converted to cholesterol by a complex cyclization reaction. The pathway of cholesterol synthesis is outlined in Figure 18-10. We have seen that the major sources of acetyl CoA are carbohydrates, fatty acids, and amino acids. The major pathways of utilization of acetyl CoA are (1) oxidation via the tricarboxylic acid cycle, (2) conversion to ketone bodies, (3) fatty acid synthesis via malonyl CoA, and (4) conversion to hydroxymethylglutaryl CoA (HMG-CoA), which in turn leads to cholesterol.

Summary

Long-chain saturated fatty acids are synthesized from acetyl CoA by a cytoplasmic complex of enzymes which employs a specific SH-containing protein, acyl carrier protein (ACP), as an acyl group carrier. Acetyl-S-ACP, which is formed from acetyl CoA, reacts with malonyl-S-ACP, formed from malonyl CoA, to yield acetoacetyl-S-ACP and free CO_2. Reduction to the D-β-hydroxy derivative and its hydration to the $\Delta^{2,3}$-unsaturated acyl-S-ACP is followed by reduction of the latter to butyryl-S-ACP at the expense of NADPH. Six more molecules of malonyl-S-ACP react successively at the carboxyl end of the growing fatty acid chain to form palmityl-S-ACP, the usual end product. Palmitic acid may be elongated to stearic acid by reaction with acetyl CoA in mitochondria or with malonyl CoA in microsomes. Palmitoleic and oleic acids are formed from palmitic and stearic acids, respectively, by action of mixed-function oxygenases, which require NADPH. Linoleic and linolenic acids, the essential fatty acids, are readily formed by plants, but not by mammals, which require them in the diet.

Triacylglycerols are formed in a sequence of reactions in which two molecules of fatty acyl CoA react with glycerol 3-phosphate to form phosphatidic acid, which is dephosphorylated and then acylated by a third molecule of fatty acyl CoA. Phosphatidic acid is also the major precursor of phosphoglycerides; it reacts with CTP to form cytidine diphosphate diacylglycerol, a carrier of phosphatidic acid. The latter then reacts with serine to yield the phospholipid phosphatidylserine, which is the precursor of both phosphatidylethanolamine and phosphatidylcholine. Cholesterol is formed from acetyl CoA via the intermediates hydroxymethylglutaryl CoA, mevalonic acid, and squalene.

References

GREENBERG, D. M. (ed.): *Metabolic Pathways*, vol. II, Academic Press, New York, 1968. Contains up-to-date reviews of fatty acid and lipid synthesis.

MASORO, E. J.: *Physiological Chemistry of Lipids in Mammals*, Saunders, Philadelphia, 1968. A short textbook.

LOWENSTEIN, J. M.: "Citrate and The Conversion of Carbohydrate into Fat," in T. W. Goodwin (ed.), *Metabolic Roles of Citrate*, Academic Press, New York, 1968, p. 61. An excellent review.

Problems

1. Write the overall equation for the biosynthesis of stearic acid in the liver of a rat, starting from cytoplasmic acetyl CoA, NADPH, and ATP.

2. Write a series of equations for the major stages in the biosynthesis of tripalmitin from D-glucose.

3. Which of the following can serve as precursors for the net synthesis of fatty acids in the rat? How many carbon atoms of each can be converted into fatty acid carbon?
   (a) Lactic acid
   (b) D-fructose
   (c) Succinic acid
   (d) L-glutamic acid
   (e) $NaHCO_3$

4. Write a series of equations for the synthesis of one molecule of phosphatidylethanolamine from one molecule each of oleic acid, palmitic acid, dihydroxyacetone phosphate, and serine.

5. A sample of D-glucose labeled in carbon atom 2 with ^{14}C isotope is injected in a rat. The animal is sacrificed one hour later and a sample of palmitic acid isolated from the liver lipids. Which carbon atoms of the palmitic acid will be most strongly labeled?

   Repeat the problem, starting with L-alanine labeled with ^{14}C in the β-carbon atom.

CHAPTER 19 BIOSYNTHESIS OF AMINO ACIDS AND NUCLEOTIDES

The biosynthesis of amino acids and nucleotides is vital for all forms of life, since amino acids are the precursors of proteins, and nucleotides the precursors of the nucleic acids. Because of the central importance of the informational macromolecules, most organisms practice strict economy in the use of amino acids and the purine and pyrimidine bases and, indeed, salvage them for reuse. Moreover, various amino acids and nucleotides are precursors of other important cell components, among them certain hormones, the porphyrin prosthetic groups of heme proteins, and several coenzymes.

Essential and Nonessential Amino Acids

Living organisms differ considerably with respect to their ability to synthesize amino acids and the forms of nitrogen which they can utilize for this purpose. Higher vertebrates are not able to synthesize all the common amino acids. For example, the albino rat and the human can make only 10 of the 20 amino acids required as building blocks for protein synthesis (Table 19-1). These are the _nonessential amino acids_ for these animals. The remainder, which are called _essential amino acids,_ must be obtained from the diet. Higher plants are more versatile; they can make all of the amino acids required for protein synthesis, starting from either ammonia or nitrate as the nitrogen source. Microorganisms differ widely in their capacity to synthesize amino acids. For example, _Escherichia coli_ cells can make all the amino acids, but the lactic

Table 19-1 The essential and nonessential amino acids (albino rat).

Essential	Nonessential
Arginine	Alanine
Histidine	Asparagine
Isoleucine	Aspartic acid
Leucine	Cysteine
Lysine	Glutamic acid
Methionine	Glutamine
Phenylalanine	Glycine
Threonine	Proline
Tryptophan	Serine
Valine	Tyrosine

acid bacteria, which cause the souring of milk, cannot, and must obtain these preformed.

The 20 different amino acids are synthesized by 20 different multienzyme sequences, some of which are exceedingly complex. As in the case of most biosynthetic routes, the pathways of amino acid synthesis are usually different from those employed in their degradation (Chapter 15).

Biosynthesis of Nonessential Amino Acids

In this section we shall examine the biosynthesis of the nonessential amino acids, those that can be synthesized by man, the rat, and most other vertebrate species. In most of these pathways an α-keto acid derived from tricarboxylic acid cycle intermediates is the precursor of the carbon skeleton of the amino acid; the amino group is usually introduced by a transamination from glutamic acid.

Glutamic Acid, Glutamine, and Proline

The biosynthetic pathways for these closely related amino acids are simple and appear to be identical in all forms of life. Glutamic acid is formed from ammonia and α-keto-glutarate, an intermediate of the tricarboxylic acid cycle, by the action of L-*glutamate dehydrogenase*. Here the required reducing power is furnished by NADPH

$$NH_3 + \alpha\text{-ketoglutarate} + NADPH + H^+ \rightleftharpoons$$
$$\text{L-glutamate} + NADP^+ + H_2O$$

This reaction is of fundamental importance in the biosynthesis of all amino acids in all species, since it is the primary pathway for the formation of α-amino groups directly from free ammonia. The glutamic acid so formed becomes the amino group donor in the biosynthesis of most other amino acids.

The amino acid glutamine is formed from glutamic acid by the action of *glutamine synthetase*, which has been discussed earlier (Chapter 15)

$$NH_3 + \text{glutamic acid} + ATP \rightleftharpoons \text{glutamine} + ADP + P_i$$

This reaction is rather complex and involves two or more intermediate steps; enzyme-bound γ-glutamyl phosphate is an intermediate in the reaction. The allosteric regulation of glutamine synthetase activity will be considered

Figure 19-1
Biosynthesis of proline. All five carbon atoms arise from glutamic acid. The end product proline is an allosteric inhibitor. The feedback inhibitory mechanism is shown by colored arrows and the point of inhibition by a colored bar.

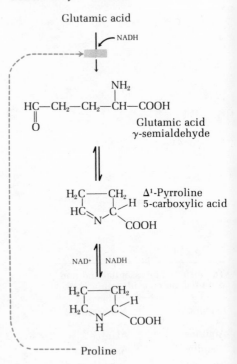

Figure 19-2
Biosynthesis of cysteine.

Methionine

ATP

S-Adenosylmethionine

—CH$_3$

S-Adenosylhomocysteine

H$_2$O
Adenosine

COOH
|
HC—NH$_2$
|
CH$_2$
| Homocysteine
CH$_2$
|
SH

Serine
Cystathionine synthetase

COOH
|
H—C—NH$_2$
|
CH$_2$
|
CH$_2$ Cystathionine
|
S
|
CH$_2$
|
CHNH$_2$
|
COOH

cystathionase

NH$_3$
CH$_3$CH$_2$COCOOH
α-Ketobutyrate

SH
|
CH$_2$
|
HCNH$_2$
|
COOH
Cysteine

below. Proline is formed from glutamic acid by the pathway shown in Figure 19-1.

Alanine, Aspartic Acid, and Asparagine

In most organisms alanine and aspartic acid arise by transamination from glutamic acid to pyruvic acid and oxaloacetic acid, respectively

Glutamic acid + pyruvic acid $\rightleftharpoons$ α-ketoglutaric acid + alanine

Glutamic acid + oxaloacetic acid $\rightleftharpoons$
α-ketoglutaric acid + aspartic acid

Here again, the carbon skeletons of these amino acids arise from the tricarboxylic acid cycle. In many organisms aspartic acid is the direct precursor of asparagine, in a reaction catalyzed by *asparagine synthetase*, which is similar to that catalyzed by glutamine synthetase

NH$_3$ + aspartic acid + ATP $\rightarrow$ asparagine + ADP + P$_i$

In other organisms there is a different pathway for asparagine synthesis

Glutamine + aspartic acid $\rightleftharpoons$ glutamic acid + asparagine

Tyrosine

Although tyrosine is a nonessential amino acid, its synthesis requires the essential amino acid phenylalanine as precursor. Tyrosine is formed from phenylalanine by a hydroxylation reaction catalyzed by *phenylalanine hydroxylase*, which we have seen (Chapter 15) also participates in the degradation of phenylalanine. This requires NADPH as coreductant. The overall reaction is

Phenylalanine + NADPH + H$^+$ + O$_2$ $\rightarrow$ tyrosine + NADP$^+$ + H$_2$O

Cysteine

Cysteine arises in mammals from methionine, which is essential, and serine, which is not. Three major stages are involved in its synthesis (Figure 19-2). In the first, methionine loses the methyl group from its sulfur atom to become *homocysteine*. This reaction, which takes place in three or more steps, requires ATP to convert methionine into an activated form, *S-adenosylmethionine* (Figure 19-3)

L-Methionine + ATP → S-adenosylmethionine + PP$_i$ + P$_i$

Its methyl group may then be enzymatically donated to a number of different methyl group acceptors, leaving S-*adenosylhomocysteine* as the demethylated product. S-Adenosylhomocysteine then undergoes hydrolysis to yield free *homocysteine*

S-Adenosylmethionine + H$_2$O → adenosine + homocysteine

In the second stage of cysteine synthesis, homocysteine reacts with serine in a reaction catalyzed by *cystathion-ine synthetase* to yield *cystathionine* (Figure 19-2)

Homocysteine + serine → cystathionine + H$_2$O

In the last step, *cystathionase*, a pyridoxal phosphate enzyme, catalyzes the cleavage of cystathionine to yield free cysteine (Figure 19-2)

Cystathionine → α-ketobutyrate + NH$_3$ + cysteine

The overall equation of cysteine synthesis is thus

L-Methionine + ATP + methyl acceptor + H$_2$O + serine →
methylated acceptor + adenosine + α-ketobutyrate
+ NH$_3$ + cysteine + PP$_i$ + P$_i$

The final result of this reaction is to bring about replacement of the —OH group of serine with an —SH group. Note that the carbon chain of cysteine comes from serine, but the sulfur atom comes from methionine (Figure 19-2).

Serine and Glycine

Since serine is the precursor of glycine, these two amino acids are considered together. The major pathway for the formation of serine in animal tissues, shown in Figure 19-4, begins with 3-phosphoglyceric acid, an intermediate of glycolysis. In the first step the α-hydroxyl group is oxidized by NAD⁺ to yield 3-*phosphohydroxypyruvic acid*. Transamination from glutamic acid yields 3-*phos-phoserine*, which undergoes hydrolysis by serine phos-phatase to yield free serine.

   The three-carbon amino acid serine is the precursor of the two-carbon glycine through removal of one carbon atom, that at the β position (Figure 19-4). This is accomplished by an enzyme which requires tetrahydrofolate,

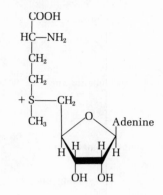

Figure 19-3
S-Adenosylmethionine

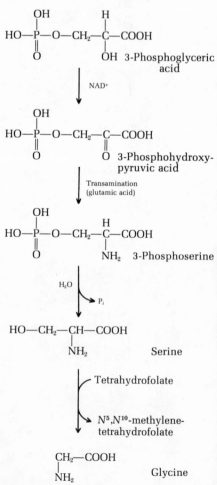

Figure 19-4
Biosynthesis of serine and glycine.

the active or coenzyme form of the vitamin *folic acid* (Chapter 8).

Tetrahydrofolate, which is often abbreviated as FH_4, serves as the acceptor of the β-carbon atom of serine as the latter is cleaved to yield glycine. The carbon atom so removed from serine forms a methylene bridge between nitrogen atoms 5 and 10 of tetrahydrofolate to yield N^5,N^{10}-*methylenetetrahydrofolate* (Figure 19-4). The overall reaction is

Serine + tetrahydrofolate $\rightarrow$

glycine + N^5,N^{10}-methylenetetrahydrofolate

This reaction completes the formation of glycine from serine. N^5,N^{10}-Methylenetetrahydrofolate is one member of the family of folic acid coenzymes, which are capable of carrying different kinds of one-carbon groups, such as methyl, hydroxymethyl, formyl, and formimino groups (Chapter 8). Tetrahydrofolate coenzymes take part in a variety of important one-carbon transfers, of which the transfer of methyl groups derived from methionine is quantitatively of greatest importance. The one-carbon unit derived from serine and carried by tetrahydrofolate can be transferred in the form of a methyl group to various acceptors, such as phosphatidylethanolamine (Chapter 18).

Biosynthesis of the Essential Amino Acids

The pathways for the synthesis of the amino acids essential in the nutrition of the albino rat and man have been deduced largely from biochemical and genetic studies on bacteria. The pathways leading to synthesis of the essential amino acids are longer (5 to 15 steps) than those leading to nonessential amino acids, most of which have fewer than 5 steps. They are also more complex, probably because various intermediates in these pathways serve as precursors of many other kinds of biomolecules. Although we shall not examine these pathways in detail, Table 19-2 lists the essential amino acids and the precursors of their carbon-skeleton. It is noteworthy that five of the essential amino acids are synthesized from nonessential amino acids: Threonine, methionine, and lysine are formed from aspartic acid, and both arginine and histidine are formed from glutamic acid. Isoleucine is formed from the essential amino acid threonine. These relationships show that the biosynthesis of various amino acids is closely interconnected.

Table 19-2 Precursors of the essential amino acids in bacteria and plants.

Amino acid	Precursor
Arginine	Glutamic acid
Histidine	ATP, ribose, glutamic acid
Isoleucine	Pyruvic acid
Leucine	Pyruvic acid
Lysine	Aspartic acid; pyruvic acid
Methionine	Aspartic acid
Phenylalanine	Phosphoenolpyruvic acid
Threonine	Aspartic acid
Tryptophan	Phosphoenolpyruvic acid
Valine	Pyruvic acid

Allosteric Regulation of Amino Acid Biosynthesis

The most important mechanism by which amino acid synthesis is controlled is through _feedback inhibition_ of the first reaction in the biosynthetic sequence by the end product of the sequence (Chapters 4 and 9). The first reaction of such a sequence, which is usually irreversible, is catalyzed by an allosteric enzyme. As an example, Figure 19-5 shows the allosteric control of the synthesis of isoleucine from threonine. The end product isoleucine is a negative modulator of the first reaction in the sequence.

Another noteworthy example is the remarkable set of allosteric controls exerted on the activity of glutamine synthetase of _E. coli_. Glutamine is a precursor or amino group donor of many metabolic products (Figure 19-6). Eight products of glutamine metabolism in _E. coli_ are now known to serve as negative feedback modulators of the activity of glutamine synthetase, which is perhaps the most complex regulatory enzyme now known.

Regulation of amino acid synthesis is a complex, integrated process, because of the many interconnecting pathways in amino acid synthesis and the basic necessity of cells to conserve nitrogen.

Biosynthesis of Porphyrins

The amino acids are precursors of many biomolecules other than proteins. Important biological functions are served by such compounds as hormones, vitamins, co-enzymes, alkaloids, porphyrins, antibiotics, pigments, and neurotransmitter substances, all of which are synthesized from amino acids. Space will not permit development of the many different biosynthetic pathways that start with the amino acids. However, the biosynthesis of porphyrins, for which glycine is a major precursor, does require special note because of the central importance of the porphyrin nucleus in the function of hemoglobin, of cytochromes, and of chlorophyll.

The porphyrins are constructed from four molecules of the monopyrrole derivative _porphobilinogen_, which is synthesized in the steps shown in Figure 19-7. This pathway was deduced largely from isotopic tracer studies. In the first reaction, glycine reacts with succinyl CoA to yield α-amino-β-ketoadipic acid, which is then decarboxylated to give δ-_aminolevulinic acid_ and carbon dioxide. Two molecules of δ-aminolevulinic acid next condense to form porphobilinogen. The enzymes catalyzing these reactions are regulatory enzymes which are

Figure 19-5
Regulation of isoleucine synthesis.

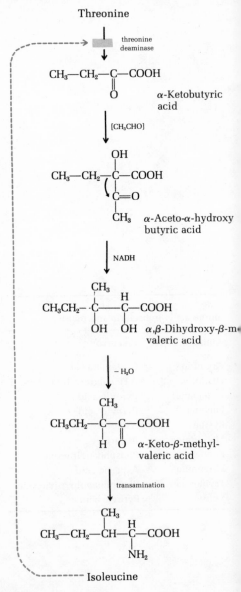

Figure 19-6
Multivalent allosteric inhibition of glutamine synthetase. Although glycine and alanine are not direct products of glutamine metabolism, they are very strong inhibitors, suggesting that the steady-state levels of glycine and alanine in the cell are critically related to glutamine synthesis and metabolism.

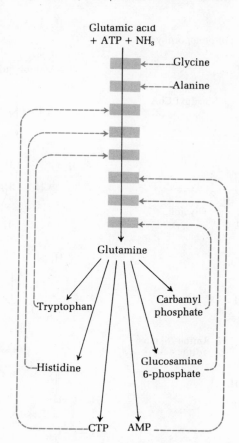

inhibited by heme, hemoglobin, and other heme proteins, the ultimate end products of this biosynthetic pathway. Four molecules of porphobilinogen serve as the precursors of *protoporphyrin*, through a series of complex reactions, some of which are still obscure. The iron atom is incorporated after the protoporphyrin molecule is completed. Figure 19-7 shows the origin of the carbon and nitrogen atoms of protoporphyrin IX from glycine and succinyl CoA.

Biosynthesis of Nucleotides

Central to the biosynthesis of the nucleic acids and nucleotides is the pathway of formation of their bases, the pyrimidines and purines. Nearly all living organisms, except for some bacteria, appear to have the capacity to synthesize these bases from very simple precursors.

Figure 19-7
Biosynthesis of protoporphyrin

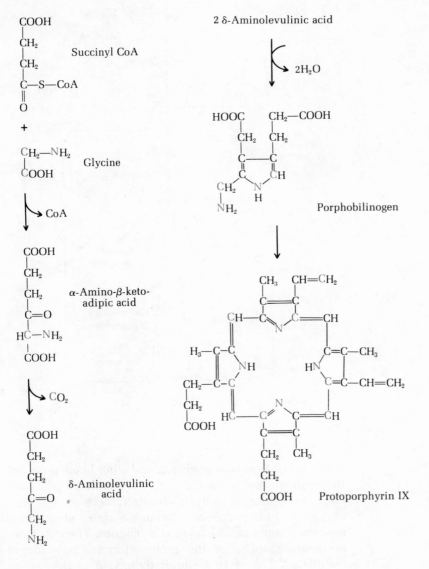

Protoporphyrin IX

Biosynthesis of Purine Nucleotides

The first important clues as to the biosynthetic origin of the purine bases came from experiments in which various isotopic metabolites were fed to animals and the sites of incorporation of the labeled atoms into the purine ring determined. Such experiments were carried out in birds, which excrete nitrogen largely in the form of uric acid, a purine derivative (Chapter 15). Figure 19-8 shows the

Figure 19-8
Origin of the purine ring system.
The three atoms in color arise from
glycine.

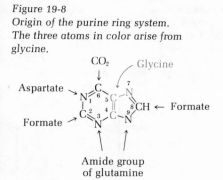

Figure 19-9
Important steps in the synthesis of
adenylic and guanylic acids. Note
that the purine ring is built stepwise
on carbon 1 of D-ribose 5-phosphate.

origin of the carbon and nitrogen atoms of the purine nucleus.

The first step in purine nucleotide synthesis begins with D-ribose 5-phosphate. It leads to the formation of an open-chain ribonucleotide, which subsequently undergoes ring closure to yield a purine nucleotide. Major steps in the pathway leading to adenylic and guanylic acids are shown in Figure 19-9. To ribose 5-phosphate is attached an amino group at position 1. Further substitution and additions at the 1-amino group are then made and the ring closed to yield the imidazole portion of the purine nucleus. Subsequently substituents are added to

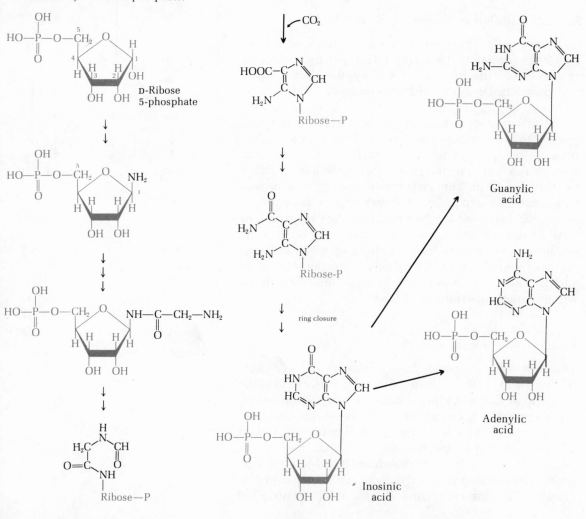

the imidazole group and another ring closure takes place to yield the two fused rings of the purine nucleus. The first intermediate to have a complete purine ring is _inosinic acid_. The conversion of inosinic acid to adenylic acid requires the insertion of an amino group derived from aspartic acid; this takes place by a rather complex set of reactions which we shall not detail here. Inosinic acid is converted to guanylic acid via intermediate reactions that require ATP. The phosphorylation of adenylic and guanylic acids to ATP and GTP, respectively, proceeds by two successive phosphokinase steps. In the case of guanylic acid the reactions are

$$GMP + ATP \rightleftharpoons GDP + ADP$$

$$GDP + ATP \rightleftharpoons GTP + ADP$$

Regulation of Purine Nucleotide Biosynthesis

In *E. coli* cells, three major feedback mechanisms cooperate in regulating the overall rate of purine nucleotide synthesis and the relative rates of synthesis of the two end products adenylic acid and guanylic acid (Figure 19-10).

The first control is exerted on the early reaction step leading to the transfer of an amino group to 5-phosphoribosyl-1-pyrophosphate. The enzyme catalyzing this reaction is a multivalent regulatory enzyme: It is inhibited by AMP and GMP. Thus, whenever either purine nucleotide accumulates to excess, the first step in its synthesis undergoes feedback inhibition. Moreover, an excess of GMP in the cell brings about allosteric inhibition of its synthesis from inosinic acid, without affecting the synthesis of adenylic acid. Conversely, an accumulation of adenylic acid can result in inhibition of its synthesis without affecting synthesis of GMP.

Synthesis of Pyrimidine Nucleotides

The first step in pyrimidine nucleotide biosynthesis (Figure 19-11) is the carbamylation of aspartic acid at the expense of carbamyl phosphate to yield N-carbamyl-aspartic acid. This reaction is catalyzed by _aspartate transcarbamylase_, one of the most thoroughly studied allosteric enzymes. By removal of water from N-carbamylaspartic acid, which is catalyzed by _dihydroorotase_, the pyrimidine ring is closed with formation of _L-dihydroorotic acid_. This compound is now oxidized to yield

Figure 19-10
Feedback control mechanisms in the biosynthesis of adenine and guanine nucleotides.

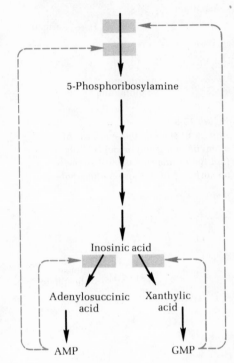

5-Phosphoribosyl-1-pyrophosphate

5-Phosphoribosylamine

Inosinic acid

Adenylosuccinic acid   Xanthylic acid

AMP   GMP

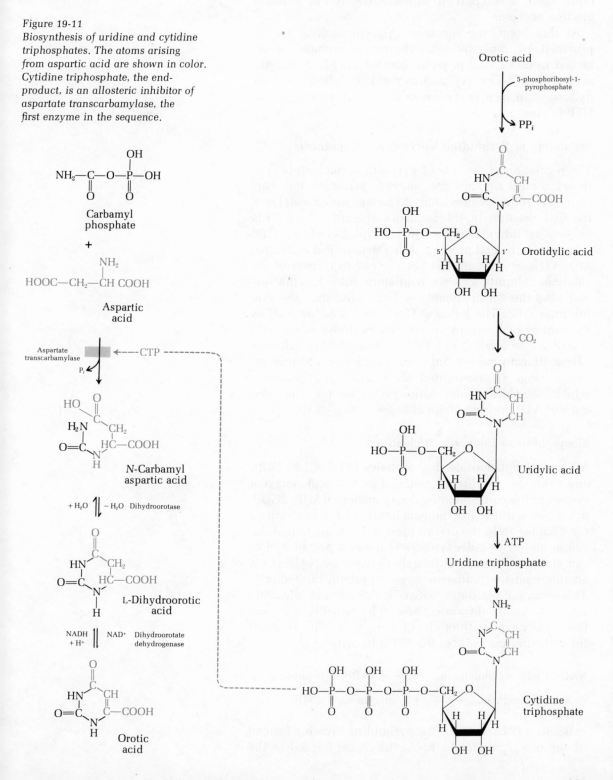

Figure 19-11
Biosynthesis of uridine and cytidine triphosphates. The atoms arising from aspartic acid are shown in color. Cytidine triphosphate, the end-product, is an allosteric inhibitor of aspartate transcarbamylase, the first enzyme in the sequence.

orotic acid, a reaction in which NAD^+ is the ultimate electron acceptor.

At this point, the D-ribose 5-phosphate side chain, provided by 5-phosphoribosyl-1-pyrophosphate, is attached to orotic acid to yield *orotidylic acid*. Orotidylic acid is then decarboxylated to yield uridylic acid. Uridylic acid in turn is the precursor of cytidylic acid via UTP (Figure 19-11).

Regulation of Pyrimidine Nucleotide Biosynthesis

The regulation of the rate of pyrimidine nucleotide synthesis occurs through the enzyme aspartate transcarbamylase (often abbreviated ATCase), which catalyzes the first reaction in the sequence (Figure 19-11). This enzyme is inhibited by cytidine triphosphate or CTP (Chapter 7), the end product of this sequence of reactions. The ATCase molecule has been found to consist of six catalytic subunits and six regulatory subunits; the former bind the substrate and the latter bind the allosteric inhibitor CTP. The entire ATCase molecule, as well as its subunits, exists in two conformations, active and inactive. The binding of CTP by the regulatory subunits causes transformation into an inactive conformation. This change is transmitted to the catalytic subunits, which also shift to the inactive conformation. The presence of ATP prevents these changes.

Biosynthesis of Deoxyribonucleotides

All four ribonucleoside diphosphates (ADP, GDP, UDP, and CDP) may be directly reduced by a specific enzyme system to the corresponding deoxy analogs dADP, dGDP, dUDP, and dCDP (for nomenclature and abbreviations see Chapter 7). In the overall process the reduction of the ribose moiety to 2-deoxyribose requires a pair of hydrogen atoms, which are ultimately donated by NADPH via an intermediate hydrogen-carrying protein, *thioredoxin*. This enzyme oscillates between dithiol and disulfide forms. Its oxidized form is reduced by NADPH in a reaction catalyzed by *thioredoxin reductase*. The reduced thioredoxin then reduces the NDPs to dNDPs:

$$NADPH + H^+ + \text{thioredoxin}_{ox} \xrightarrow{\text{reductase}} NADP^+ + \text{thioredoxin}_{red}$$

$$\text{Thioredoxin}_{red} + NDP \rightarrow \text{thioredoxin}_{ox} + dNDP$$

Because DNA contains the pyrimidine thymine instead of the uracil present in RNA, the dUDP formed in the reaction with thioredoxin is converted into the corresponding deoxythymidylic acid via methylation of dUMP by N^5,N^{10}-methylenetetrahydrofolate. To complete the synthesis of the deoxyribonucleoside 5'-triphosphates, which are direct precursors in DNA biosynthesis, the following phosphotransferase reactions occur:

$$ATP + dADP \rightarrow ADP + dATP$$

$$ATP + dCDP \rightarrow ADP + dCTP$$

$$ATP + dTDP \rightarrow ADP + dTTP$$

$$ATP + dGDP \rightarrow ADP + dGTP$$

Nitrogen Fixation

Most organisms are unable to use molecular nitrogen (N_2) from the atmosphere. However, certain microorganisms, among them the blue-green algae and some soil bacteria, are capable of "fixing" atmospheric nitrogen. Moreover, leguminous plants, such as peas, beans, clover, alfalfa, and soybeans, can also fix atmospheric nitrogen, in a process which requires the cooperative action of the host plant and certain symbiotic bacteria present in their root nodules. This type of nitrogen fixation is called sym-biotic nitrogen fixation. The nitrogen-fixing enzymes are actually located in the bacteria, but the plant supplies some essential components that the bacterium lacks.

Because the first stable product of nitrogen fixation is ammonia (NH_3), the overall process is believed to consist of the reduction of molecular nitrogen to ammonia (Figure 19-12). However, the mechanism by which this process occurs has been very difficult to study and not all the steps in it have been elucidated. In some nitrogen-fixing organisms molecular hydrogen itself may be the hydrogen donor, but other reducing agents may serve this role in other species. It is generally accepted that the nonheme-iron protein ferredoxin is an intermediate carrier of reducing equivalents during reduction of nitrogen. It has a molecular weight of 6000 and contains seven iron atoms as well as an equal number of acid-labile sulfur atoms. Ferredoxin transmits its reducing equivalents to a sequence of two metalloenzymes, the first an iron protein and the second a protein containing both iron and molybdenum.

ATP is an obligatory component of the nitrogen-fixing system, but the detailed nature of its function is not

Figure 19-12
The pathway of nitrogen fixation.

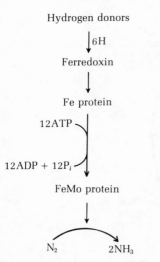

Hydrogen donors

$\downarrow$6H

Ferredoxin

$\downarrow$

Fe protein

12ATP

12ADP + 12P$_i$

FeMo protein

$\downarrow$

N$_2$ 2NH$_3$

clear. Since two molecules of ATP are required for each electron transported to nitrogen, a total of 12 ATPs are required to reduce one molecule of N_2 to 2 NH_3, which requires input of 6 H atoms (or 6 electrons). The overall equation is therefore

$$N_2 + 3H_2 + 12ATP \rightarrow 2NH_3 + 12ADP + 12P_i$$

The ammonia so formed can then be utilized in the synthesis of amino acids and other nitrogenous products.

The Nitrogen Cycle

Figure 19-13 shows the nitrogen cycle in the biosphere. Ammonia generated by nitrogen fixation can be converted into amino acids in plants, which are then utilized by animals to build their own proteins. On the death of the animals the degradation of their proteins returns ammonia to the soil, where nitrifying bacteria convert it to nitrite (NO_2^-) and nitrate (NO_3^-) successively. Plants and many bacteria can convert nitrate into ammonia again. Although reduced nitrogen in the form of ammonia or amino acids is the form utilized in most living organisms, there are some soil bacteria that derive their energy by oxidizing ammonia with the formation of nitrite and ultimately nitrate. Because these organisms are extremely abundant and active, nearly all ammonia reaching the soil ultimately becomes oxidized to nitrate, a process known as *nitrification*.

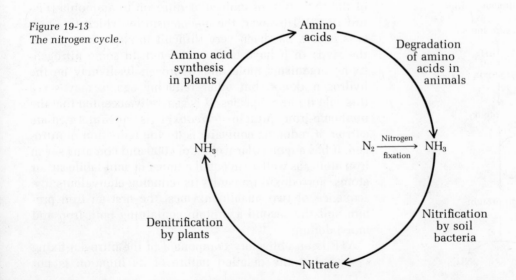

Figure 19-13
The nitrogen cycle.

Summary

Man and the albino rat can synthesize ten of the twenty protein amino acids. The remainder, which are required in the diet and are called essential amino acids, are synthesized by plants and bacteria. Among the nonessential amino acids glutamic acid is formed by reductive amination of α-ketoglutarate. It is the direct precursor of glutamine and proline. Alanine and aspartic acid are formed by transamination to pyruvic and oxaloacetic acids, respectively. Tyrosine is formed by hydroxylation of phenylalanine. Cysteine is formed from methionine by a more complex series of reactions in which S-adenosylmethionine and cystathionine are the most significant intermediates. The carbon chain of serine is derived from 3-phosphoglycerate. Serine is the precursor of glycine; the β-carbon atom of serine is transferred to tetrahydrofolate.

The pathways of synthesis of the essential amino acids in bacteria and plants are more complex and longer. They are formed from certain of the nonessential amino acids and other metabolites.

The biosynthetic pathways leading to the amino acids are subject to allosteric or end-product inhibition; the regulatory enzyme is usually the first in the sequence. Amino acids are precursors of many other important biomolecules; the porphyrin ring of heme proteins is derived from glycine and succinyl CoA. The purine ring system of purine nucleotides is built by formation of an open-chain ribonucleotide from ribose 5-phosphate, followed by two ring-closure steps; inosinic acid is an important intermediate. The pyrimidine ring is formed from aspartic acid, CO_2, and ammonia, followed by attachment of ribose 5-phosphate, to yield the pyrimidine ribonucleotides.

Formation of ammonia by fixation of molecular nitrogen in legume root nodules, nitrification of ammonia to form nitrate by soil organisms, and the denitrification of nitrate by higher plants complete the nitrogen cycle.

References

GREENBERG, D. M. (ed.): *Metabolic Pathways,* vols. I, II, III, 2nd ed., Academic Press, New York, 1967–1969. Excellent papers in review form.

LEHNINGER, A. L.: *Biochemistry,* Worth Publishers, New York, 1970. All the amino acid biosynthetic pathways are shown in Chapter 24.

MEISTER, A.: *Biochemistry of the Amino Acids,* vols. I and II, 2d ed., Academic Press, New York, 1965. Comprehensive review of amino acid biosynthesis and precursor functions, particularly in Vol. II.

POSTGATE, J. (ed.): *Chemistry and Biochemistry of Nitrogen Fixation*, Plenum Press, New York, 1971.

Problems

1. Write a balanced equation for the synthesis of glycine starting from D-glucose as sole carbon source and NH_3 as nitrogen source.

2. Methionine labeled with ^{14}C isotope in the α-carbon atom is injected into a rat. One hour later the animal is sacrificed and a sample of cysteine isolated from the liver proteins. Which carbon atom(s) of cysteine will be labeled?

3. Repeat Problem 2, starting from serine labeled in the α-carbon atom with ^{14}C.

4. Write a balanced equation for the synthesis of proline from α-ketoglutarate.

   If the α-carbonyl carbon atom of α-ketoglutarate is labeled with ^{14}C, where would the label appear in the proline formed?

CHAPTER 20

**ABSORPTION, TRANSPORT, AND
THE REGULATION OF METABOLISM
IN MAMMALS**

In this chapter we shall outline the major biochemical events involved in the metabolism of the intact mammalian organism. We shall examine the absorption of nutrients from the intestine, their distribution and transport to the various organs, the transport of oxygen to the tissues, and the removal of carbon dioxide and various wastes. We shall also outline the metabolic interplay among various organs and the endocrine regulation of metabolism in mammals. Much of this information has come from study of metabolism in the human.

Biochemistry of Digestion and Absorption

The three major bulk nutrients of mammals—carbohydrates, proteins, and lipids—undergo enzymatic degradation to their building blocks during the process of digestion in the gastrointestinal tract.

Polysaccharides

The polysaccharides starch and glycogen are hydrolyzed completely to form free glucose. This process begins in the mouth during mastication, through the action of amylase secreted in the salivary fluid. Amylase is also secreted in the pancreatic juice into the small intestine, where the degradation of starch, glycogen, and other digestible polysaccharides is completed. Disaccharides are also hydrolyzed by digestive enzymes secreted into the small intestine. Ingested sucrose is hydrolyzed to glucose and fructose by invertase; similarly, maltose and lactose are hydrolyzed to their constituent monosaccharides.

However, for lack of the appropriate digestive enzymes, some polysaccharides cannot be digested by man and certain other mammals to yield simple sugars. The major example is cellulose, which forms a large portion of ingested plant polysaccharides, but which cannot be degraded and thus passes through the intestinal tract of man and other vertebrates more or less intact. Cellulose can be digested by ruminant animals, such as the cow; the ruminants possess an extra stomach, the rumen, which contains large numbers of symbiotic bacteria capable of secreting an enzyme, *cellulase*, which catalyzes hydrolysis of cellulose into simple sugars and thus makes cellulose available to the cow as a nutrient.

Only simple sugars or monosaccharides can be absorbed by the epithelial cells lining the small intestine. Such sugars are transported across the membrane at the expense of energy ultimately furnished by hydrolysis of ATP. Once these sugars have entered the epithelial cells they undergo further enzymatic reactions (Chapters 11, 17), the major end product being D-glucose, which enters the bloodstream and is brought to the liver, the main distributing organ.

Proteins

Ingested proteins also undergo enzymatic hydrolysis to their constituent amino acids during the course of digestion in the gastrointestinal tract. This process begins in the acid gastric juice of the stomach through the action of *pepsin*, a proteolytic enzyme having an optimum pH of 1–2, about the pH of the gastric juice. At this low pH most proteins undergo denaturation or unfolding (Chapter 3), which makes their peptide bonds more accessible to enzymatic degradation. Pepsin is reasonably specific for those peptide bonds involving the amino acids tyrosine, phenylalanine, and tryptophan, as well as leucine (Table 20-1). Thus the long polypeptide chains of proteins are incompletely degraded by pepsin in the stomach to a mixture of smaller oligopeptide fragments.

Pepsin itself is secreted by the chief cells of the gastric mucous membrane in the form of an enzymatically inactive precursor, or zymogen, called *pepsinogen*, which has a molecular weight of about 40,000. In the presence of the hydrochloric acid of the gastric juice and the active pepsin already present, 42 amino acid residues are removed

Table 20-1 The activation and specificity of the proteolytic enzymes in the gastrointestinal tract. The specificity is given in terms of the amino acid furnishing the carbonyl group of the peptide bond.

	Specificity
Stomach (gastric juice)	
Pepsinogen $\xrightarrow{\text{HCl, pepsin}}$ pepsin	Tyr, Phe, Trp, Leu
Small intestine (pancreatic juice)	
Trypsinogen $\xrightarrow{\text{enterokinase}}$ trypsin	Lys, Arg
Chymotrypsinogen $\xrightarrow{\text{trypsin}}$ chymotrypsin	Phe, Tyr, Trp
Procarboxypeptidase → carboxypeptidase	Carboxy-terminal residues

from the NH_2-terminal end of the polypeptide chain of pepsinogen to yield free, enzymatically active pepsin (molecular weight 33,000) and a mixture of small peptides.

In the small intestine the digestion of proteins continues. Three major proteolytic enzymes are secreted into the small intestine via the pancreatic juice in the form of enzymatically inactive zymogens, *trypsinogen*, *chymotrypsinogen*, and *procarboxypeptidase*. Trypsinogen is converted into its active form *trypsin* by removal of a hexapeptide from the NH_2-terminal end of the trypsinogen molecule by the action of *enterokinase*, another enzyme secreted into the intestinal juice. Trypsin hydrolyzes those peptide bonds of the mixture of oligopeptides entering the intestine which involve the carbonyl groups of lysine and arginine. Chymotrypsinogen consists of a single polypeptide chain. In the presence of trypsin, chymotrypsinogen is activated to form *chymotrypsin*. In this process the single long polypeptide chain of chymotrypsinogen is broken at two points by removal of dipeptides. The three resulting segments of the original polypeptide chain are held together in chymotrypsin by —S—S— cross-linkages. Chymotrypsin hydrolyzes those peptide bonds involving phenylalanine, tyrosine, and tryptophan. Procarboxypeptidase is converted into *carboxypeptidase*, which removes COOH-terminal residues from peptides. Another major proteolytic enzyme in the small intestine is an aminopeptidase which can hydrolyze the NH_2-terminal residues from many peptides. The proteolytic enzymes of the digestive tract are summarized in Table 20-1.

By the combined sequential action of these proteolytic enzymes, most ingested proteins are ultimately hydrolyzed to free amino acids, which are transported across the membrane of the epithelial cells lining the small intestine at the expense of ATP energy. The free amino acids are then transported in the blood to the liver and other organs.

Lipids

Ingested lipids also must undergo at least partial enzymatic degradation prior to absorption. The enzyme lipase (Chapter 6) is secreted by the pancreas into the small intestine. It catalyzes the hydrolysis of triacylglycerols by removal of one, two, or all three fatty acid residues from the glycerol moiety. Phospholipases (Chapter 6) can hydrolyze phospholipids to yield fatty acids, glycerol and other alcohols, and phosphoric acid.

Since the acylglycerols are not truly water-soluble, they are first emulsified into a fine dispersion in the intestine. The fat droplets in this emulsion contain triacylglycerols, as well as free fatty acids, and mono- and diacylglycerols formed by the action of lipases. After passing into the intestinal cells, the absorbed fat appears in the lymph draining the intestine in the form of a milky fluid called *chyle*, which consists of a suspension of *chylomicrons*, small micelles (Chapter 1) or droplets of emulsified triacylglycerols about 1 micron (one-thousandth of a millimeter) in diameter. The chylomicrons pass into the bloodstream and are later reprocessed in the liver (see below) for transport to the peripheral tissues.

The emulsification and absorption of lipids in the small intestine is facilitated by the action of derivatives of the *bile acids*. In the human, four bile acids have been found, of which *cholic acid* (Figure 20-1) is by far the most abundant. Cholic acid is secreted in the bile into the small intestine in the form of two water-soluble derivatives, *glycocholic acid* and *taurocholic acid*. The sodium salts of such conjugated bile acids, which are often called *bile salts*, are powerful detergents. They aid in the formation and stabilization of micelles of lipids.

Active Transport Systems in Intestinal Absorption

The absorption of simple sugars and amino acids from the small intestine into the blood requires metabolic energy in the form of ATP. If respiration is inhibited or uncoupled glucose and amino acids are no longer absorbed. The transport of these simple molecules across the cell membrane into the epithelial cells lining the intestine is a special case of a more general and funda-

Figure 20-1
Cholic acid and glycocholic acid.

Cholic acid

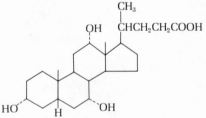

Glycocholic acid

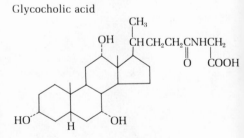

mental property of all cells, namely, the ability to maintain their internal ionic and metabolite composition constant despite fluctuations in the surrounding aqueous environment, by transporting specific metabolites across the cell membrane at the expense of energy.

In order to maintain their internal constancy cells possess specific *transport systems* in their membranes which can transport specific ions or metabolites either inward or outward, depending on the solute and the circumstances. Cell membranes, because of their nonpolar hydrocarbon inner core (Chapter 6), are rather impermeable to most ionic or polar solutes, and thus allow only those solutes to pass for which there are specific transport systems. Transport systems are very selective in their specificity. For example, the sugar transport system of the small intestine can transport D-glucose, D-fructose, and D-mannose effectively, but does not transmit disaccharides, L-glucose, nor some D-pentoses. Amino acids in turn are transported by a series of four systems, one for small neutral amino acids, one for large neutral amino acids, one for basic amino acids, and another for acidic ones.

Transport systems consist of one or more specific protein molecules bound within the membrane, one of which can combine with the specific substrate reversibly, in much the same way as an enzyme, and then carry it across the membrane and "unload" it on the other side. Transport systems can be inhibited by specific inhibitors and also show "saturation" in rate when the substrate concentration is raised beyond a given level.

Another important attribute of many membrane transport systems is that they can transport their specific substrates across the membrane *against* a gradient of concentration, a process which requires input of free energy. Normally, solute molecules tend to diffuse in such a direction as to equalize their concentration throughout the solution, and thus maximize entropy in the system, in accordance with the second law of thermodynamics (Chapter 10). To move a solute against or up a concentration gradient, which is called *active transport*, requires input of free energy, according to the equation

$$\Delta G = 2.303 \; RT \log \frac{C_2}{C_1}$$

where R is the gas constant (1.98 cal mole^{-1} degree^{-1}), T the absolute temperature, C_1 the concentration of solute in the zone from which it is pumped to a higher concentration C_2. It requires 1.34 kcal free energy input to transport one mole of solute up a 10-fold gradient of concentration at 25°C.

The energy required by active transport systems comes from their capacity to bring about the hydrolysis of ATP to yield ADP and phosphate. Indeed, there appears to be a specific ratio between the number of solute molecules transported and the number of ATP molecules hydrolyzed.

Nearly all cell types of vertebrates contain active transport systems for the cations Na^+, K^+, and Ca^{2+}, as well as for glucose and for amino acids. The epithelial cells of the small intestine have extremely active membrane transport systems for amino acids and for glucose, which can transport these molecules from the intestine into the cells and thus into the circulating blood when their concentration in the intestine is much lower than in the blood. There is considerable evidence that inward transport of amino acids requires simultaneous inward transport of Na^+, but the mechanism of this effect is not known with certainty. In any case, the intestinal epithelium utilizes a large part of the ATP generated from glycolysis and respiration for the purpose of transporting solutes across the membrane during absorption of nutrients. Further information on active transport systems is given later in this chapter.

Composition of the Blood

The blood is the vehicle by which nutrients are transported to the liver and other organs and by which waste products are returned to the lungs and kidneys for excretion. The blood is also the vehicle for transport of oxygen from the lungs to the tissues and for the removal of the CO_2 generated during the respiratory metabolism of the tissues. Moreover, hormones are transported from the endocrine glands via the blood to their specific target organs, in their function as chemical messengers. The blood is thus very complex in its chemical composition and it plays a vital role in the metabolic interplay and regulation of the various tissues of the mammalian organism. The major inorganic ions of blood plasma, namely, Na^+, Ca^{2+}, Mg^{2+}, Cl^-, HCO_3^- and phosphate, also play important roles in the regulation of tissue activity.

The blood in the vascular system of a human amounts to about 5 to 6 liters. Nearly one-half of the volume of blood is due to its cells, which consist largely of red blood cells (erythrocytes) and much smaller amounts of white blood cells (leukocytes), and blood platelets. The liquid portion of the blood is the _blood plasma_, of which about 10 percent consists of dissolved solutes. Nearly three-fourths of the latter, or some 7 percent of the blood plasma, consists of the _plasma proteins_. Inorganic salts make up about 1 percent of the solutes. The remaining 2 percent consists of organic nutrients passing between various organs and organic waste products to be excreted in the urine. Table 20-2 shows the distribution of the major nutrients and waste products in normal human blood, together with an indication of their general function in metabolism.

The concentration of the molecular and ionic components of the blood plasma are maintained at characteristic levels by various regulatory systems. Some of the components vary in concentration, depending on the nature of the nutrient intake and the time at which they are measured. Certain components of the blood are critical in the function of specific organs. We shall see that the blood glucose level, for example, is extremely critical in

Table 20-2 The major organic components of normal human blood plasma and their function

	Concentration range (mg per 100 ml)	Function
Proteins (total)	5800–8000	
Serum albumin	3000–4500	Osmotic regulation; transport of fatty acids
α-globulins	700–1500	Transport of lipids, copper, thyroid hormone
β-globulins	600–1100	Transport of lipids, iron, and other metals
γ-globulins	700–1500	Antibodies
Fibrinogen	300	Participates in blood clotting
Lipids (total)	400–700	
Triacylglycerols	100–250	Fuel en route to storage
Phospholipids	150–250	Precursors of membrane structure
Cholesterol and its esters	150–250	Precursors of membrane structure
Free fatty acids	10–30	Immediate fuel for muscles
Glucose	70–90	Transport form of carbohydrate from liver to peripheral tissues
Amino acids	35–65	Precursors in protein synthesis in tissues
Urea	20–30	Excretory product of nitrogen from amino acid catabolism
Uric acid	2–6	Excretion product of purine metabolism

the function of the brain and when it falls below a certain level, the normal function of the brain is disturbed. Measurements of the concentration of certain components of the blood plasma are extremely important in medicine, since they can yield insight into the nature of pathological metabolic disturbances.

The Central Role of the Liver in Distribution of Nutrients

After being absorbed from the intestinal tract most of the incoming nutrients pass directly to the liver, the major distributing center in the metabolism of the mammal. In the liver the incoming sugars, amino acids, and lipids are processed and apportioned for distribution to the peripheral organs.

Sugars

The mixture of free sugars entering the liver is processed enzymatically to convert various hexoses, such as fructose, galactose, and mannose, into D-glucose 6-phosphate (Chapters 11 and 17). There are at least five major pathways glucose 6-phosphate can take: (1) Some of it will be dephosphorylated by glucose 6-phosphatase (Chapter 17) and the glucose so formed will pass into the systemic blood, to be transported to the other organs for consumption. (2) Some will undergo oxidative degradation to CO_2 via the glycolytic sequence and the tricarboxylic acid cycle (Chapters 11 and 12), to serve as a source of energy in the liver. (3) Some will serve as a source of reducing power (NADPH) via the pentose phosphate pathway (Chapter 12). (4) Some will be converted into the storage polysaccharide glycogen by the action of glycogen synthetase (Chapter 17), a process we shall examine in more detail below. (5) Some of the glucose 6-phosphate not needed for other purposes will be degraded into acetyl CoA (Chapters 11 and 12), which will then be converted into fatty acids and thence into fat (Chapter 18). The fat may then be transported to fat tissues under the skin or in the abdominal cavity for longer-term storage. These relationships are shown in Figure 20-2.

Amino acids

The amino acids entering the liver following absorption from the intestinal tract also have several important fates:

Figure 20-2
The fate of glucose 6-phosphate in the liver.

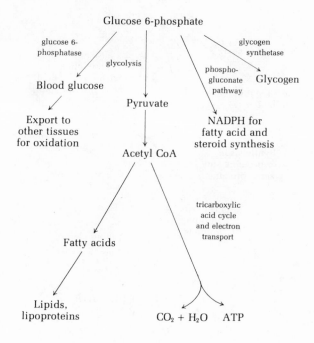

(1) Some of the amino acid influx is allowed to pass through the liver into the bloodstream, and out to other organs for utilization as building blocks in tissue protein synthesis. (2) Some of the amino acid influx is utilized by the liver for the synthesis, not only of the intrinsic proteins of liver cells, but also of the plasma proteins of the blood, which are then exported. (3) Some of the amino acids in excess of needs for protein synthesis may be converted into pyruvate and thence to glucose and glycogen, in the process of gluconeogenesis (Chapter 17). (4) Some of the amino acids may be converted into acetyl CoA (Chapter 15), which may be burned as energy source in the tricarboxylic acid cycle or converted into lipids (Chapter 18). The ammonia released on degradation of amino acids is then converted into the waste product urea by the liver (Chapter 15). (5) Some of the amino acids may be converted in the liver into various specialized products, such as porphyrins (Chapter 19). These relationships are shown in Figure 20-3.

Lipids

The lipids entering the liver also have several different fates, as outlined in Figure 20-4: (1) Some of the incoming

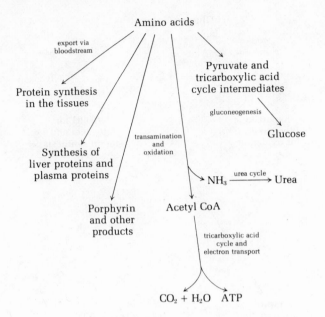

Figure 20-3
The fate of amino acids in the liver.

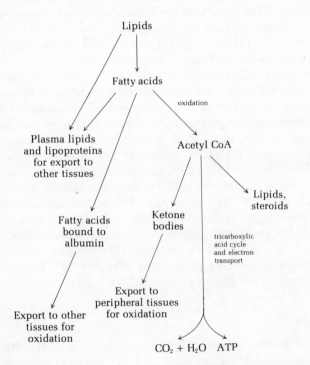

Figure 20-4
The fate of lipids in the liver.

Table 20-3   The composition of the human plasma lipoproteins. The composition of chylomicrons is also shown for comparison.

	Percent composition			
	Protein	Triacyl-glycerol	Cholesterol and its esters	Phospholipids
α-Lipoproteins (high density)	33–57	6–11	15–27	20–29
β-Lipoproteins (low density)	7–21	10-52	21-46	19–22
Chylomicrons	1	87	4	8

lipids, largely in the form of triacylglycerols, will be reprocessed and used in the synthesis of lipoproteins, in which form lipids are transported to peripheral tissues, particularly to the fat cells. (2) Some of the lipids will be degraded to yield free fatty acids, which are bound to serum albumin and pass to peripheral tissues, particularly the heart and skeletal muscle, which can utilize free fatty acids. (3) Some of the free fatty acids will be directly oxidized by the liver to yield acetyl CoA, most of which will be oxidized in the tricarboxylic acid cycle. The remainder will be converted into the ketone bodies aceto-acetic acid and D-β-hydroxybutyric acid, which are circulated via the blood to peripheral tissues as fuel for the tricarboxylic acid cycle (Chapter 14). (4) Some of the acetyl CoA derived from fatty acids (and from glucose) will be used as a building block for the synthesis of cholesterol and other steroids (Chapter 18).

Metabolic Characteristics of the Major Organs

Although all cells of eukaryotic organisms are capable of catalyzing the main pathways of metabolism, the various tissues and organs of mammals are specialized in function and participate in a harmonious metabolic interplay, in which the principle of maximum economy and regulation are paramount. We have already seen that the liver plays a central processing and distributing role and furnishes all other organs with a proper mix of nutrients via the bloodstream. We shall now consider some of the metabolic characteristics of some other major organs and how they use their ATP energy supply.

Skeletal Muscle and its Contractile System

The muscle mass accounts for over 50 percent of the total metabolizing tissue in the resting human and a much greater fraction during great muscular activity. Skeletal muscle metabolism is primarily specialized to yield ATP, the immediate fuel for muscular work, from glycolysis and respiration. Skeletal muscle utilizes glucose, free fatty acids, and ketone bodies as fuel. In resting muscle the basic fuels are fatty acids and ketone bodies, which are converted into acetyl CoA (Chapter 14) and then enter the tricarboxylic acid cycle for combustion. However, in actively working muscle, blood glucose is utilized in addition to fatty acids. The glucose is phosphorylated and is degraded by the glycolytic sequence to pyruvate. Although some of the pyruvate is oxidized to acetyl CoA, most will be reduced to lactate in very active muscles, the lactate then diffusing into the bloodstream. Very active muscles use the anaerobic glycolysis pathway to provide extra ATP energy, supplementing the basal ATP production resulting from the oxidation of fatty acids by the tricarboxylic acid cycle. The lactic acid so formed returns to the liver where it is converted through gluconeogenesis into blood glucose again, for recirculation to the working muscles. Thus glucose and lactic acid cycle between actively working muscle and the liver in a metabolic partnership. This cycle is called the Cori cycle after C. Cori and G. Cori, who first described its function (Figure 20-5).

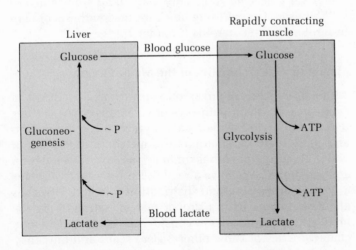

Figure 20-5

The metabolic cooperation between contracting skeletal muscle and the liver. To supplement the ATP produced from oxidation of fatty acids, pyruvate and ketone bodies also bring about glycolysis to yield extra ATP, thus forming lactate. The lactate is carried to the liver where it is converted again into blood glucose at the expense of energy.

Skeletal muscle contains significant amounts of glycogen, up to 1 percent. While this is much less than the glycogen in the liver, which fluctuates in the range 5–10 percent, the muscle glycogen is essential for furnishing energy to the contracting muscle in emergencies, as we shall see later in this chapter. Because muscle glycogen cannot be converted into blood glucose, since muscle lacks glucose 6-phosphatase, all of the muscle glycogen is available for muscle activity. Muscles also contain considerable amounts of phosphocreatine (Chapter 10), a reservoir of high-energy phosphate groups which is utilized during extreme activity to replenish ATP.

The contractile system of skeletal muscle consists of parallel *myofibrils*, each made up of protein filaments. There are two types of *myofilaments*, thick and thin filaments, which are arranged in hexagonal array about each other (Figure 20-6). The thick filaments consist of bundles of the protein *myosin*, a long slender molecule, having two helically twisted polypeptide chains of mol wt 225,000 (~1800 residues). Each myosin molecule has a complex "head"; these heads protrude at regular intervals along the thick filament. The thin filaments consist of bead-like strings of the monomeric protein *G actin* (globular actin); each monomeric unit has a molecular weight of about 60,000. During energy-dependent contraction the thin filaments slide further into the space between the thick filaments (Figure 20-6). The basic unit of the contractile system is the *sarcomere*, which includes a set of thick filaments and two sets of thin filaments penetrating into the zone of thick filaments. The entire sarcomere is about 2.5 μ long.

When contraction is initiated by a nerve impulse to the muscle cell membrane, the impulse travels to all the individual myofibrils via a system of tubular membranes called *transverse tubules* or the *T-system*. The incoming nerve impulse causes the sarcoplasmic reticulum to release Ca^{2+} into the cytoplasm. The Ca^{2+} then activates the contraction by stimulating the "heads" of the myosin molecules to break and remake cross-bridges to the thin filaments, simultaneously bringing about translocation of the thin (actin) filaments, along the thick ones. At the same time, ATP is hydrolyzed by the heads of the myosin molecules to yield ADP and phosphate. For each cross-bridge between a myosin head and the actin filament to be made and broken, one molecule of ATP is hydrolyzed. In this way ATP energy is used to bring about translocation and thus contraction. The muscle is induced to relax again when the nerve impulses cease and Ca^{2+}

Figure 20-6
The structure of muscle fibers and the sliding filament model of contraction. Muscle cells contain parallel bundles of muscle fibers, surrounded by sarcoplasmic reticulum and bathed by the sarcoplasm.

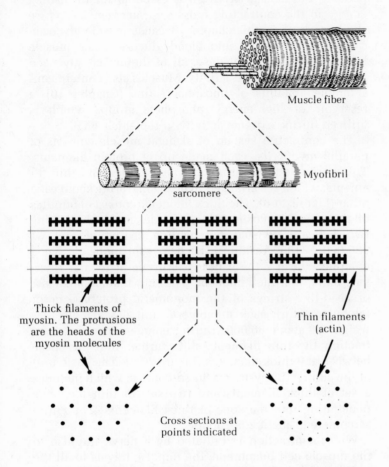

Muscle fiber

Myofibril

sarcomere

Thick filaments of myosin. The protrusions are the heads of the myosin molecules

Thin filaments (actin)

Cross sections at points indicated

Sliding filament model of contraction. The thin actin filaments (color) slide into the spaces between the thick myosin filaments, causing shortening of the sarcomeres.

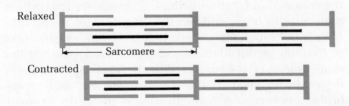

Relaxed

Sarcomere

Contracted

returns to the sarcoplasmic reticulum, where it is bound in an ATP-dependent process.

The actomyosin system is basic to the function not only of skeletal but also smooth muscle and heart muscle. It is also involved in other types of motile systems, for example, of amoebae and other protozoa.

Heart Muscle

Heart muscle differs in its metabolism from skeletal muscle in that it is continuously active and has a completely aerobic metabolism at all times. Thus, lactic acid is not produced as an end product of glycolysis, and the tricarboxylic acid cycle is the main source of ATP. Heart muscle cells have numerous mitochondria and utilize a mixture of blood glucose, free fatty acids, and ketone bodies, which are oxidized via the tricarboxylic acid cycle. Like skeletal muscle, heart muscle also can store small amounts of high-energy phosphate bonds in the form of phosphocreatine, but otherwise does not store either glycogen or fat in any significant amounts.

Brain

The metabolism of the brain is remarkable in that this organ in normal adult mammals has a stringent requirement for glucose as its major fuel. The brain has a very active respiratory metabolism and utilizes almost 20 percent of the total oxygen consumption in a resting human. Because the brain contains very little glycogen, it depends on the incoming glucose from the blood on a minute-to-minute basis. If the blood glucose should fall below certain critical levels for even short periods of time, symptoms of brain disturbance, sometimes severe and irreversible, may appear.

Glucose is utilized by the brain via the glycolytic sequence and the tricarboxylic acid cycle. ATP generated by oxidative phosphorylation in the brain is used to support the capacity of the brain and nerves to transmit nerve impulses and to provide energy for the biosynthesis of protein, a very active process in this organ.

The brain has a rather high concentration of certain amino acids, some of which are synthesized in the brain from other precursors. Particularly abundant are glutamic acid, glutamine, aspartic acid, glycine, and γ-aminobutyric acid (Chapter 2), some of which serve as chemical messengers or modulators in transmitting impulses across nerve synapses.

Although the brain has no capacity to utilize free fatty acids or lipids from the blood, it can utilize the ketone body β-hydroxybutyric acid, which is formed from the oxidation of fatty acids in the liver. During starvation large amounts of β-hydroxybutyric acid are generated in the liver at the expense of body fat and it then becomes a major fuel for the brain, which oxidizes it to CO_2 via the tricarboxylic acid. The brain of infant animals also utilizes blood ketone bodies actively.

Kidneys

The kidneys have a very active respiratory metabolism and have considerable metabolic flexibility. They can utilize blood glucose, ketone bodies, free fatty acids, and amino acids as sources of fuel, degrading these ultimately via the tricarboxylic acid cycle and making available ATP energy as a result of oxidative phosphorylation. Most of this energy is utilized in the formation of urine, which takes place in a two-stage process. In the first, the blood plasma is filtered through microscopic structures called glomeruli in the cortex (outer layer) of the kidney. These "filters" allow all components of the blood plasma except the proteins and lipids to pass freely into ducts called renal tubules. As the filtrate passes down the tubules, Na^+, Cl^-, and water are reabsorbed back into the blood passing through capillaries surrounding the tubules. As a result the glomerular filtrate undergoes concentration as it proceeds down the tubules. Each ml of the final urine is formed by the concentration of from 50 to 100 ml of the glomerular filtrate. The solute composition of normal human urine is shown in Table 20-4.

Some components of urine, particularly glucose and bicarbonate, are normally present in much lower concentration than in blood. This group of solutes is reabsorbed from the glomerular filtrate into the blood, against a concentration gradient. A second group of solutes, including ammonia and H^+, are present in relatively high concentration in urine compared to blood; they are actively secreted from the blood into the tubules, also against a concentration gradient. A third group of substances, including urea and creatinine (the end-product of phosphocreatine degradation), shows inert behavior; they are neither absorbed or secreted against gradients. Na^+ is a special case; it is reabsorbed from the glomerular filtrate into the blood plasma by active transport in the first portion of the tubule, but some of it later

Table 20-4 Major components of human urine. The 24-hour volume and composition of urine vary widely depending on fluid intake and diet. The data shown are for an average 24-hour specimen of total volume 1200 ml.

Component	Grams per 24 hours	Approximate urine/plasma concentration ratio
Glucose	<0.05	<0.05
Amino acids	0.80	1.0
Ammonia	0.80	100
Urea	25	70
Creatinine	1.5	70
Uric acid	0.7	20
H^+	pH 5–8	Up to 300
Na^+	3.0	1.0
K^+	1.7	15
Ca^{2+}	0.2	5
Mg^{2+}	0.15	2
Cl^-	6.3	1.5
Phosphate	1.2 g P	25
Sulfate	1.4 g S	50
Bicarbonate	0–3	0–2

passes back into the urine again by secondary exchange with other cations.

The transport of Na^+ and K^+ is especially conspicuous in the kidney, which must help regulate the concentration of these vital cations in the blood. Virtually all mammalian cells contain a relatively high concentration of K^+ and low concentration of Na^+, whereas the blood plasma and other extracellular fluids have a high concentration of Na^+ and low K^+ (Figure 20-7). The plasma membranes of most cells contain an ATP-dependent system for transporting K^+ inward and Na^+ out. This transport system, often called the Na^+- and K^+-trans-

Figure 20-7
The electrolyte composition of intracellular fluid, blood plasma, gastric juice, and pancreatic juice. The left portion of each bar graph shows the cation composition; the right portion the anion composition. The solid black zones represent the sums of the minor components. Note the large differences in Na^+ and K^+ content of intracellular fluid and blood plasma, as well as the large gradient of H^+ ions between gastric juice and blood plasma. Such concentration gradients are generated by ATP-consuming active transport systems.

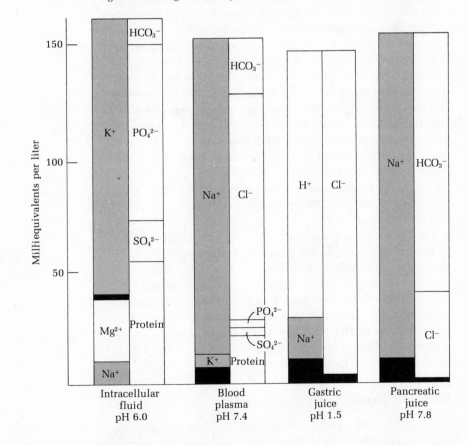

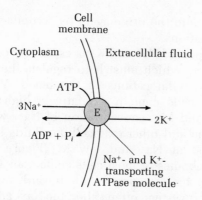

Cell
membrane

Cytoplasm   Extracellular fluid

The two stages in Na^+ and K^+ transport:

1. $Na^+_{inside} + ATP + E \rightarrow [Na^+ \cdot E \sim P] + ADP$

2. $[Na^+ \cdot E \sim P] + K^+_{outside} \rightarrow Na^+_{outside} + K^+_{inside} + P_i + E$

Figure 20-8
*The action of the Na^+, K^+-transporting
ATPase in the plasma membrane of
animal cells. The drug ouabain
inhibits both the transport of Na^+ and
K^+ and the hydrolysis of ATP.*

porting ATPase, has been studied in great detail. It appears to consist of a large protein molecule with several subunits exceeding a total of 500,000 in molecular weight. The complex is tightly bound in the cell membrane in such a way that it can accept ATP only from the inside or cytoplasmic compartment for hydrolysis; the ADP and phosphate formed return to the cytoplasm. However, for ATP hydrolysis to occur K^+ must move into the cell and Na^+ out. This process occurs in two stages. In the first, the ATP donates its terminal phosphate group to phosphorylate the transport enzyme, a reaction which requires binding of Na^+ from the cytoplasm. In the second stage, the phosphoenzyme undergoes hydrolysis to form the free transport enzyme and phosphate; this stage causes delivery of Na^+ to the outside and K^+ to the inside. Three Na^+ ions are transported out and two K^+ ions in for each molecule of ATP hydrolyzed. This transport process is shown in Figure 20-8.

Through the action of various active transport systems in the kidney, including the Na^+, K^+-transporting ATPase, the urine is formed in such a way as to excrete those substances whose concentration in the blood must be lowered and to reabsorb those substances from the urine required to maintain the composition of the blood constant. Over three-fourths of the ATP generated by respiration in the kidneys is utilized in the formation of urine by the action of active transport mechanisms.

Metabolic Regulation by Hormones

We have already seen (Chapters 4, 9) that metabolism may be regulated (1) by the intrinsic kinetic properties of enzymes, that is, their affinity for substrates and their optimum pH in relationship to tissue concentrations of substrates and tissue pH; (2) by allosteric control mechanisms, of which we have seen numerous instances; and (3) by induction or repression of enzyme synthesis, a form of regulation to be discussed further in Part IV of this book (see Chapter 23).

In higher animals another type of metabolic regulation takes place: regulation by hormones, chemical messengers secreted by various glands of the endocrine system. The anterior pituitary gland of the brain secretes a series of trophic, or "master", hormones which can stimulate the various endocrine glands to produce their characteristic hormones. For example, the thyrotrophic and adrenocorticotrophic hormones, polypeptides secreted by the anterior pituitary gland, pass via the blood to the thyroid gland and the adrenal cortex and there stimulate these tissues to produce their hormones, thyroxine and the adrenal cortical steroids, respectively. Other target glands stimulated by the anterior pituitary are the gonads and the islet cells of the pancreas.

A complex network of regulatory checks and balances characterizes the function of the endocrine system in regulating many different activities of the mammalian organism. This subject lies within the domain of the field of physiology. However, in the last few years some profound advances have been made in our understanding of the molecular basis of the action of some of the hormones, which brings them into the domain of biochemistry.

We shall concern ourselves in the following discussion with only a few hormones, each of which can produce profound effects on the regulation of carbohydrate and fat metabolism, namely, the hormones epinephrine, glucagon, and insulin. These three hormones regulate the release of the stored fuels glycogen and fat in times of metabolic need.

Regulation of Glycogen Storage and Release

We have seen (Chapters 11 and 17) that the synthesis and breakdown of glycogen takes place by the action of two distinct enzymes, glycogen synthetase and glycogen phosphorylase, respectively. They are allosteric enzymes which are sensitive to different sets of regulatory modulators. Superimposed on the allosteric regulation of glycogen phosphorylase is another level of regulation by the hormones epinephrine and glucagon, which promote glycogen breakdown in skeletal muscle and in the liver, respectively.

Although skeletal muscle contains a much lower amount of glycogen than the liver, muscle glycogen is essential to the mammal because its breakdown by phosphorylase is an immediate source of glucose 6-phosphate for glycolysis to lactic acid, providing energy for sudden and extreme activity. The breakdown of muscle glycogen catalyzed by phosphorylase is triggered by the hormone _epinephrine_ (Figure 20-9), also known as adrenaline, which is secreted by the adrenal medulla. When an animal is suddenly frightened and must fight or flee, epinephrine is immediately secreted into the blood from the adrenal medulla, although only in trace concentrations of the order of 10^{-9} M. It is carried by the bloodstream to the muscles. There the epinephrine molecule becomes bound to specific receptor sites in the muscle cell membrane. Once these sites are occupied, an enzyme called _adenyl cyclase_, which is attached to the inner surface of the membrane, is converted from a resting, inactive state to a state of maximal activity. Adenyl cyclase catalyzes the removal of pyrophosphate from ATP in such a way as to yield _cyclic 3',5'-adenylic acid_ (cyclic AMP) (Figure 20-9)

$$\text{ATP} \rightarrow \text{cyclic 3',5'-adenylic acid} + \text{PP}_i$$

Like ordinary adenylic acid, cyclic AMP contains a single phosphate group, but this forms a diester linkage between the 5' and 3' positions of the D-ribose ring. Cyclic AMP is normally present in vanishingly small amounts in resting muscle, but stimulation of adenyl cyclase by epinephrine quickly increases the amount of cyclic AMP in the muscle one hundred-fold or more. This is the first step in an amplification cascade (Figure 20-10) which ultimately results in the stimulation of phosphorylase to break down glycogen to yield large amounts of glucose 1-phosphate and from the latter, glucose 6-phosphate.

The cyclic AMP now makes possible the next step in this amplification of the incoming hormone signal. Cyclic AMP is a specific activator of the enzyme _phosphorylase b kinase kinase_ which catalyzes the phosphorylation of an inactive form of the enzyme _phosphorylase b kinase_, at the expense of ATP

ATP + inactive phosphorylase b kinase $\xrightarrow{\substack{\text{phosphorylase b} \\ \text{kinase kinase}}}$

ADP + active phosphorylase b kinase

This is accomplished by transfer of the terminal phosphate group of ATP to the hydroxyl group of a specific

Figure 20-9
The structure of epinephrine and 3',5'-cyclic adenylic acid (cyclic AMP).

Epinephrine

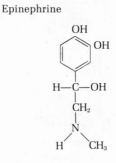

3',5'-Cyclic adenylic acid

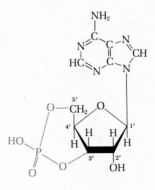

Figure 20-10
Stimulation of glycogen breakdown and glycolysis in muscle following
secretion of epinephrine by the adrenal medulla. The signal provided
by 10^{-9} M epinephrine is amplified over 5 million-fold by the
cascade effect produced as a result of the activation of each
enzyme by a product of the preceding reaction.

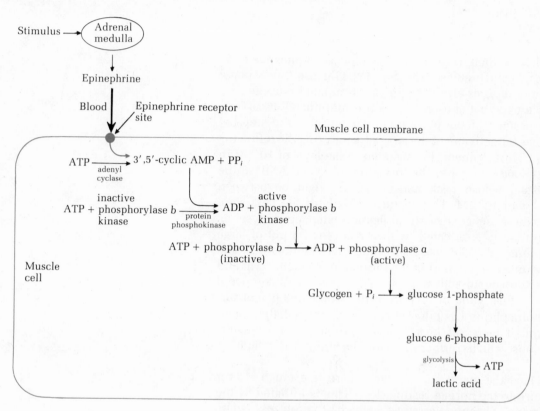

serine residue in the polypeptide chain of the inactive form of phosphorylase *b* kinase. Once this enzyme is brought to its active phosphorylated form it makes possible the next step in the amplification of the incoming epinephrine signal.

The active form of the enzyme phosphorylase *b* kinase now carries out its catalytic function and promotes the phosphorylation by ATP of the inactive form of phosphorylase, namely, phosphorylase *b* (Chapter 11), to yield the active form phosphorylase *a*

$$4\text{ATP} + 2 \text{ phosphorylase } b \xrightarrow{\substack{\text{phosphorylase } b \\ \text{kinase}}}$$

$$\text{phosphorylase } a + 4\text{ADP}$$

In this reaction the hydroxyl groups of specific serine residues in each of the two subunits of phosphorylase *b* become phosphorylated and two molecules of phosphorylated phosphorylase *b* now combine to form a single molecule of phosphorylase *a*, which has four subunits (Chapter 11).

Finally, in the last stage of amplification, phosphorylase *a* catalyzes the breakdown of glycogen to glucose 1-phosphate

$$\text{Glycogen}_n + P_i \rightarrow \text{glycogen}_{n-1} + \text{glucose 1-phosphate}$$

which is then converted into glucose 6-phosphate by phosphoglucomutase (Chapter 11). Glucose 6-phosphate then undergoes glycolysis in the stimulated muscle.

We note that at each step in this amplification cascade the product of the preceding enzyme-catalyzed step becomes the activator of the enzyme catalyzing the following step. Thus, epinephrine secreted at the level of 10^{-9} M in the blood stimulates the formation of cyclic AMP in the muscle, whose peak concentration might be between 10^{-8} and 10^{-7} M. This in turn stimulates the next step. By these successive amplification steps glycolysis is enormously stimulated, with a resulting output of large amounts of ATP used for muscular contraction. The amounts of ATP used in the formation of cyclic AMP and the subsequent activation steps are very small compared to the final yield of ATP from the stimulated glycolysis.

Epinephrine has other effects on the mammalian organism. It increases heart rate and dilates blood vessels, actions which further prepare the animal to "fight or flee."

Once the emergency is over, there is a much slower return to the normal resting state. This is initiated by the enzyme *phosphodiesterase*, which hydrolyzes cyclic AMP to yield the inactive, ordinary form of adenylic acid

$$\text{Cyclic AMP} + H_2O \xrightarrow{\text{phospho-diesterase}} \text{adenylic acid}$$

Once the cyclic AMP level subsides, the activated enzymes revert to their inactive forms.

Cyclic AMP has been aptly termed a "second messenger" or "intracellular hormone," in consonance with the concept that hormones such as epinephrine are chemical messengers from one organ to another. Cyclic AMP also plays a messenger role in the action of other hormones; another example is given in the next section.

Action of Glucagon on Liver Phosphorylase

Glucagon is a polypeptide hormone of 21 amino acid residues which is secreted by the β- cells of the islet tissue of the pancreas. This hormone is secreted into the blood whenever the blood glucose concentration falls below a critical level. The glucagon acts as a chemical messenger to the liver. It becomes bound to specific receptor sites on the surface of liver cells and, like epinephrine, activates the adenyl cyclase attached to the inner surface of the membrane. Cyclic AMP is then formed. In a series of amplification steps resembling, but not identical with, those in skeletal muscle, liver phosphorylase is ultimately stimulated to catalyze breakdown of glycogen to glucose 1-phosphate, which in turn is transformed to glucose 6-phosphate. The latter undergoes hydrolysis to free glucose which passes into the blood. As the blood glucose concentration increases to its normal level, the secretion of glucagon is halted and the system reverts to its normal state.

Regulation of Fat Storage and Release

Cyclic AMP also plays a central role in the endocrine control of fat storage and release. Both glucagon and epinephrine serve as chemical messengers to the cells of adipose or fat tissue. These cells are gorged with large amounts of triacylglycerols. When these hormones are bound by specific receptor sites on the fat cell membrane, they stimulate the formation of cyclic AMP in the cell. The cyclic AMP in turn stimulates a protein phosphokinase which phosphorylates the inactive form of the enzyme *lipase* at the expense of ATP, to yield its active form. Lipase catalyzes hydrolysis of the triacylglycerols present in the fat cells. The free fatty acids so produced are yielded to the bloodstream, where they combine with serum albumin. The fatty acids pass to other tissues as fuel to meet the demand for ATP energy signaled by the secretion of glucagon and epinephrine.

The Action of Insulin

The hormone insulin (Chapter 3), which is also secreted by the islet tissue of the pancreas, but by a different cell type, has profound effects on the metabolism of carbohydrate, fat, and protein. These are shown most dramatically in the human disease *diabetes mellitus*, in which there is a failure of insulin formation or secretion. In this disease, there is a very high blood glucose concentration, excessive oxidation of fatty acids leading to high levels of the ketone bodies in the blood, and a cessation of fatty acid synthesis from acetyl CoA. Administration of insulin alleviates these symptoms.

The precise mode of action of insulin is not yet known. However, recent research shows that the secretion of insulin is triggered whenever the blood glucose concentration rises to abnormally high levels. The insulin passes via the blood to the peripheral organs, where it promotes the uptake of blood glucose and its conversion to glycogen, promotes the uptake of amino acids, and promotes the synthesis of fatty acids from acetyl CoA.

In the fat cells insulin causes an *inhibition* of adenyl cyclase and thus counteracts the effects of glucagon and epinephrine. However, the action of insulin is not yet completely understood in molecular terms.

Intermediary metabolism is also regulated by other hormones, but the biochemical basis of their action is not fully understood. Thyroxine, for example, increases the rate of oxidative metabolism of animals but its mode of action is still under study.

Transport of Oxygen

Oxygen from the atmosphere is required as the ultimate electron acceptor for the aerobic metabolism of mammals and it must be provided at a high rate. Whereas oxygen dissolved in an aqueous medium can readily pass into single-celled organisms by simple physical diffusion at a high enough rate to satisfy metabolic needs, this is not the case in higher organisms, such as mammals, which have a very large metabolizing mass of tissue in relation to the surface area exposed to dissolved oxygen. A normal human daily requires over 500 liters of pure oxygen gas, equivalent to the oxygen content of about 2500 liters of air. It is the function of the respiratory protein hemoglobin of the red blood cells to transport oxygen from the lungs to all portions of the body very rapidly and efficiently, via the bloodstream.

Mature erythrocytes of mammals are relatively small, nonnucleated cells that primarily depend on breakdown of glucose by glycolysis for energy; they have no respiratory metabolism. The cytoplasm of erythrocytes contains about 35 percent by weight of hemoglobin, which represents about 90 percent of the total cell protein. Hemoglobin (Chapter 3) contains four polypeptide chains of about 150 amino acid residues and four heme groups (Chapter 13), each of which can bind one molecule of molecular oxygen reversibly. Because of this property, whole mammalian blood, when fully oxygenated, can carry about 21 ml of gaseous oxygen per 100 ml, whereas blood plasma alone, in the absence of the red cells, can absorb by physical solution only about 0.3 ml of oxygen per 100 ml.

At any given pH the amount of oxygen bound by hemoglobin is determined by the partial pressure of oxygen in the gas phase with which the hemoglobin solution (or the blood) is in equilibrium. Figure 20-11 shows that a plot of percent oxygenation of hemoglobin versus oxygen pressure has a sigmoid shape. This relationship indicates that initial binding of the first molecules of oxygen increases the affinity of the hemoglobin molecule for binding further amounts of oxygen. As the partial pressure of oxygen is increased further a plateau is reached in which each of the hemoglobin molecules contains the limit of four molecules of oxygen. At this point the hemoglobin is spoken of as being saturated with oxygen.

This sigmoid relationship is due to the fact that the hemoglobin molecule contains two α and two β polypeptide chains (each carrying a heme group), which are arranged into two $\alpha\beta$ subunits. When a solution of hemoglobin is exposed to a limited amount of oxygen each of the two $\alpha\beta$ subunits combines reversibly but loosely with one molecule of oxygen

$$\begin{matrix}\textcircled{α}\textcircled{β} \\ \textcircled{β}\textcircled{α}\end{matrix} \;+\; 2O_2 \;\rightleftharpoons\; \begin{matrix}O_2-\boxed{\alpha}\textcircled{β} \\ \textcircled{β}\boxed{\alpha}-O_2\end{matrix}$$

Binding of one molecule of oxygen to one of the two hemes of each $\alpha\beta$ subunit causes the second heme to undergo a large increase in affinity for oxygen, symbolized by the change from circles to squares

$$\begin{matrix}O_2-\boxed{\alpha}\textcircled{β} \\ \textcircled{β}\boxed{\alpha}-O_2\end{matrix} \;\rightleftharpoons\; \begin{matrix}O_2-\boxed{\alpha}\boxed{\beta} \\ \boxed{\beta}\boxed{\alpha}-O_2\end{matrix}$$

This form now combines with two more molecules of oxygen with very high affinity, to yield fully oxygenated hemoglobin, called *oxyhemoglobin*

$$\begin{matrix}O_2-\boxed{\alpha}\boxed{\beta} \\ \boxed{\beta}\boxed{\alpha}-O_2\end{matrix} \;+\; 2O_2 \;\rightleftharpoons\; \begin{matrix}O_2-\boxed{\alpha}\boxed{\beta}-O_2 \\ O_2-\boxed{\beta}\boxed{\alpha}-O_2\end{matrix}$$

Binding of one molecule of oxygen to the $\alpha\beta$ subunit is believed to result in a conformational change in the oxygenated chain, which is transmitted to the nonoxygenated chain to endow it with greater affinity for oxygen.

Figure 20-11
The effect of the partial pressure of oxygen and the effect of pH on the oxygenation of hemoglobin.

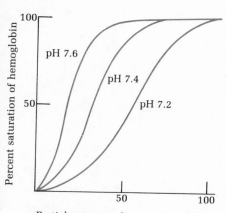

Partial pressure of oxygen (mm Hg)

This property of hemoglobin has served as a model for the study of allosteric enzymes (Chapter 4), some of which show increases in affinity for their substrates after binding the modulator or effector molecule. Biologically, the sigmoid oxygenation curve for hemoglobin represents an adaptation to make possible high-affinity loading of the hemoglobin molecule with oxygen in the lungs and efficient unloading of oxygen in the tissues, where the oxygen partial pressure is about half that in the lungs.

The position of the oxygen–hemoglobin equilibrium is also affected by the pH (Figure 20-11). The higher the pH of the solution at a given partial pressure of oxygen, the greater the percent saturation with oxygen at equilibrium. This reversible effect is due to the fact that when hemoglobin is oxygenated it ionizes to set free a proton according to the equation

$$HHb^+ + O_2 \rightleftharpoons HbO_2 + H^+$$

in which HHb^+ is the protonated form of hemoglobin. Since this reaction is freely reversible, increasing the hydrogen ion concentration (decreasing the pH) will cause oxygenated hemoglobin to lose some of its oxygen, as the reaction equilibrium is pushed to the left. Conversely, decreasing the hydrogen ion concentration (raising the pH) will increase the amount of oxygen bound, as the equation is pulled to the right by removal of H^+.

The partial pressure of oxygen and the pH are the two most important factors regulating the function of hemoglobin in the transport of oxygen. In the lungs, where the partial pressure of oxygen is high, about 100 mm Hg, and the pH relatively high, hemoglobin will tend to become almost maximally saturated with oxygen, about 96 percent (see Figure 20-11). In the interior of peripheral tissues, where the oxygen tension is low (about 50 mm Hg) and the pH also low, the hemoglobin binds oxygen less strongly. Therefore the hemoglobin will unload some of its oxygen to the respiring cell mass until the hemoglobin is only about 65 percent saturated. In normal blood, then, the hemoglobin cycles between 65 and 95 percent saturation with oxygen in its continuous passage between the lungs and the tissues.

Transport of CO_2

The blood also can transport large amounts of carbon dioxide, the end product of aerobic metabolism, from the tissues, where it is formed as the end product of oxidation of fuels, to the lungs, where it is excreted in the expired air. Venous blood leaving the tissues is rich in CO_2; it contains about 600 ml of CO_2 per 100 ml. On the contrary, arterial blood, after leaving the lungs, contains only 50 ml of CO_2 per 100 ml. About two-thirds of the blood CO_2 is present in the plasma and about one-third in the red blood cells. However, as we shall see, nearly all the CO_2 of blood must pass through the red blood cells during CO_2 transport between the tissues and the lungs. In both the plasma and the red blood cells the total CO_2 is present in two forms, some as dissolved CO_2 but most as the bicarbonate ion HCO_3^-. Since dissolved CO_2 is hydrated to yield carbonic acid (H_2CO_3), the mixture of CO_2 and bicarbonate together constitutes a buffer system (Chapter 1), in which the carbonic acid formed from CO_2 is the proton donor and the bicarbonate ion the proton acceptor.

The following sequence of events occurs in CO_2 transport from the tissues to the lungs (Figure 20-12). Dissolved CO_2, the direct product of the tricarboxylic acid cycle oxidations in the tissues, diffuses through the cell membranes into the blood plasma and then into the erythrocyte. Within the erythrocyte the CO_2 becomes hydrated to yield free carbonic acid in the reaction

$$CO_2 + H_2O \rightleftharpoons H_2CO_3$$

In the absence of a catalyst this reaction is relatively slow and does not proceed fast enough to keep pace with the production of CO_2 by the respiring tissues. However, erythrocytes contain an enzyme, *carbonic anhydrase*, which greatly accelerates this reaction. Once carbonic acid is formed in this enzyme-catalyzed process it ionizes to yield bicarbonate

$$H_2CO_3 \rightleftharpoons H^+ + HCO_3^-$$

The bicarbonate ions so formed then pass out of the erythrocyte into the blood plasma, in exchange for chloride ions (Cl^-). The H^+ ion resulting from the ionization of the carbonic acid in the erythrocyte is then absorbed by the oxyhemoglobin in the red blood cell, causing it to lose its oxygen in the reverse of the reaction described earlier

$$H^+ + HbO_2 \rightarrow HHb^+ + O_2$$

Thus the H^+ ion formed by the dissociation of carbonic acid in the red blood cell promotes the unloading of oxygen by hemoglobin in the tissues.

When the deoxygenated but CO_2-rich venous blood now returns to the lungs the reverse of this cycle takes place.

Figure 20-12
*Role of red blood cells in transport of oxygen and carbon dioxide
between lungs and tissues.*

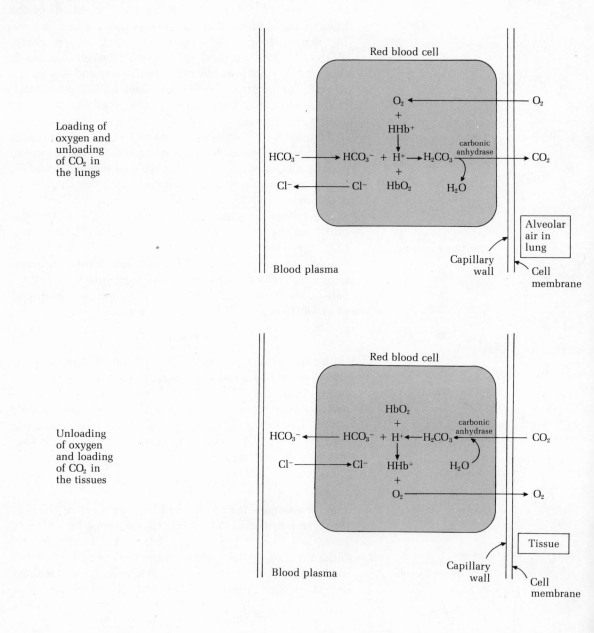

Loading of
oxygen and
unloading
of CO_2 in
the lungs

Unloading
of oxygen
and loading
of CO_2 in
the tissues

Oxygen taken up by the hemoglobin causes the oxyhemo-globin to lose an H^+ ion.

$$HHb^+ + O_2 \rightarrow H^+ + HbO_2$$

The H^+ ion now promotes the formation of carbonic acid from HCO_3^- in the erythrocyte

$$H^+ + HCO_3^- \rightarrow H_2CO_3$$

The carbonic acid is then dehydrated by carbonic anhy-drase to yield dissolved CO_2, which passes out of the erythrocyte through the blood plasma into the air space of the lungs. In this way the transport of oxygen and trans-port of CO_2 mutually promote each other via the action of hemoglobin, a molecule beautifully adapted for its special transport function.

Summary

The three major classes of energy-yielding nutrients undergo enzymatic hydrolysis to their building-block components during digestion in the gastrointestinal tract. Most polysac-charides and disaccharides are degraded to their monosaccha-ride components. Digestion of proteins begins in the stomach by the action of pepsin, secreted in the form of pepsinogen. The inactive zymogens—trypsinogen, chymotrypsinogen, and procarboxypeptidase—secreted in the pancreatic juice, are converted in the intestine into the corresponding active en-zymes—trypsin, chymotrypsin, and carboxypeptidase—which carry out sequential hydrolysis of the peptide fragments of ingested proteins to yield ultimately free amino acids. Lipids are partially hydrolyzed by lipases and phospholipases to yield mono- and diacylglycerols, which together with triacylglycerols are emulsified with the aid of bile salts. Sugars and amino acids are transported into the epithelial cells lining the intestine against a gradient of concentration, through the action of ATP-dependent active transport systems which are specific for their substrates.

The liver participates in the processing and distribution of glucose, amino acids, and lipids for transport via the blood to the peripheral tissues. The liver is active in conversion of glu-cose via acetyl CoA into fatty acids, lipids, and lipoproteins, the transport form of lipids in the blood. The liver is also active in synthesis of the plasma proteins, and is the site of formation of urea, the end-product of amino acid catabolism. The brain, skeletal muscles, heart, and kidneys have characteristic met-abolic patterns and engage in metabolic interplay with the liver. Muscles of various types contain as contractile systems parallel thick filaments, composed of the protein myosin, and thin filaments of actin. These filaments slide along each other and produce contraction at the expense of ATP, which is hydrolyzed during the making and breaking of translocating crossbridges between the filaments.

The hormone epinephrine, secreted by the adrenal medulla, prepares the mammal for emergencies by stimulating glycogen breakdown in the skeletal muscles. Trace amounts of epinephrine stimulate adenyl cyclase of the muscle membrane to produce cyclic AMP from ATP, a reaction which initiates a sequence of enzyme amplification steps ultimately resulting in rapid conversion of muscle glycogen to glucose 1-phosphate. Glucagon, secreted by the pancreas when the blood glucose level becomes low, initiates a similar cycle of events in the liver, which leads to the conversion of liver glycogen into blood glucose. Insulin, also secreted by the islet tissue of the pancreas, is formed when the blood glucose level becomes elevated; it promotes the utilization of glucose by the tissues and its conversion into fat.

Oxygen is transported from the lungs to the tissues by the hemoglobin of red blood cells. The amount of oxygen bound by hemoglobin depends on its partial pressure, in such a way that binding of the first molecules of oxygen enhances the binding of subsequent molecules, an effect that is due to conformational changes induced in the four subunits of the hemoglobin molecule. The binding of oxygen by hemoglobin is diminished at high pH and increased by low pH values; this effect causes more efficient loading of hemoglobin in the lungs and unloading in the tissues. Red blood cells also participate in the transport of CO_2 from the tissues to the lungs because of their content of carbonic anhydrase, which catalyzes hydration of CO_2 and dehydration of H_2CO_3.

References

WHITE, A., P. HANDLER, and E. L. SMITH: *Principles of Biochemistry*, Fourth Edition, McGraw-Hill, New York, 1968. Textbook with detailed treatment of biochemical aspects of blood and urine and the function of the lungs and kidneys.

PART IV REPLICATION, TRANSCRIPTION,
AND TRANSLATION OF
GENETIC INFORMATION

PART **IV** **REPLICATION, TRANSCRIPTION,**
AND TRANSLATION OF
GENETIC INFORMATION

In this the last part of this book we shall consider the biochemical basis of some of the most fundamental and central questions that are posed by the genetic continuity and the evolution of living organisms. What is the molecular nature of the genetic material and its functional units—the chromosomes, genes, coding symbols, and mutational units? How is genetic information replicated with such fidelity? How is it transcribed and used elsewhere in the cell? How is genetic information ultimately translated into the amino acid sequence of protein molecules?

The last fifteen years have brought an intellectual revolution in molecular genetics that many consider comparable to that which commenced with Darwin's theory of the origin of species. All fields of biology have been profoundly influenced by these new developments. They have brought penetrating new insight into some of the most fundamental problems in cell structure and function and have led to a more comprehensive and a more widely applicable conceptual framework for the science of biochemistry.

Today's knowledge of the biochemical basis of genetics arose from advances in three different fields: genetics, biochemistry, and molecular physics. In the field of genetics, experimental progress was immensely accelerated by the use of x-rays and other mutagenic agents, which greatly increase the rate of spontaneous mutation, and by the use of organisms with a very short life cycle, such as molds, bacteria, and viruses. In addition, methods were developed for the rapid selection of mutants and for construction of genetic maps of chromosomes. Such maps have revealed the sequence of various genes in chromosomes and have shown that a gene has a large number of sites at which it may undergo mutation. But the most pervasive development in the field of genetics, the development which initiated the confluence of genetics and biochemistry, was the idea that each enzyme in a cell is determined by a specific gene, the "one gene–one enzyme" hypothesis enunciated by G. Beadle and E. Tatum in 1940. Another was the discovery by O. Avery and his colleagues two years later that DNA contains genetic information.

In the field of biochemistry new chromatographic methods led to quantitative analysis of the amino acid composition of proteins and determination of the sequence in which the amino acids occur. Such methods were also instrumental in establishing the base composition and the covalent structure of nucleic acids. The isotopic tracer technique also played a vital role, since it made possible direct experimental approaches to the enzymatic biosynthesis of nucleic acids and proteins.

In the field of molecular physics, the application of x-ray diffraction analysis to the structure of fibrous proteins by Astbury and Pauling and globular proteins by Kendrew and Perutz yielded the concept that each protein molecule has a specific conformation of precise dimensions, which determines its biological function. Then came the application of the powerful x-ray method to the analysis of the three-dimensional structure of DNA. In 1953 J. Watson and F. Crick postulated a double-helical structure for DNA which not only accounted for its characteristic x-ray diffraction pattern, but also suggested a simple mechanism by which genetic information can be precisely transferred from parent to daughter cells. This hypothesis quickly resulted in a remarkable confluence of ideas and experimental approaches from genetics, biochemistry, and molecular physics.

The Watson-Crick hypothesis was soon built upon and extended to yield what Crick termed the central dogma of molecular genetics, which states that genetic information normally flows from DNA to RNA to proteins, a relationship often symbolized as "DNA → RNA → protein." The central dogma defined three major processes in the preservation and transmission of genetic information. The first is replication, the copying of DNA to form identical daughter DNA molecules. The second is transcription, the process in which the genetic message in DNA is transcribed into the form of RNA, to be carried to the ribosomes. The third is _translation_, the process in which the genetic message is decoded and used in the synthesis of proteins on the ribosomes. The central dogma was supported not only by the discovery of messenger RNA, which carries genetic information from DNA to the ribosomes (the sites of protein synthesis), but also by the demonstration that the sequence of nucleotides in a gene bears a linear correspondence to the sequence of amino acids in the protein it specifies. Moreover, in one of the greatest achievements in modern science, the genetic "dictionary" of triplet code words for the various amino acids has been deduced.

CHAPTER **21** **DEOXYRIBONUCLEIC ACID**
AND THE STRUCTURE OF
CHROMOSOMES AND GENES

In this chapter we shall try to answer three major questions. What is the evidence that deoxyribonucleic acid (DNA) is the primary form in which genetic information is stored? What is the molecular basis for the capacity of DNA to store information? What are the dimensions of some of the functional units of the genetic material, such as chromosomes, genes, and the units of mutation?

The covalent structure of DNA has already been examined in Chapter 7, which should first be reviewed.

Evidence for DNA as Genetic Material

Although DNA was first discovered in cell nuclei over 100 years ago, it was not directly identified as bearing genetic information until 1943, when Avery and his colleagues discovered that DNA extracted from a virulent (disease-causing) strain of the bacterium *Pneumococcus* permanently transformed a non-virulent strain of this organism into a virulent one. Evidence that DNA is the bearer of genetic information has also come from other sources. For one thing, the amount and composition of DNA in any given species of cell or organism is remarkably constant and cannot be altered by environmental circumstances or by changes in the nutrition or metabolism of the cell. Moreover, the amount of DNA per cell appears to be in proportion to the complexity of the cell and thus to the amount of genetic information it contains.

Data in Table 21-1 show that the higher the organism in the evolutionary scale, the greater its content of DNA. The amount of DNA present in DNA-containing viruses, which have only a few genes and thus relatively little genetic information, is correspondingly very small. Another point of evidence is that the germ cells of higher animals, which are haploid, that is, possess only one set of chromosomes, contain only half the amount of DNA found in somatic cells of the same species, which are diploid and have two sets of chromosomes. In any given species of higher organism, the amount of DNA per diploid cell is approximately constant from one cell type to another.

Experiments on the replication of bacterial viruses also point to DNA as genetic material. During the infection of a susceptible bacterial cell by a DNA-containing virus, it is the DNA portion of the virus particle which enters the host cell and which carries the information to specify the synthesis of progeny virus particles.

Another important piece of evidence supporting the concept that DNA is the bearer of genetic information was the discovery that the base composition of DNA is related to the species of origin in a very specific manner. Table 21-2 shows the amounts of the four bases in DNA specimens isolated from different organisms. These and other data have shown the following statements to be correct:

1. The base composition of DNA varies from one species to another.

Table 21-1 Approximate DNA content of some cells

Species	DNA, picograms† per cell	Number of nucleotide pairs, millions
Mammals	6	5500
Amphibia	7	6500
Fishes	2	2000
Reptiles	5	4500
Birds	2	2000
Crustaceans	3	2800
Mollusks	1.2	1100
Sponges	0.1	100
Higher plants	2.5	2300
Fungi	0.02–0.17	20
Bacteria	0.002–0.06	2

† One picogram = 1.0 $\mu\mu$g = 10^{-12} grams. The figures given for eukaryotic organisms are for somatic (diploid) cells.

Table 21-2 Base equivalences in DNA

	Base composition, mole percent				Base ratios			Asymmetry ratio
	A	G	C	T	A/T	G/C	Purines / Pyrimidines	$\dfrac{A+T}{G+C}$
Animals								
Man	30.9	19.9	19.8	29.4	1.05	1.00	1.04	1.52
Sheep	29.3	21.4	21.0	28.3	1.03	1.02	1.03	1.36
Hen	28.8	20.5	21.5	29.2	1.02	0.95	0.97	1.38
Turtle	29.7	22.0	21.3	27.9	1.05	1.03	1.00	1.31
Salmon	29.7	20.8	20.4	29.1	1.02	1.02	1.02	1.43
Sea urchin	32.8	17.7	17.3	32.1	1.02	1.02	1.02	1.58
Locust	29.3	20.5	20.7	29.3	1.00	1.00	1.00	1.41
Plants								
Wheat germ	27.3	22.7	22.8	27.1	1.01	1.00	1.00	1.19
Yeast	31.3	18.7	17.1	32.9	0.95	1.09	1.00	1.79
Aspergillus niger (mold)	25.0	25.1	25.0	24.9	1.00	1.00	1.00	1.00
Bacteria								
E. coli	24.7	26.0	25.7	23.6	1.04	1.01	1.03	0.93
Staphylococcus aureus	30.8	21.0	19.0	29.2	1.05	1.11	1.07	1.50

2. DNA specimens isolated from different tissues of the same species have the same base composition.

3. The base composition of DNA in a given species does not change with age, nutritional state, or change in environment.

4. The number of adenine residues in all DNAs is equal to the number of thymine residues (that is, A = T). Moreover, the number of guanine residues is always equal to the number of cytosine residues (G = C). Thus the sum of the purine residues equals the sum of the pyrimidine residues; that is, A + G = C + T.

The Watson-Crick Model of DNA Structure

The relationships outlined in the preceding section among the purine and pyrimidine bases in DNA from different species raised the intriguing possibility that there is a level of structural organization of DNA molecules which is compatible with certain base equivalences and incompatible with others. It had long been suspected that DNA has a specific three-dimensional conformation. For example, solutions of native DNA are highly viscous, suggesting that the DNA molecule is long and rigid, rather than compact and folded. Moreover, heating of freshly isolated DNA produces a decrease in viscosity and other physical properties, without breaking the covalent bonds in the DNA backbone. But the most convincing and illuminating experimental evidence came from the application of the powerful method of x-ray diffraction analysis to the structure of DNA. This approach showed that native DNA has two periodicities in its native structure, a major one of 3.4 Å and a secondary one of 34 Å. The problem was then to formulate a three-dimensional conformation of the DNA molecule which could account for these periodicities in DNA structure, and for the equivalence of certain bases.

In 1953 Watson and Crick postulated a three-dimensional model of DNA structure which accounted for both the x-ray data and the base equivalences. This model accounted for many of the observations on the chemical and physical properties of DNA and also suggested a mechanism by which genetic information may be accurately replicated. The Watson-Crick model of DNA structure is shown in Figure 21-1. It consists of two right-handed helical polynucleotide chains coiled around the same axis to form a double helix. The two chains or strands are

Figure 21-1
Structure of DNA.

Space-filling model

Skeleton model

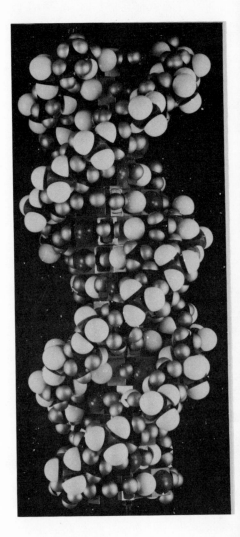

antiparallel; that is, their 3',5'-internucleotide phos-phodiester bridges (see Chapter 7) run in opposite direc-tions. The purine and pyrimidine bases of each strand are stacked on the inside of the double helix, with their planes parallel to each other and perpendicular to the long axis of the double helix. The bases of one strand are paired in the same planes with the bases of the other strand. The pairing of the bases from the two strands is such that only certain base pairs fit inside this structure in such a manner that they can hydrogen-bond to each other. The allowed pairs are A-T and G-C, which are precisely the base pairs showing exact equivalence in DNA. The manner in which they form hydrogen bonds is shown in Figure 21-2. Other pairings of bases do not

Figure 21-2
The hydrogen-bonded base pairs
adenine-thymine and cytosine-
guanine. The former has two hydrogen bonds, the latter three. Cytosine-
guanine pairs are slightly closer
together, more compact, and thus
slightly more dense than adenine-
thymine pairs.

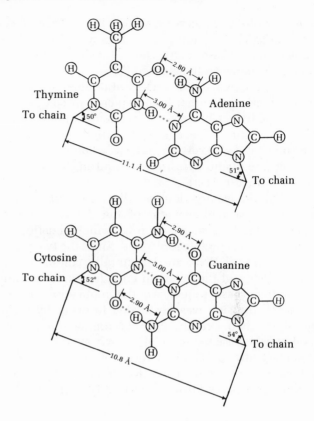

fit such a structure. The pair A-G would be too large to fit inside a helix having these dimensions, and the pair C-T would be too far apart in such a helix to form stable hydrogen bonds with each other. Furthermore, hydrogen bonding in A-G and C-T pairs would be very weak. Thus, the Watson-Crick double helix involves not only the maximum possible number of hydrogen-bonded base pairs but also those pairs giving maximum stability.

To account for the 3.4 Å periodicity observed by x-ray methods, Watson and Crick postulated that the vertically stacked bases are 3.4 Å apart. Since there are exactly 10 nucleotide residues in each complete turn of the double helix, the secondary repeat distance of 34 Å is also accounted for. These repeat distances are possible only if the purines and pyrimidines are paired in a helical structure in the manner postulated.

The hydrophobic, relatively insoluble bases are closely stacked within the double helix, shielded from water, whereas the hydrophilic sugar residues and the electrically charged secondary phosphate groups are located on the periphery, exposed to water. The double helix is thus stabilized not only by hydrogen-bonding within complementary base pairs, but also by hydrophobic interactions between the stacked bases.

The two polynucleotide chains of double-helical DNA are not identical in either base sequence or composition. Instead, the two chains are complementary to each other; wherever adenine appears in one chain thymine is found in the other and vice versa. Similarly, wherever guanine is found in one chain, cytosine is found in the other, and vice versa (Figure 21-3).

The complementary double-helical structure of DNA leads to the second element of the Watson-Crick hypothesis, namely, a mechanism by which genetic information can be accurately replicated. Since the two strands of double-helical DNA are structurally complementary to each other and thus contain complementary information in their base sequences, the replication of DNA during cell division was postulated to occur by separation of the two strands, each becoming the template specifying the base sequence of a newly synthesized complementary strand. The end result of such a process is the formation of two daughter double-helical DNAs, each containing one strand from the parent DNA (Figure 21-4).

Some Physical Properties of Native DNA

Native double-helical DNA may be separated from disrupted cells by gentle extraction with dilute salt solutions, followed by precipitation with cold alcohol, in which it is insoluble. It may be purified by chromatography.

Denaturation

The native double-helical form of DNA is quite stable at pH 7.0 and ordinary temperatures. However, it readily undergoes unwinding into random, disordered coils when it is subjected to extremes of pH, temperatures above 70°-80°C, or when it is exposed to high concentrations of alcohols, urea, and certain other substances. Since these agencies are identical to those producing denaturation or unfolding of globular proteins (Chapter 3) it has been concluded that native double-helical DNA undergoes a similar denaturation process. Native DNA is stabilized by two sets of forces: hydrogen bonding and hydrophobic bonding. When either or both sets of forces are interrupted, the native, double-helical structure

Figure 21-3
Complementary antiparallel strands of DNA. Note that their base composition is not identical.

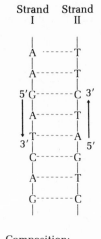

Composition:
Strand I $A_4G_2C_1T_1$
Strand II $A_1G_1C_2T_4$

Figure 21-4

Replication of DNA as suggested by Watson and Crick. The complementary strands are separated and each forms the template for synthesis of a complementary daughter strand. [From James D. Watson, Molecular Biology of the Gene, copyright © 1970, J. D. Watson; W. A. Benjamin, Inc., Menlo Park, Cal.]

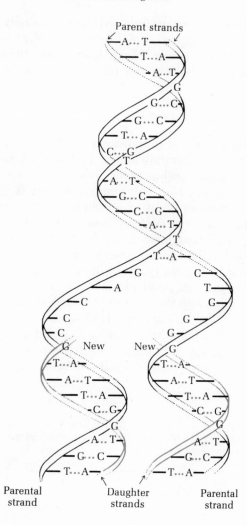

undergoes unwinding into irregular random coils. During such denaturation of DNA no covalent bonds in the backbone are broken.

There are two stages in the denaturation of duplex DNA molecules. In the first stage, the two strands are partially unwound but remain united by at least a short segment of double helix with its complementary bases still in register. In the second phase, the two strands completely separate from each other. Heat denaturation of homogeneous DNA is easily reversible if the unwinding process has not gone past the first stage. As long as a double-helical segment of 12 or more residues unites the two strands, with base pairing in register, the unwound segments of the two strands will spontaneously rewind to reform a complete duplex when the temperature is lowered. They literally "snap back" into their native conformation, which is the form having the least free energy.

The Melting Point of DNA

Unlike many globular proteins, which denature gradually over a wide temperature range, native DNA molecules usually denature within a very small increment of temperature (Figure 21-5). This sharp transition is similar to the sharp melting point of a simple organic crystal. In fact, denaturation of DNA by heat is often designated as *melting*. DNA specimens from different cell types have characteristically different melting points, defined as the temperature at the midpoint of the melting curve (Figure 21-6). The melting point (designated T_m) increases in a linear fashion with the content of G-C base pairs, because the triply hydrogen-bonded G-C base pairs are more compact and stable than the A-T pairs. The higher the content of G-C pairs, the more stable the structure and the more heat energy required to disrupt it (Figure 12-6). Careful

Figure 21-5
The melting curve of a bacterial DNA. T_m is the temperature at the midpoint of the curve (82.5°C).

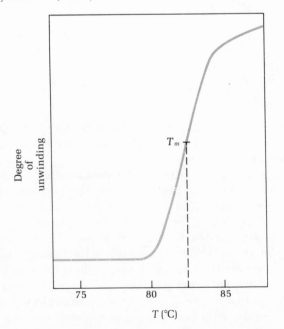

Figure 21-6
A plot of T_m for 40 different DNA
specimens from plant, animal and
viral sources against GC content. All
samples were measured under identical conditions.

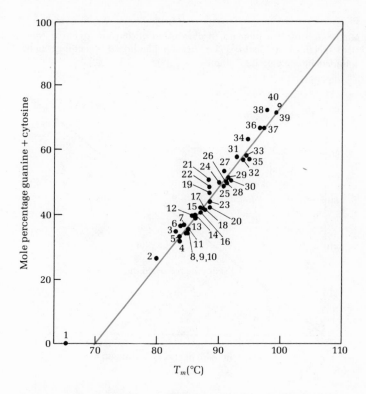

Figure 21-6
A plot of T_m for 40 different DNA
specimens from plant, animal and
viral sources against GC content. All
samples were measured under identical conditions.

determination of the melting point of a DNA specimen, under fixed conditions of pH and ionic strength, can give a remarkably accurate estimate of its base composition.

The Hyperchromic Effect

As was pointed out in Chapter 7, nucleotides and nucleic acids strongly absorb ultraviolet light at 260 nm. When native DNA is denatured there is a dramatic increase in light absorption at 260 nm; this is termed the *hyper-chromic effect*.

In native double-helical DNA molecules electronic interactions between the stacked bases diminish the amount of light each purine or pyrimidine base can absorb. But when the double-helical structure is dis-ordered the bases "unstack" and in this less hindered form they absorb as much light as they would were they present as free nucleotides.

The percent increase in light absorption on heating DNA is directly related to the content of A-T base pairs. The base composition of a DNA sample can thus be esti-mated by carrying out spectrophotometric measurements of the magnitude of the hyperchromic effect accompany-ing heating (Figure 21-7).

Figure 21-7
Maximum absolute increase in absorption at 260 nm given by some
DNA specimens on heating. The largest hyperchromic effect is given by
complementary poly dAT strands.

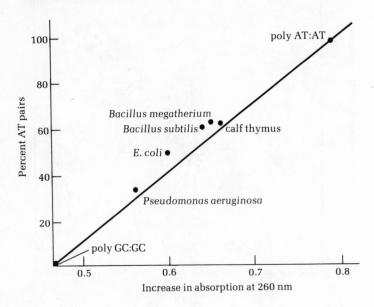

The Size of DNA Molecules in Chromosomes

Native DNA molecules in bacteria and animal cells are
so large that they are not readily isolated in intact form,
since they are easily broken by shear forces during extraction and handling. Even simple stirring or pipetting
of DNA solutions can cause fragmentation of DNA. The
single chromosomes of DNA-containing bacterial viruses
and of bacteria are now known to consist of single, very
large double-helical DNA molecules, which are often
circular (Figure 21-8). The chromosome of the E. coli
bacteriophage lambda (λ) (Chapter 7) is a circular duplex
DNA molecule of molecular weight 32 million; it is about
17.2 μ long. The chromosome of the E. coli cell itself
consists of a single, enormous double-stranded DNA
molecule with a molecular weight of about 2,800,000,000,
and a contour length that exceeds 1200 μ or 1.2 mm. The
circular E. coli chromosome contains about 4.2 million
base pairs.

Eukaryotic cells contain several or many chromosomes,
depending on the species. Each contains one or more

Figure 21-8
Electron micrographs of two viral
DNA molecules.

DNA of bacteriophage T₂ (linear form). Its contour length is 55.9 μ
and its molecular weight is 130 million.

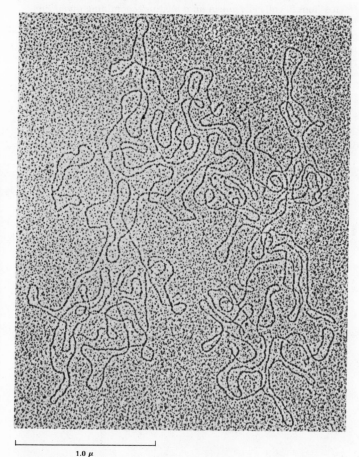

DNA of bacteriophage lambda (λ). Its
molecular weight is 32 million and
its length is 17.2 μ.

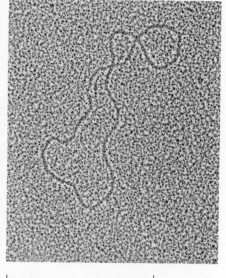

5.0 μ

1.0 μ

very large DNA molecules. The total length of all the
DNA in a single human cell has been calculated to be
about 2 meters. This is equivalent to 5.5×10^9 base pairs.
DNA is also present in certain cell organelles, particularly the mitochondria and chloroplasts.

Genes

From early genetic experiments and more recent biochemical approaches, it is now possible to specify the
size of a gene, which may be defined as that segment of a
chromosome that codes for a single polypeptide chain
of a protein. A gene may have anywhere from 300 to 6000
or more nucleotide pairs (molecular weights from about
100,000 to 2,000,000), corresponding to polypeptide

Figure 21-9
Radioautograph of a DNA molecule of E. Coli. The DNA was extracted from E. coli cells grown on a medium containing thymidine labeled with the radioactive hydrogen isotope tritium. It was spread on a sensitive photographic plate and the radioactive "track" of the molecule detected microscopically. This DNA molecule is undergoing replication.

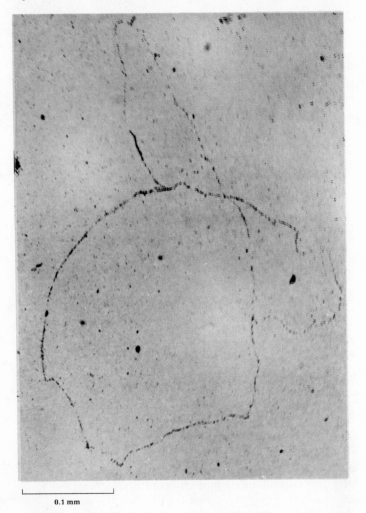

0.1 mm

chains having from 100 to 1800 amino acid residues, the extremes of polypeptide chain length in most proteins. The size of the gene which determines any given protein may be calculated simply by multiplying the number of amino acid residues in the protein by 3.0. This follows from the fact (see Chapter 23) that a sequence of three nucleotides in DNA, called a codon, is required to code for a single amino acid residue. Since some proteins or enzymes have two or more different polypeptide chains, each coded by a different gene, it is clear that more than one gene may be required to code for such proteins.

Some genes are smaller than those specifying proteins. Transfer RNA molecules have only 75 to 90 mononucleotide units (Chapter 7). They are formed enzymatically by complementary transcription from a segment of DNA (see Chapter 22). The genes specifying tRNAs are thus only about 75–90 nucleotides long.

The exact location of a gene specifying a given enzyme or protein in the chromosome of a virus or bacterial cell can be deduced by genetic mapping methods. Often the genes specifying the individual enzymes of a multienzyme system are located adjacent to each other in the chromosome and may be transcribed and translated as a group. Such a cluster of related genes is called an *operon* (see Chapter 23).

The single chromosomes of DNA viruses may contain as few as 5 or as many as 200 genes. The single chromosome of an *E. coli* cell is believed to contain at least 4000 different genes.

Mutations

Many agencies are known which cause heritable mutations—chemical or physical changes in DNA which result in the synthesis of proteins with altered amino acid sequences. Often such defective proteins lack their normal biological activity, which may have lethal consequences to the organism. Mutations may be caused by radiant energy in the form of x-ray or ultraviolet rays, or by chemical agents capable of chemically combining with or modifying a purine or pyrimidine base. An example is nitrous acid, which can convert an amino group into a hydroxyl group. Some mutagenic agents are capable of deleting or inserting bases.

For many years it was thought that the gene itself was the unit of mutation, that is, each gene contained only one mutable site. However, it has been proved that each gene has many mutable sites; indeed the smallest mutable site in a gene is a single nucleotide. In some mutations there is replacement of one purine base by the other (A for G or G for A) or one pyrimidine base by the other (C for T or T for C). Some mutations involve replacement of a purine by a pyrimidine or vice versa. Still other mutations are caused by deletion of a nucleotide or by insertion of an extra one. Sometimes several nucleotides may be deleted. Mutant proteins have already been described in Chapter 3.

Summary

That DNA bears genetic information was first shown by experiments in which DNA isolated from one strain of the bacterium *Pneumococcus* transformed another strain in a heritable manner. All somatic (diploid) cells of a given species of higher organism contain the same amount of DNA, which is not modified by age, diet, or environmental circumstances. The base composition of DNA is specific for each species. The total amount of DNA per cell is greater, the higher the position of the organism in the evolutionary scale. Whatever the species, the number of adenine residues equals the number of thymine residues; similarly, the guanine residues equal the cytosine residues.

From x-ray analysis of DNA and from the base equivalences, Watson and Crick postulated and others have confirmed that native DNA consists of two antiparallel chains in a double-helical arrangement, with the complementary bases A-T and G-C paired by hydrogen bonding within the helix. The base pairs are closely stacked perpendicular to the long axis, 3.4 Å apart. The double helix is about 20 Å in diameter. This structure accounts for accurate replication of the two strands, since it can form only from precisely complementary strands. A chromosome consists of a single, double-helical DNA molecule which is often an endless, closed loop; the *E. coli* chromosome has a molecular weight of nearly 3 billion.

The buoyant density of DNA in cesium chloride gradients increases with its G-C base-pair content. DNA undergoes unwinding or melting with no rupture of covalent bonds. During this process light absorption at 260 nm increases (the hyperchromic effect). The magnitude of the hyperchromic effect increases with the A-T content, whereas the melting temperature (T_M) increases with the G-C content.

A gene is a segment of chromosomal DNA which codes for a single polypeptide chain or a single tRNA molecule; it contains anywhere from about 75 to 6000 or more nucleotide pairs. Genes may undergo mutation on exposure to x-rays or to certain chemical reagents, which produce an alteration in nucleotide sequence, leading to synthesis of altered proteins which may be functionally defective. The smallest mutational unit is a single nucleotide pair.

The central dogma of molecular biology is that genetic information flows in the sequence DNA → RNA → protein. Replication is the process by which DNA is copied to form identical daughter DNA molecules; transcription is the process by which the genetic message in DNA is transcribed in the form of mRNA; and translation is the process by which the genetic message is decoded and used to determine the amino acid sequence in protein synthesis.

References

DAVIDSON, J. N.: *The Biochemistry of The Nucleic Acids,* Methuen, London, 6th edition, 1969. A popular and useful introduction (350 pages).

LEHNINGER, A. L.: *Biochemistry,* Worth Publishers, New York, 1970. A more detailed textbook treatment in Chapters 28 and 29.

WATSON, J. D.: *Molecular Biology of the Gene,* W. A. Benjamin, Menlo Park, Cal., 1965. An excellent elementary account of the principles of molecular genetics.

Problems

1. Write the base sequence of the complementary strand of double helical DNA in which one strand is
   (5′)ATGCCGTATGCATTC(3′)

2. In a sample of DNA isolated from a given species of organism, adenine made up 21 percent of the total purine and pyrimidine bases. Predict the relative proportions of adenine, guanine, thymine, and cytosine in the sample.

3. Calculate the weight in grams of a double helical DNA molecule stretching from the earth to the moon (~200,000 miles). (The DNA double helix weighs about 1×10^{-18} grams per 1000 nucleotide pairs. There are 1.6×10^{13} Å per mile.)

4. Calculate the weight in grams and the length in miles of all the DNA in the human infant, which has 2×10^{12} cells. Compare with your answer to Problem 3.

5. How many nucleotide pairs are present in the genes for the following:
   (a) Ribonuclease (124 amino acid residues)
   (b) Hemoglobin α chains
   (c) Chymotrypsinogen

6. How long (in Å) are each of the genes in Problem 5?

CHAPTER **22** **REPLICATION AND**

TRANSCRIPTION OF DNA

In this chapter we shall examine the enzymatic mechanisms by which DNA serves as a template for its own replication and by which DNA is transcribed to yield messenger RNA. We shall see that the basic features of the Watson-Crick hypothesis and the central dogma of molecular genetics are verified by present knowledge of the mechanisms of replication and transcription.

The Mechanics of DNA Replication

A significant feature of the Watson-Crick hypothesis is its postulate that the two strands of double-helical DNA are complementary to each other and that each strand is used as a template for the replication of complementary daughter strands. In this way two daughter duplex DNA molecules identical to the parent DNA would be formed, each of which would contain one intact strand from the parental DNA. Ingeniously contrived experiments carried out by M. Meselson and F. Stahl in 1957–1958 conclusively proved that in intact, living E. coli cells, DNA is replicated in the manner postulated by Watson and Crick. They grew E. coli cells for several generations in a medium containing the nearly pure "heavy" isotope ^{15}N instead of the ^{14}N. All nitrogenous components of these cells, including their DNA, thus contained ^{15}N instead of ^{14}N. Since DNA containing ^{15}N has a significantly greater density than normal ^{14}N DNA, it is possible to separate and measure the heavy (^{15}N) and light (^{14}N) forms present in a sample of DNA by centrifuging it in a medium of 6 M cesium chloride. The ^{15}N DNA sediments to a lower equilibrium position in the tube than the ^{14}N DNA (Figure 22-1).

Meselson and Stahl transferred some "heavy" E. coli cells grown on the ^{15}N medium to a fresh medium in which the NH_4Cl contained the normal ^{14}N isotope. After the number of cells had doubled, the DNA was isolated from the cells and its density analyzed by the centrifugal method. The DNA of these cells formed but a single band midway in density between the normal or "light" DNA and the "heavy" DNA of cells grown exclusively on ^{15}N. This result is what might be expected if the double-helical DNA of the daughter cells contained one ^{14}N strand and one ^{15}N strand. After two generations, the DNA isolated

Figure 22-1. (below and right). The Meselson-Stahl experiment.

Two possible mechanisms of replication of heavy (^{15}N) DNA. Newly replicated strands are shown in color.

Conservative replication of heavy DNA. Each of the two heavy strands of parent DNA is replicated, without strand separation, to yield original heavy DNA and newly synthesized light DNA. F_2 generation yields one heavy DNA, no hybrids, and three light DNAs.

Semiconservative replication of heavy DNA. In first daughters, each duplex contains one of parent heavy strands. F_2 generation yields two hybrid DNAs and two light DNAs.

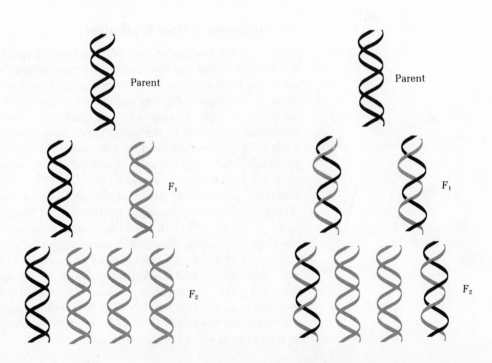

Parent

F_1

F_2

Parent

F_1

F_2

exhibited two bands, one having a density equal to that of normal light DNA and the other equal to that of the hybrid DNA observed after one generation. These results, which are shown schematically in Figure 22-1, are exactly those postulated by Watson and Crick. This experiment was particularly convincing because it was carried out on intact dividing cells, without use of inhibitors or other injurious agents.

DNA Polymerase I

The enzymatic mechanisms by which DNA is replicated were opened to direct biochemical investigation by the important research of A. Kornberg and his colleagues

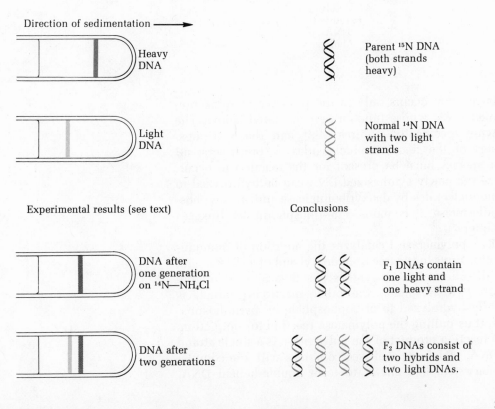

Identification of heavy and light DNA in centrifuge

Direction of sedimentation ⟶

Heavy DNA

Parent ^{15}N DNA (both strands heavy)

Light DNA

Normal ^{14}N DNA with two light strands

Experimental results (see text)

Conclusions

DNA after one generation on ^{14}N—NH$_4$Cl

F$_1$ DNAs contain one light and one heavy strand

DNA after two generations

F$_2$ DNAs consist of two hybrids and two light DNAs.

beginning in 1956. They incubated crude extracts of *E. coli* cells with mixtures of dATP, dGTP, dCTP, and dTTP, labeled with ^{32}P in the α-phosphate group, that is, that esterified to the 5'-hydroxyl group of deoxyribose (Figure 22-2). After incubation they isolated the DNA from the extract and found it to contain some ^{32}P isotope, which they showed to be present in the phosphodiester bridges of newly formed DNA.

After painstaking purification Kornberg and his colleagues ultimately isolated from *E. coli* extracts small quantities of an enzyme, now called *DNA polymerase I*, which catalyzes the synthesis of the internucleotide linkages of DNA starting from a mixture of dATP, dGTP, dCTP, and dTTP. The overall reaction was found to be

Figure 22-2

Deoxyribonucleoside 5'-triphosphate labeled with ^{32}P in the α position.

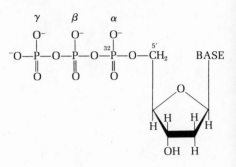

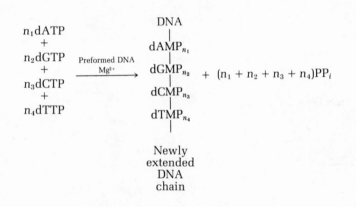

This reaction occurs only in the presence of some preformed DNA, whose function is considered below. The enzyme specifically requires Mg^{2+} and the 5'-triphosphates of the deoxyribonucleosides as precursors; all four species must be present for the reaction to occur. Since the newly synthesized DNA can be hydrolyzed to mononucleotides by deoxyribonuclease and spleen phosphodiesterase, it contains 3',5'-phosphodiester linkages (Chapter 7).

DNA polymerase I catalyzes the addition of mononucleotide units to the free 3'-hydroxyl end of a DNA chain; the direction of DNA synthesis is thus 5' → 3' (Figure 22-3). Pyrophosphate, the other reaction product, is rapidly hydrolyzed to orthophosphate by pyrophosphatase, thus pulling the polymerase reaction to completion. The reaction product of the polymerase is a single strand of DNA, which we shall see combines with the complementary template strand to form double-helical DNA.

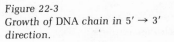

Figure 22-3
Growth of DNA chain in 5' → 3'
direction.

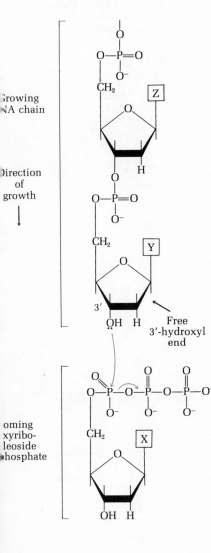

Growing
DNA chain

Direction
of
growth

3'

OH H Free
3'-hydroxyl
end

Incoming
deoxyribo-
nucleoside
5'-phosphate

OH H

DNA polymerase I activity has been detected in many animal, plant, and bacterial cells. In eukaryotic cells it is found primarily in the nucleus.

The Role of Preformed DNA

DNA polymerase requires the presence of some preexisting DNA for activity. In the absence of the latter, the purified enzyme will make no polymer at all, except under very special conditions. RNA will not substitute for preformed DNA.

Kornberg and his colleagues found that intact native double-helical DNA is completely inactive in supporting DNA synthesis by this enzyme. However, denatured DNAs, in which the two strands are largely separated, yield maximal activity. Moreover, denatured DNAs from many different sources (animal, plant, bacterial, and viral) were found to support the activity of E. coli DNA polymerase. In most such experiments the amount of newly synthesized DNA did not exceed the amount of preformed DNA added. These facts strongly suggested that the polymerase requires a single strand of DNA primarily as a template for synthesis of a complementary chain.

That preformed DNA functions as a template was more conclusively supported by comparison of the base composition of the newly synthesized DNA with that of the template DNA added to the system (Table 22-1). DNAs isolated from various species having widely different proportions of A-T and G-C pairs were heated and added to the purified E. coli enzyme. The DNA product of the polymerase always showed a base composition nearly identical to that of the added DNA, as would be expected if each of the two strands of the denatured template DNA

Table 22-1 Similarity of base composition of template and product DNA. In these experiments the template DNA was heat-denatured and both strands were therefore replicated.

DNA	A	T	G	C
Mycobacterium phlei				
Template	0.65	0.66	1.35	1.34
Product	0.65	0.65	1.34	1.37
E. coli				
Template	1.00	0.97	0.98	1.05
Product	1.04	1.00	0.97	0.98
Calf thymus				
Template	1.14	1.05	0.90	0.85
Product	1.12	1.08	0.85	0.85

were replicated by the enzyme to form two complementary strands.

Kornberg and his colleagues also developed a method called *nearest-neighbor base-frequency analysis* to determine with more certainty whether the new strand of DNA formed by the polymerase is complementary to the added DNA. This method makes possible determination of the frequency with which any two mononucleotides occur as adjacent or "nearest neighbors" in a given DNA chain. Figure 22-4 shows that the four bases of DNA can occur in 16 different nearest-neighbor pairs. Obviously, if two DNA chains have identical nearest-neighbor base frequencies, the probability must be extremely high that they have identical base sequences. By this approach Kornberg and his colleagues provided overwhelming statistical evidence that the DNA formed by action of DNA polymerase is complementary to the template DNA in the manner predicted by the Watson-Crick hypothesis; that is, A always pairs with T and G always pairs with C. Their data also proved that the newly synthesized strand, which is always made in the $5' \rightarrow 3'$ direction, runs in the opposite direction from the template strand, in accordance with the Watson-Crick hypothesis

Recent investigations of Kornberg and his colleagues show that preformed DNA may serve a second function, namely as a *primer*. DNA polymerase functions best when the preformed DNA furnishes not only a template strand but also a priming strand, to which it can add mononucleotide residues in response to the template. The template strand and the priming strand are complementary to each other (Figure 22-5).

The fact that DNA polymerase I can make the internucleotide linkages of DNA in only the $5' \rightarrow 3'$ direction raises an important question. We have seen that in intact cells, both strands of duplex DNA are replicated. When the "growing point" moves along the chromosome, the replication process must make two chains, one in the $5' \rightarrow 3'$ and one in the $3' \rightarrow 5'$ direction. Thus a dilemma arises since the Kornberg polymerase can make linkages only in the $5' \rightarrow 3'$ direction. Is a second polymerase required to replicate the other chain in the $3' \rightarrow 5'$ direction? Or is some entirely different enzyme required, one that can replicate in both directions? These questions have prompted much research. One key has been provided by the discovery of an enzyme called *DNA ligase*. When this enzyme acts in conjunction with DNA polymerase I *both* strands of double-helical DNA can be replicated.

Figure 22-4
The 16 possible nearest-neighbor mononucleotide residues in DNA. The small p indicates the phosphate group forming the phosphodiester bridge between each pair of nearest neighbors.

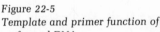

$5' \rightarrow 3'$

ApA
GpA
CpA
TpA

ApG
GpG
CpG
TpG

ApC
GpC
CpC
TpC

ApT
GpT
CpT
TpT

Figure 22-5
Template and primer function of preformed DNA.

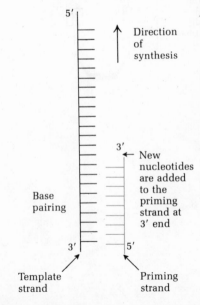

5'

Direction of synthesis

3'

← New nucleotides are added to the priming strand at 3' end

Base pairing

3'   5'

Template strand

Priming strand

Figure 22-6
Action of DNA ligase in repairing a single-strand nick (above) and in closing a circle (below). The ends to be joined must be held by complementary base pairing to a template strand.

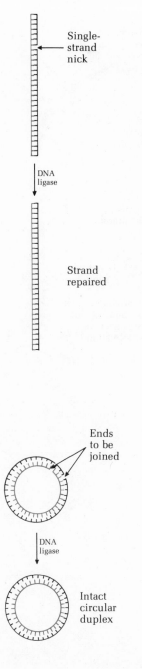

DNA Ligase

Nearly simultaneously, several investigators discovered in E. coli extracts an enzyme, distinct from DNA polymerase, that can join the ends of two DNA chains; it cannot, however, replicate a whole chain. This enzyme is called DNA ligase.

The DNA ligase reaction is rather complex and takes place in two or more steps. Its overall effect is to join the 5'-phosphate end of one segment of the DNA to the 3'-hydroxyl end of the other DNA segment (Figure 22-6).

DNA ligase can link the ends of linear DNA molecules to yield circular forms, it can catalyze the repair of breaks or nicks in one chain of double-stranded DNA, it can join genes together during genetic recombination, and it can cooperate with DNA polymerase in the replication of both strands of antiparallel DNA, as we shall now see.

Enzymatic Synthesis of φX174 DNA

For years many attempts have been made to determine whether a biologically active form of DNA can be synthesized by pure enzymes in the test tube. This objective was achieved in 1968 by M. Goulian, A. Kornberg, and R. Sinsheimer. They carried out the enzymatic synthesis of the circular, biologically active double-stranded DNA of E. coli bacteriophage φX174 by the combined action of DNA polymerase and DNA ligase. Kornberg and his colleagues started by using one strand of natural φX174 DNA as a template to synthesize a complementary strand by the action of DNA polymerase (Figure 22-7). They then joined the ends of the new strand by the action of DNA ligase to form a circle. Next, they used the resulting circular strand as a template to make a new complementary strand, again by the action of DNA polymerase. This new strand was found to be identical with the original strand of φX174 DNA they used as starting template. The synthetic strand was as infectious as the original natural φX174 DNA strand which they began.

These elegant experiments established that DNA polymerase can synthesize a biologically active DNA molecule, the first ever created in the test tube by the action of highly purified enzymes acting on highly purified substrates with a pure template of known biological activity. This experiment also proved that DNA polymerase can make both strands of a viral DNA when the process is carried out in two steps.

Figure 22-7
Enzymatic synthesis of biologically active DNA.

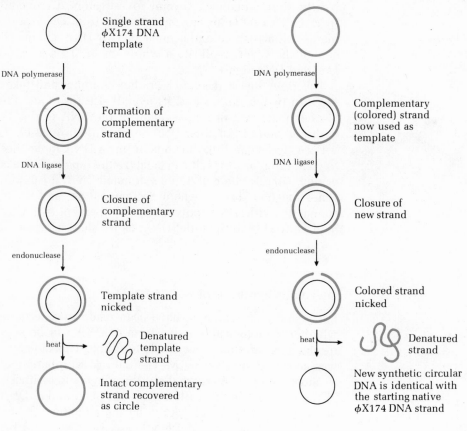

The Replication of Double-Stranded DNA In Vivo

Despite the successful synthesis of single-strand ϕX174 DNA, a nagging question remained. How can both strands of antiparallel duplex DNA molecules be replicated *simultaneously*, as is known to occur in the intact cell (Figure 22-8), so that two chains, one running in the $5' \rightarrow 3'$ and one in the $3' \rightarrow 5'$ direction, are made at the same time? We have seen that isolated, pure DNA polymerase cannot replicate native double-stranded DNA by itself and that it can build DNA linkages only in the $5' \rightarrow 3'$ direction. An interesting hypothesis for the replication of both strands of DNA has been constructed. This mechanism requires three enzymes: DNA polymerase, DNA ligase, and an endonuclease. Replication begins by introduction of a "nick" in one strand by action of the

Figure 22-8

The course of replication of the E. coli chromosome. Electron microscopic investigations showed that during replication of the circular DNA double helix the two new daughter strands (in color) begin their growth at a fixed point (gray dot) and continue in the direction of the arrows until complete daughter circles are formed, each complementary to a parent strand. The new daughter double helixes then separate. Redrawn from "The Bacterial Chromosome" by J. Cairns. Copyright © January 1966 by Scientific American, Inc. All rights reserved.

endonuclease (Figure 22-9). DNA polymerase then binds at this nick and begins to extend the 3'-hydroxyl end of the nicked strand by adding successive mononucleotide units complementary to the intact strand, while the 5' end of the nicked strand "peels" away from the duplex structure. After a given length of the intact strand has been replicated, the DNA polymerase is postulated to "jump" from the intact template strand to the nicked strand at the fork where the two strands are separated. The enzyme is then postulated to replicate the nicked strand in the 5' → 3' direction. The enzyme continues until the entire loose end of the nicked strand is replicated and then leaves. The endonuclease then cleaves the newly formed strand at the fork, leaving a short fragment of complementary DNA duplexed with the 5' end of the original nicked template strand.

This process is then repeated as another DNA polymerase molecule attaches to the 5' end at the fork and replicates another short segment of the intact template strand, jumps to the other template strand, and replicates the latter in the other direction. The short DNA segments so made are then "stitched" together by DNA ligase. This mechanism accounts for the fact that many short segments of DNA are known to accumulate in E. coli cells at the time their DNA is undergoing replication.

DNA Polymerase II

Despite the ability of purified Kornberg DNA polymerase to make biologically active DNA, some investigators have held the view that this enzyme functions biologically in the repair of spontaneous nicks or breaks in DNA rather than in its replication. Some support for this view has come from the discovery that certain mutants of E. coli appear not to contain the Kornberg DNA polymerase, yet are still able to replicate their DNA normally. In such mutants another DNA polymerase has been found, designated DNA polymerase II, which appears to be quite different from the Kornberg enzyme, now called DNA polymerase I.

DNA polymerase II also can replicate DNA from a mixture of deoxyribonucleotide 5'-triphosphates as precursor. It differs from DNA polymerase I in being rather firmly attached to the E. coli cell membrane and not easily extractable. It also differs in its sensitivity to various inhibitors. It is not yet known which of these DNA polymerases is responsible for normal DNA replication.

Figure 22-9
Hypothesis of Kornberg for the repli-
cation of both strands of antiparallel
duplex DNA by DNA polymerase.

DNA polymerase binds
to nick on strand b.

5′ Nick in
3′ b strand

a b

The newly formed strand
is nicked at the fork by
an endonuclease.

5′
5′
3′

Strand a is replicated
while nicked strand b
is peeled back.

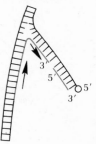

3′
5′

DNA polymerase now
returns and resumes
replication of strand a
at 3′ end. At the fork,
it jumps to strand b and
replicates it until earlier
fragment is reached.

3′
5′
5′
3′

DNA polymerase jumps
from strand a to strand b
and replicates the latter
in the 5′ → 3′ direction.

5′
3′

DNA ligase joins the two
fragments complementary
to strand b. Endonuclease
nicks new strand at the
fork and a new cycle
begins. In this fashion
both strands are repli-
cated in short lengths,
with the polymerase
replicating always in the
5′ → 3′ direction.

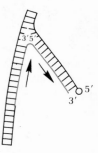

3′ 5′
5′
3′

Messenger RNA

Early investigators had found that the onset of protein
synthesis in intact cells is sometimes accompanied by a
simultaneous increase in the amount or rate of turnover
of cytoplasmic RNA. The function of this RNA fraction of
high turnover rate remained obscure for some time. However, in 1961 F. Jacob and J. Monod, on the basis of these
and other observations, proposed that the rapidly labeled
RNA formed during or preceding protein synthesis is a
species of RNA which serves as a messenger carrying
genetic information from the DNA of the chromosomes to
the surface of the ribosomes. They postulated that mes-

senger RNA is formed enzymatically in such a way that
it has a base sequence complementary to that of one
strand of DNA. The messenger RNA molecule was pre-
sumed to contain the complete message for specifying one
or more polypeptide chains. It was postulated to bind to
the ribosomes and to serve as the "working" template
for protein synthesis.

Messenger RNA is difficult to isolate, not only because
it has a short half-life, but also because it makes up only
a few percent of the total cellular RNA. Moreover, since
each protein or group of proteins requires its own specific messenger RNA molecule, isolation of one pure
messenger RNA from a pool containing only small amounts
of hundreds of different mRNAs is an extremely difficult
task.

We shall now examine the enzymatic mechanisms
involved in the synthesis of messenger RNA, the process
called _transcription_.

DNA-Directed RNA Polymerase

The discovery of DNA polymerase and its dependence
on a template began an active search for the enzymes that
participate in the transcription of DNA. In 1959 a DNA-
dependent enzyme capable of forming an RNA polymer
from ribonucleoside 5'-triphosphates was identified. The
RNA formed by this enzyme was found to be complementary to the DNA template strand. This enzyme, called
RNA polymerase, is very similar in its action to DNA
polymerase. The reaction requires the appropriate nucleoside 5'-triphosphates (ATP, GTP, UTP, and CTP) as
well as Mg^{2+}. It proceeds with elimination of pyrophosphate as follows:

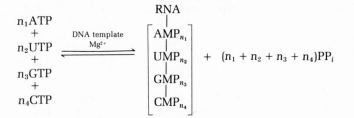

All four ribonucleoside 5'-triphosphates are required
simultaneously. The polymer formed was found to possess 3',5'-phosphodiester bridges and was hydrolyzed
by ribonuclease. RNA polymerase adds mononucleotide
units to the 3'-hydroxyl end of the RNA chain and thus
builds RNA in the 5' → 3' direction. RNA polymerase

is most active with a double-stranded DNA as template and shows much less, though significant activity with single-stranded or denatured DNA. The RNA formed in response to various DNA templates has a base composition complementary to that of the template strand of DNA; in the RNA so formed, uracil residues are inserted complementary to adenine residues of the DNA template.

Nearest-neighbor base-frequency analysis shows that the base sequences of template and product are perfectly complementary to each other and that the newly formed RNA chain has the opposite polarity of the template DNA chain. When highly purified RNA polymerase of E. coli is presented with completely intact duplex DNA templates, only one of the two strands is transcribed, as is true in the intact cell. Most available evidence suggests that in E. coli there is but a single DNA-directed RNA polymerase which can make mRNAs, tRNAs, and rRNAs.

The formation of RNA by DNA-directed RNA polymerase preparations is specifically inhibited by the antibiotic actinomycin D (Figure 22-10), which binds to DNA by hydrogen-bonding to guanine residues. Actinomycin D also inhibits RNA synthesis in intact cells and has become a very important diagnostic tool for detection of processes dependent on transcription of DNA. RNA polymerase is a very complex enzyme; it contains six subunit polypeptide chains ranging from 10,000 to 160,000 in molecular weight. Its total molecular weight is about 500,000. One of its subunits, referred to as sigma (σ), has

Figure 22-10
Structure of actinomycin D, an inhibitor of transcription of DNA. Sarcosine is N-methylglycine. The linkages between sarcosine, L-proline, and D-valine are peptide bonds.

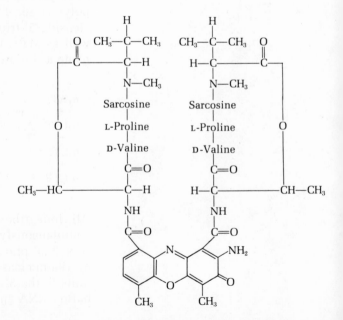

the function of recognizing certain "start" signals on the DNA undergoing replication, thus making possible transcription of a gene at the proper point along the DNA template strand. Transcription stops at certain specific points in DNA which can be recognized by the enzyme.

RNA-Directed RNA Polymerase (RNA Replicase)

Some bacterial viruses contain RNA, including the *E. coli* bacteriophages f2, MS2, and Qβ. These RNA viruses induce the formation in the host cell of an RNA polymerase specific for the viral RNA as template. This enzyme has been called *RNA-directed RNA polymerase* or *RNA replicase*.

RNA replicase isolated from virus-infected cells catalyzes the formation of RNA from the ribonucleoside 5'-triphosphates with elimination of pyrophosphate; the reaction is formally similar to that of DNA-directed RNA polymerase. The RNA replicase requires RNA as template and will not function with DNA. However, contrary to DNA and RNA polymerases, RNA replicase is template-specific. It can employ as template only the RNA of the infecting virus; the RNAs of the host cell do not support replication. This finding obviously explains how RNA viruses are preferentially replicated in the host cell, which contains many other types of RNA.

Polynucleotide Phosphorylase

This enzyme does not function in normal biological synthesis of informational RNA. However, it can make an RNA-like polymer. It is apparently found only in bacteria. Polynucleotide phosphorylase catalyzes the reaction

$$n\text{NDP} \underset{\substack{\text{RNA primer}}}{\overset{\substack{\text{Mg}^{2+}}}{\rightleftharpoons}} \underset{\text{ribopolynucleotide}}{(\text{NMP})_n} + n\text{P}_i$$

It acts on the 5'-diphosphates of ribonucleosides, singly or in combination; it does not act on the homologous 5'-triphosphates or on deoxyribonucleoside 5'-di- or triphosphates. Mg^{2+} is required for its action. The RNA-like polymer it forms contains 3',5'-phosphodiester linkages which can be attacked by ribonuclease. The reaction is reversible and can be pushed in the direction of breakdown of the polyribonucleotide by increasing the phosphate concentration.

The polynucleotide phosphorylase reaction does not utilize a template and therefore does not form a polymer having a specific base sequence. It does require a priming strand of RNA, which merely furnishes a free 3'-hydroxyl terminus to which additional nucleotide residues may be added. The base composition of the polynucleotide formed merely reflects the composition of the nucleoside diphosphates present; thus, if only ADP is present, polyadenylic acid will be formed. One possible biological function of this enzyme is to degrade mRNA and recover the monomeric units as nucleoside 5'-diphosphates. This enzyme has been utilized to synthesize RNA polymers whose use as synthetic mRNAs helped in identifying the nucleotide triplets coding for specific amino acids (see Chapter 23).

RNA-Directed DNA Polymerase
(Reverse Transcriptase)

Recently it has been discovered that several RNA viruses that infect animal cells contain an enzymatic activity capable of forming DNA from a mixture of deoxyribonucleoside 5'-triphosphates. Among these are the Rous sarcoma virus, which can transmit a cancer in chickens. The viral enzyme requires a template nucleic acid. However, the unexpected and extraordinary finding has been made that this type of viral DNA polymerase requires RNA rather than DNA as a template. The most active template is the RNA of the virus particle itself, although other RNAs, including synthetic ribopolynucleotides, are also active. The DNA formed by the enzyme has been proved to have a base sequence complementary to that of the template RNA strand.

Although the existence of such an enzyme had first been postulated by H. Temin some years ago, it had seemed unlikely that such an enzyme would be found, in view of the central dogma of molecular biology, which holds that information flows from DNA to RNA to proteins. Over the course of many years, no case of genetic information flowing in the opposite direction has been found. However, a number of laboratories have now proved beyond a doubt the existence of RNA-directed DNA polymerase, not only in a number of RNA viruses but also in some animal cells.

There is no reason to abandon the classical central dogma DNA→RNA→protein, but perhaps it may have to be modified, at least in special cases, to DNA⇌RNA→protein.

The discovery of this enzyme is significant from another point of view. It has been implicated in the process by which normal animal cells undergo malignant transformation into cancer cells. Many kinds of cancer in experimental animals can be transmitted by specific RNA viruses, some of which have been found to contain RNA-directed DNA polymerase. It has been postulated that this enzyme of the cancer virus is capable of directing the synthesis of DNA complementary to the viral RNA in the host cell, and that the DNA formed, which may constitute or contain a "cancer gene," may then become inserted into the normal host cell chromosomes. In this way the host cell is believed to be genetically transformed into a cancer cell, whose progeny are new cancer cells.

Summary

Double-stranded DNA undergoes semiconservative replication in bacterial cells, so that daughter cells contain one strand of parental DNA and a new complementary strand. There is ordinarily a single growing point in the circular DNA molecule and both chains are replicated simultaneously, one in the $5' \rightarrow 3'$ and the other in the $3' \rightarrow 5'$ direction. DNA polymerase I of E. coli can catalyze synthesis of DNA from the four deoxyribonucleoside 5'-triphosphates in the presence of Mg^{2+}, with elimination of pyrophosphate. The reaction requires preexisting single-strand DNA as template; the chain growth is in the $5' \rightarrow 3'$ direction. The enzyme forms a strand of DNA complementary to the single-stranded template, as shown by analysis of base composition. The enzyme DNA ligase can join the ends of two DNA chains, or join the ends of a single chain to make a circular DNA. Sequential action of DNA polymerase I and DNA ligase has yielded biologically active circular DNA of E. coli bacteriophage ϕX174. To account for the fact that both strands of antiparallel DNA are replicated simultaneously it has been proposed that DNA polymerase makes short hairpin segments, first along one strand in the $5' \rightarrow 3'$ direction and then back along the other strand, also in the $5' \rightarrow 3'$ direction. The newly formed segment is cleaved at the fork and the polymerase returns to the template strand to replicate another segment. The resulting fragments are then "stitched" together by DNA ligase. A second type of polymerase, DNA polymerase II, is believed to be concerned with DNA replication, whereas DNA polymerase I may be concerned primarily with repair of nicks.

RNAs are synthesized by RNA polymerase from ribonucleoside 5'-triphosphates. Double-stranded DNA is required as template, but only one strand of DNA is transcribed. Initiation of RNA chains requires recognition of the starting point by sigma factor. RNA-directed RNA polymerase is formed by cells infected by RNA-viruses; they will accept as template only intact homologous viral RNA. Polynucleotide phosphorylase can reversibly form RNA-like polymers having no specific sequences by elimination of phosphate from ribonucleoside 5'-diphos-phates; it adds or removes mononucleotides at the 3'-hydroxyl end of a primer chain. RNA-directed DNA polymerase, also called reverse transcriptase, an enzyme found in some animal viruses, can form DNA from an RNA template.

References

INGRAM, V. M.: *The Biosynthesis of Macromolecules*, W. A. Benjamin, Menlo Park, Cal., 2nd ed., 1971. A readable short paper-back.

LEHNINGER, A. L.: *Biochemistry*, Worth Publishers, New York, 1970. More detailed textbook treatment in Chapters 29–32.

LEWIN, B. M.: *The Molecular Basis of Gene Expression*, Wiley-Interscience, London, 1970.

WATSON, J. D.: *Molecular Biology of the Gene*, W. A. Benjamin, Menlo Park, Cal., 2nd ed., 1971

Problems

1. Write the base sequence of a segment of DNA replicated by DNA polymerase from the DNA template sequence.

    (5')AGCTTGCAACGTTGCATTAG(3')

2. Write the overall equation for the synthesis of this DNA segment, starting from the required deoxyribonucleoside 5'-triphosphates and assuming the presence of pyrophos-phatase.

3. Write the base sequence of a segment of messenger RNA transcribed by RNA polymerase from the DNA template sequence in Problem 1.

4. Write the overall equation for the enzymatic synthesis of the messenger RNA of Problem 3, starting from the required ribonucleoside 5'-triphosphates and assuming the presence of excess pyrophosphatase.

5. The generation time of *E. coli* cells in a culture medium giving maximal rate of cell division is 20 min at 37°. The chromosome of DNA has a molecular weight of 2.8 billion and contains 4,200,000 base pairs. Calculate the rate of synthesis of DNA in such *E. coli* cells in terms of nucleotide residues per sec per cell.

CHAPTER 23 BIOSYNTHESIS OF PROTEINS

We come now to consider a final major question posed by the genetic continuity of living organisms. How is the genetic information contained in the nucleotide sequence of messenger RNA translated so that amino acids are assembled to form a polypeptide chain having a specific amino acid sequence? We shall also see how the triplet code words corresponding to the various amino acids were identified and how protein synthesis is regulated.

Ribosomes as the Site of Protein Synthesis

The modern era of research on protein synthesis began with the important investigations of P. Zamecnik and his colleagues from 1950 onwards. They injected radioactive amino acids into rats. At different time intervals after the injection, the liver was removed, homogenized, and fractionated by centrifugation. The various intracellular fractions collected in this manner were examined for the presence of radioactive protein. When hours or days were allowed to elapse before removal and fractionation of the liver, then all the intracellular fractions contained labeled protein. However, if the liver was fractionated very shortly after the injection of this labeled amino acid, only the ribonucleoprotein particles of the cytoplasm (Figure 23-1), which are today known as ribosomes (Chapter 7), contained radioactive protein. It was tentatively concluded that ribosomes are the site of synthesis of proteins from amino acids (Chapter 11).

Figure 23-1
Ribosomes in a eukaryotic cell. They are found in two major locations
(1) along the surface of the endoplasmic reticulum vesicles and
(2) in free form in the cytosol, as shown in this electron micrograph
of a pancreatic cell.

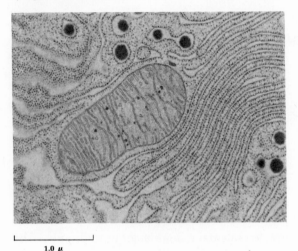

1.0 μ

Subsequent research has revealed that freshly isolated
ribosomes have the capacity for synthesizing polypeptide
chains from a mixture of amino acids if they are properly
supplemented with other factors. The necessary auxiliary
factors include (1) soluble enzymes from the cytoplasm
called aminoacyl-tRNA synthetases, (2) a mixture of
transfer RNAs (Chapter 7), also normally found in the
cytoplasm, (3) a messenger RNA (Chapters 7, 22), (4)
ATP, (5) GTP, and (6) Mg^{2+}. Once these necessary cofactors were identified very rapid progress was made in
study of the mechanism of protein synthesis.

Direction of Chain Growth during Polypeptide Synthesis

Although it had long been assumed that the polypeptide
chain is elongated during its synthesis by successive
addition of single amino acid residues to one end, whether
the chain was built starting from the amino-terminal or
the carboxyl-terminal residue was an open question. The
answer came from cleverly designed isotope tracer experiments. Radioactive leucine was added to reticulocytes,
immature red blood cells which actively synthesize hemoglobin. This amino acid was chosen because it occurs

Figure 23-2
Direction of growth of polypeptide chain. The colored zones show the portions of hemoglobin chains containing labeled leucine added at time zero. At 4 minutes only a few residues at the carboxyl end were labeled, indicating that the chain grows by addition at the COOH end.

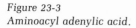
Direction of growth of chain

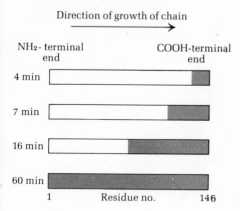

NH₂-terminal end   COOH-terminal end

4 min

7 min

16 min

60 min

1   Residue no.   146

Figure 23-3
Aminoacyl adenylic acid.

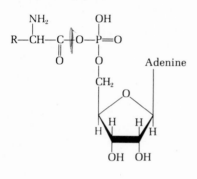

frequently along both the polypeptide chains of hemoglobin. By determining the relative amounts of radioactivity in the leucine residues along the polypeptide chains at different time intervals, it was found that in those polypeptide chains that had just been started at the time when isotopic leucine was added, all the leucine residues were labeled, whereas those chains that were nearly complete at the time leucine was added contained radioactive leucine residues only at the COOH-terminal end (Figure 23-2). It was therefore concluded that polypeptide chains are constructed beginning with the amino-terminal amino acid, whose carboxyl group combines with the amino group of the incoming amino acid to form a peptide bond. Serial addition of new amino acid residues at the free COOH-terminal end of the growing polypeptide chain continues until the chain is complete.

We shall now examine the individual steps in protein synthesis.

Activation of Amino Acids

In the first stage of protein synthesis, the 20 different amino acids are "activated." They are esterified to their corresponding transfer RNAs (Chapter 7) by the action of enzymes known as the *aminoacyl-tRNA synthetases*, each of which is specific for one amino acid and for its corresponding tRNA. Thus 20 different activating enzymes are required to activate the 20 amino acids found in proteins. This stage occurs in the cell cytoplasm. The reaction catalyzed is

$$\text{Amino acid} + \text{tRNA} + \text{ATP} \xrightleftharpoons{\text{Mg}^{2+}}$$
$$\text{aminoacyl-tRNA} + \text{AMP} + \text{PP}_i$$

The activation reaction occurs in two separate steps on the enzyme catalytic site. In the first, an enzyme-bound intermediate, *aminoacyl adenylic acid* (Figure 23-3), is formed by reaction of ATP and the amino acid. The carboxyl group of the amino acid is bound in anhydride linkage with the 5′-phosphate group of the AMP, with displacement of pyrophosphate

$$\text{ATP} + \text{amino acid} \rightleftharpoons$$
$$\text{[aminoacyl adenylic acid]} + \text{pyrophosphate}$$

The second step consists of the transfer of the aminoacyl group from AMP to the corresponding specific transfer RNA

$$\text{[Aminoacyl adenylic acid]} + \text{tRNA} \rightleftharpoons$$
$$\text{aminoacyl-tRNA} + \text{adenylic acid}$$

The aminoacyl group is transferred to the free 2'-hydroxyl group of the terminal AMP residue at one end of the tRNA molecule (Figure 23-4). The ester linkage between the amino acid and the tRNA is a high-energy bond, for which the $\Delta G^{\circ\prime}$ of hydrolysis is approximately −7.0 kcal. The inorganic pyrophosphate formed in the activation reaction is ultimately hydrolyzed to orthophosphate by pyrophosphatase; thus two high-energy phosphate bonds are ultimately split for each amino acid molecule activated. The overall reaction for amino acid activation in the cell would thus be essentially irreversible

$$\text{Amino acid} + \text{tRNA} + \text{ATP} \xrightarrow{\text{Mg}^{2+}}$$
$$\text{aminoacyl-tRNA} + \text{AMP} + 2P_i \qquad \Delta G^{\circ\prime} = -7.0 \text{ kcal}$$

The aminoacyl-tRNA synthetases are very highly specific for both the amino acid and the corresponding tRNA. Thus the aminoacyl-tRNA synthetases must possess three very specific binding sites, one for the amino acid substrate, the second for the corresponding tRNA, and a third for binding the ATP molecule required in the reaction.

Structure of tRNAs and the Specificity of the Activating Enzymes

Transfer RNAs contain from 75 to 90 nucleotide units in a single chain of molecular weight 23,000 to 30,000. The base sequences of many tRNA molecules are now known (Chapter 7). A striking feature is the presence of many minor bases in addition to the normal bases A, U, G, and C. Over 30 different minor bases have been discovered in tRNAs, most of which are methylated forms of the normal bases. The surprising finding has been made that, although the tRNAs for different amino acids have different base sequences, all of them are potentially capable of existing in the same cloverleaf conformation (Chapter 7) if the chain is arranged in such a way as to yield maximal base pairing (Figure 23-4).

Base-sequence studies have revealed the presence of one nucleotide triplet in the polynucleotide chain that is different in all tRNAs examined; it is thought to represent the anticodon of the tRNA, namely, the specific nucleotide triplet complementary to the codon triplets in mRNA. In addition to the anticodon site, tRNA molecules must contain at least one other specific binding site, the enzyme recognition site, at which it is bound to the corresponding activating enzyme.

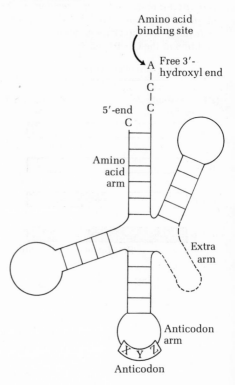

Figure 23-4
Cloverleaf structure of tRNAs. This configuration assumes maximal base pairing. Some tRNAs have an extra arm as shown.

Some amino acids have more than one specific tRNA. Yeast cells contain five different tRNAs that react specifically with the leucine-activating enzyme, five tRNAs for serine, four for glycine, and four for lysine.

The Adapter Role of tRNA

Once an amino acid is esterified to its corresponding tRNA, it makes no contribution to the specificity of the aminoacyl-tRNA, since its aminoacyl group is not recognized by either the ribosome or the mRNA template. Rather, the specificity of the aminoacyl-tRNA is furnished by the anticodon site of the tRNA portion alone. This was conclusively proved in clever experiments in which enzymatically formed cysteinyl-tRNA$_{Cys}$ was chemically converted into alanyl-tRNA$_{Cys}$. This "hybrid" aminoacyl-tRNA, which contains the anticodon for cysteine but actually carries alanine, was then incubated with a cell-free system capable of protein synthesis. The newly synthesized polypeptide chains were then examined and it was found that the labeled alanine was incorporated into the cysteine positions in significantly large amounts.

This experiment provided proof for the hypothesis, first postulated by Crick in 1958, that tRNA is a molecular "adapter" into which the amino acid is plugged so that it can be adapted to the nucleotide triplet language of the genetic code.

After completion of the amino acid activation reaction, the charged tRNA, symbolized as in the example Ala-tRNA$_{Ala}$, is ready to be incorporated into the growing polypeptide chains on the surface of the ribosome.

The Initiating Amino Acid

For some time it has been thought that the codon for the first or NH_2-terminal amino acid residue of a polypeptide chain must have some distinctive characteristic enabling the ribosome to recognize it as the starting point for the growth of a polypeptide chain. It is now clear that in *E. coli* (and perhaps all other prokaryotic cells) the synthesis of all proteins begins with the amino acid derivative N-formylmethionine. It enters as N-formylmethionyl-tRNA (symbolized as fMet-tRNA) (Figure 23-5), which is formed by the enzymatic transfer of a formyl group from N^{10}-formyltetrahydrofolate (Chapters 7, 15) to one species of methionyl-tRNA designated Met-tRNA$_{fMet}$

N^{10}-formyltetrahydrofolate + Met-tRNA$_{fMet}$ →

tetrahydrofolate + fMet-tRNA$_{fMet}$

Figure 23-5
N-Formylmethionine.

H
|
C=O
|
NH
|
CH$_3$SCH$_2$CH$_2$CHCOOH

There are two species of methionyl-tRNA; that species capable of accepting the N-formyl group is designated methionyl-tRNA$_{fMet}$. The enzyme catalyzing this reaction does not formylate free methionine or methionine attached to the other species of methionyl-tRNA, which is designated methionyl-tRNA$_{Met}$.

The blocking of the amino group of methionine by the N-formyl group not only prevents it from entering into a peptide linkage but also appears to allow fMet-tRNA to be bound at a site on the ribosome which does not accept free Met-tRNA or any other free aminoacyl-tRNA. The synthesis of all proteins of E. coli is believed to begin with the amino-terminal sequence N-formylmethionyl-alanylserine. The N-formyl group does not, however, appear in the finished protein; it is removed by enzymatic cleavage.

Initiation of Polypeptide Chains

Attention must now be focused on the structure and role of ribosomes (Chapter 7). It will be recalled that ribosomes of E. coli have two subunits; the smaller is called the 30S and the larger the 50S subunit. Each subunit contains a specific rRNA and a number of polypeptide chain subunits (Chapter 7; Figure 7-11). It has been discovered that during protein synthesis there is continuous dissociation of complete or intact 70S ribosomes into their 50S and 30S subunits and continuous reassociation of the subunits to form complete 70S ribosomes. More detailed study has shown that the intact 70S ribosome must first dissociate into 50S and 30S subunits before either mRNA or the initiating aminoacyl-tRNA can be bound and protein synthesis initiated. The free 30S subunit then binds both the initiating codon of the mRNA and fMet-tRNA at a specific site called the _peptidyl site_ to form what is called the _initiation complex_ (Figure 23-6). In this complex the anticodon triplet of the fMet-tRNA hydrogen-bonds with its complementary codon triplet in the mRNA, specifically, the codon which specifies N-formylmethionine, the initiating amino acid. The initiation complex then associates tightly with the 50S ribosome subunit to form the functional, intact 70S ribosome (Figure 23-6). The initiation of the polypeptide chain also involves specific protein initiation factors, symbolized as F_1, F_2, and F_3, which are present in the 30S ribosomal subunit. Presumably this elaborate initiation process is required to ensure that ribosomes do not start making a polypeptide chain at the middle. The com-

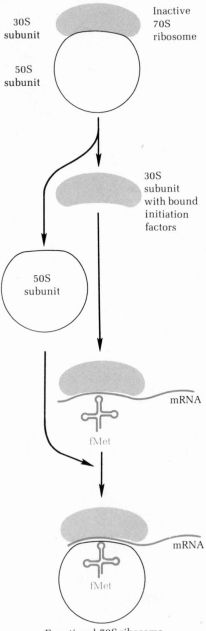

Figure 23-6
Formation of initiation complex and complete 70S ribosome, ready for chain elongation.

30S subunit

Inactive 70S ribosome

50S subunit

30S subunit with bound initiation factors

50S subunit

mRNA

fMet

mRNA

fMet

Functional 70S ribosome

plete functional ribosome, with its initiating fMet-tRNA
and mRNA in place, is now ready for elongation.

The Elongation Cycle

Step 1

Three major steps occur in chain elongation (Figure 23-7).
In the first step, the incoming aminoacyl-tRNA binds to
the next codon of the mRNA and to what is called the
aminoacyl site of the complete 70S ribosomal complex.
This binding process requires GTP and a specific cytoplasmic protein called T factor, which in turn has two
subunits, T_U, and T_S.

Step 2

In the second step, the peptide bond is formed by reaction
of the amino group of the newly bound aminoacyl-tRNA
with the esterified carboxyl group of the COOH-terminal
amino acid residue of the peptidyl-tRNA (Figures 23-7,
23-8), thus displacing the tRNA from the preceding amino
acid. The "empty" tRNA remains temporarily on the
peptidyl site and the elongated peptidyl-tRNA is now
bound to the ribosome at its aminoacyl site. Neither ATP
nor GTP is required for the formation of the new peptide
bond, which is presumably made at the expense of the
bond energy of the ester linkage between the amino acid
and its tRNA. However, an enzyme called *peptidyl transferase*, which is part of the 50S subunit of the ribosome,
is required to catalyze this reaction (Figure 23-8).

Step 3

In the third step of the elongation cycle the peptidyl-tRNA
is physically shifted or translocated from the aminoacyl to
the peptidyl site, thus "bumping" the empty tRNA from
the peptidyl site. This translocation reaction is believed
to be the result of a conformational change in the ribosome, which occurs at the expense of the hydrolysis of
GTP. A specific protein called *G factor* is required for
this step. Simultaneous with the translocation of the
peptidyl-tRNA from the aminoacyl site to the peptidyl
site there is a translocation of the mRNA, which moves
along the ribosome by one codon (Figure 23-7).   The
ribosome probably undergoes complex geometrically
specific changes in shape during each bond-making cycle.
The mRNA appears to "track" through a groove or tunnel

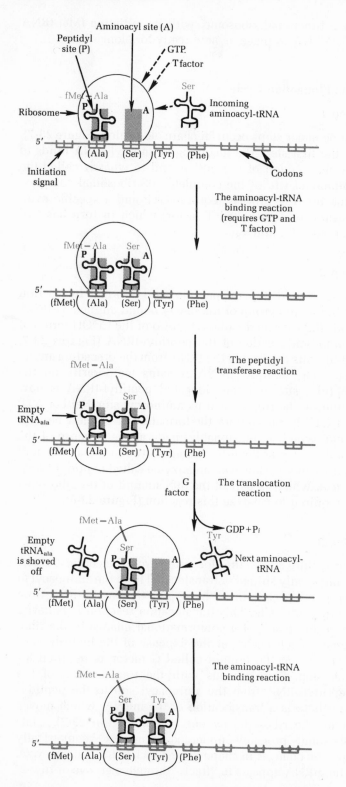

Peptidyl site (P)
Aminoacyl site (A)
GTP
T factor
Ribosome
fMet–Ala
P
A
Ser
Incoming
aminoacyl-tRNA
5′
(Ala) (Ser) (Tyr) (Phe)
Codons
Initiation
signal

The aminoacyl-tRNA
binding reaction
(requires GTP and
T factor)

fMet–Ala
P
Ser
A
5′
(fMet) (Ala) (Ser) (Tyr) (Phe)

The peptidyl
transferase reaction

fMet–Ala
P
Ser
A
Empty
tRNA_ala
5′
(fMet) (Ala) (Ser) (Tyr) (Phe)

G
factor
The translocation
reaction

fMet–Ala
Empty
tRNA_ala
is shoved
off
Ser
P
A
GDP + P_i
Tyr
Next aminoacyl-
tRNA
5′
(fMet) (Ala) (Ser) (Tyr) (Phe)

The aminoacyl-tRNA
binding reaction

fMet–Ala
Ser Tyr
P A
5′
(fMet) (Ala) (Ser) (Tyr) (Phe)

Figure 23-7
Steps in elongation. After the initiation complex is formed, the fMet-tRNA is on the peptidyl site. The next aminoacyl-tRNA is then bound to the aminoacyl site, a process that requires GTP and factor T. In the peptidyl transferase reaction the new peptide bond is formed; the lengthened peptidyl-tRNA is now on the aminoacyl or A site and the empty aminoacyl-tRNA on the peptidyl or P site. In the translocation reaction, which requires the G factor, GTP is hydrolyzed to furnish the energy required to translocate simultaneously the peptidyl-tRNA from the A site to the P site and the mRNA by one codon. The cycle then repeats with the next incoming aminoacyl-tRNA.

Figure 23-8
The peptidyl transferase reaction.

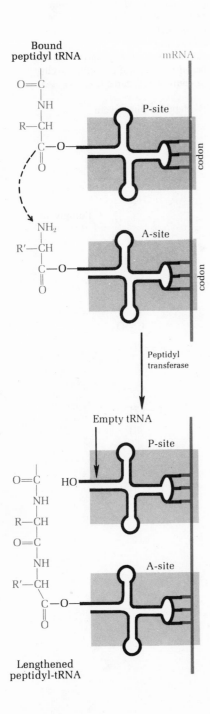

between the ribosomal subunits. The elongation cycle is repeated many times until the last amino acid is inserted.

Termination of the Polypeptide Chain

The completion of a polypeptide chain and its detachment from the ribosome require special steps which are not yet fully understood. After the last, or COOH-terminal, amino acid residue has been added to a polypeptide chain on the ribosome, the polypeptide chain is still covalently attached to the last tRNA, which is in turn noncovalently attached to the ribosome by its codon–anticodon interaction with the mRNA. The release of the polypeptide chain from the ribosome, when a termination codon is reached, is promoted by a specific *protein release factor*, which is bound to the ribosome and promotes the hydrolysis of the ester linkage between the polypeptide and the tRNA, thus releasing both from the ribosome. The 70S ribosome then "runs off" the mRNA in free form. It may start a new polypeptide chain after it undergoes dissociation into its 50S and 30S subunits, a reaction which appears to require one of the specific protein initiation factors (Figure 23-6).

As we have seen in Chapter 3, a protein is not biologically active unless it is in its specific folded configuration, which is determined by its amino acid sequence. As the polypeptide chain undergoes elongation on the ribosome, it is believed that the growing chain spontaneously assumes its minimum-energy native configuration, in a residue-by-residue manner. When it is finally complete and is released from the ribosome, the polypeptide chain has attained its biologically active conformation. In this way the linear or one-dimensional genetic code brought by the messenger RNA is converted into the specific three-dimensional structure of the newly-synthesized polypeptide chain.

Inhibitors of Protein Synthesis

Protein synthesis is inhibited by the antibiotic *puromycin*, which has a structure very similar to that of the terminal AMP residue of an aminoacyl-tRNA (Figure 23-9). Puromycin interrupts peptide-chain elongation by virtue of its capacity to replace an entering aminoacyl-tRNA with formation of a *peptidyl-puromycin* derivative. No new amino acid residues can be added to this peptidyl-puromycin because its amide linkage is substituted. Peptide bond formation is also inhibited by _chloramphenicol_ in bacteria and by _cycloheximide_ in eukaryotic cells (Figure 23-10).

Polyribosomes

When ribosomes are carefully isolated from tissues active in protein synthesis, they are often obtained in clusters containing from as few as 3 or 4 to as many as 100 individual ribosomes, depending on the tissue. Such clusters, which are called _polyribosomes_, or _polysomes_, have been examined with the electron microscope and studied chemically. Since they can be fragmented into single individual ribosomes by the action of ribonuclease, polyribosomes are apparently held together by a strand of RNA. A connecting fiber between adjacent ribosomes can actually be seen in electron micrographs. It has been deduced that the RNA connecting strand is actually mRNA, which is being "read" by several ribosomes simultaneously (Figure 23-11), spaced some distance apart. Each individual ribosome of a polyribosome can make a complete polypeptide chain and does not require the presence of the other ribosomes. Such an arrangement therefore increases the efficiency of utilization of the mRNA template, since several polypeptide chains can be made from it simultaneously.

The Energy Requirement in Protein Synthesis

Two high-energy phosphate bonds are utilized in the formation of each molecule of aminoacyl-tRNA during the activation reaction (see above). In addition, at least one molecule of GTP is cleaved to GDP and phosphate during the translocation reaction. Therefore a total of at

Figure 23-9
Action of puromycin. Puromycin resembles aminoacyl tRNA and can react with peptidyl RNA to yield peptidyl-puromycin (below). The chain cannot be elongated further and is discharged from the ribosome as a free peptidyl-puromycin. A number of peptidyl-puromycins varying in chain length have been isolated.

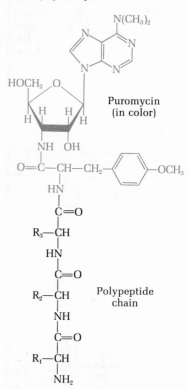

Peptidyl-puromycin

Figure 23-10
Other inhibitors of protein synthesis.

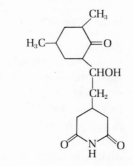

Cycloheximide

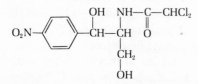

Chloramphenicol

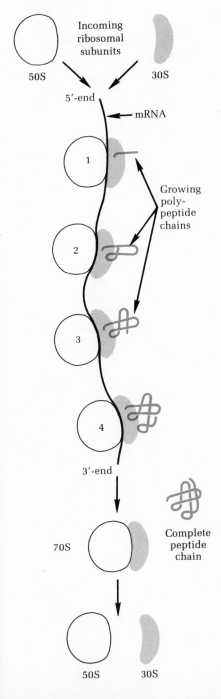

Figure 23-11
Formation and function of a polyribosome. The individual ribosomes function independently of
each other, each forming a polypeptide chain as it moves along the
mRNA molecule.

least three high-energy bonds is ultimately required for the synthesis of each peptide bond of the completed protein. This represents an exceeding large thermodynamic "push" in the direction of synthesis, since a total of at least $7.3 \times 3 = 21.9$ kcal is invested to generate a peptide bond whose standard free energy of hydrolysis is about -5.0 kcal. The net $\Delta G^{\circ\prime}$ for peptide-bond synthesis is thus about -16.9 kcal, making peptide-bond synthesis at the expense of phosphate-bond energy overwhelmingly exergonic and essentially irreversible. Although this energy expenditure may appear to be wasteful, very likely it is one of the important factors making possible nearly perfect fidelity in the biological translation of the genetic message of mRNA into the amino acid sequence of proteins.

The Genetic Code

We have seen that amino acids are coded for by specific triplets of nucleotides in DNA called codons. We shall now consider exactly how the four-letter "language" of DNA, which contains four different bases (A, G, C, and T) is translated into the 20-letter language of proteins via the formation of messenger RNA.

It had long appeared likely that at least three nucleotide residues of DNA were required to code for each amino acid, since the four code letters of DNA (A, G, C, and T) arranged in groups of two can yield only $4^2 = 16$ different combinations but arranged in groups of three can yield $4^3 = 64$ different combinations. However, a major question was whether the coding units for the amino acids are separated by symbols representing punctuation, that is, commas. Important genetic experiments in the late 1950s conclusively proved not only that the genetic code words for amino acids must be triplets of nucleotides but also that the genetic code for proteins is "commaless," that is, there are no punctuation symbols between successive amino acid residues.

In 1961 M. Nirenberg and H. Matthaei reported that when the synthetic polyribonucleotide polyuridylic acid was incubated with E. coli extracts, GTP, and a mixture of radioactive amino acids, the polyuridylic acid (designated poly U) acted as an artificial messenger RNA and promoted the synthesis of a radioactive polypeptide, which was identified as containing only phenylalanine. They concluded that the triplet UUU must code for phenylalanine. Very soon it was found that the synthetic messenger polycytidylic acid codes for formation of a polyproline and the messenger polyadenylic acid for polylysine. Thus the triplet CCC must code for proline and the triplet AAA for lysine.

By use of different artificial messenger RNAs, the bases present in the triplets coding for all the amino acids were soon identified. However, no information as to the *sequence* of bases in each coding triplet became available until Nirenberg made a second important discovery. In 1964 he and P. Leder found that isolated *E. coli* ribosomes will bind a specific aminoacyl-tRNA if at the same time the corresponding synthetic polynucleotide messenger is present. For example, ribosomes incubated with poly-uridylic acid simultaneously bind phenylalanyl-tRNA$_{Phe}$ but no other aminoacyl-tRNA. Because GTP, which is required for peptide-bond formation, is not present in such a test, polyphenylalanine is not formed. Nirenberg and Leder then found that the shortest polynucleotide that would promote specific binding of phenylalanyl-tRNA$_{Phe}$ was the trinucleotide UUU. Such experiments thus showed that simple trinucleotides suffice as synthetic messengers to specify the binding of specific aminoacyl-tRNAs.

With this binding assay, Nirenberg and two other groups led by S. Ochoa and H. Khorana ultimately established the sequence of nucleotides in the triplet code words for each of the amino acids. The complete code-word "dictionary" for the amino acids is given in Figure 23-12.

General Features of the Genetic Code

The genetic code for the amino acids is degenerate; that is, there is more than one code word for most of the amino acids (Figure 23-12). In fact, all of the amino acids except tryptophan and methionine have more than one specific codon. The term "degenerate" should not be taken to mean imperfect, for there is no code word that specifies more than one amino acid.

It is striking that the degeneracy is not uniform. Thus, there are six code words for leucine and for serine whereas other amino acids, such as glutamic acid, tyrosine, and histidine have only two codons and tryptophan but one (UGG).

In most cases the degeneracy involves only the third base in the codon (Figure 23-12). For example, alanine is coded for by the triplets GCU, GCC, GCA, and GCG; that is, the first two bases GC are common to all the codons for alanine. In fact, nearly all of the amino acid codons consist of triplets symbolized by XY_G^A or XY_C^U. Evidently

Figure 23-12
The genetic code-word dictionary.
The third nucleotide of each codon
(in color) is less specific than the first
two. The codons read in the 5′ → 3′
direction. For example, pUpUpA =
leucine. The three nonsense codons
are in color.

	U		C		A		G	
U	UUU	Phe	UCU	Ser	UAU	Tyr	UGU	Cys
	UUC	Phe	UCC	Ser	UAC	Tyr	UGC	Cys
	UUA	Leu	UCA	Ser	UAA	Ochre	UGA	
	UUG	Leu	UCG	Ser	UAG	Amber	UGG	Trp
C	CUU	Leu	CCU	Pro	CAU	His	CGU	Arg
	CUC	Leu	CCC	Pro	CAC	His	CGC	Arg
	CUA	Leu	CCA	Pro	CAA	Gln	CGA	Arg
	CUG	Leu	CCG	Pro	CAG	Gln	CGG	Arg
A	AUU	Ile	ACU	Thr	AAU	Asn	AGU	Ser
	AUC	Ile	ACC	Thr	AAC	Asn	AGC	Ser
	AUA	Ile	ACA	Thr	AAA	Lys	AGA	Arg
	AUG	Met	ACG	Thr	AAG	Lys	AGG	Arg
G	GUU	Val	GCU	Ala	GAU	Asp	GGU	Gly
	GUC	Val	GCC	Ala	GAC	Asp	GGC	Gly
	GUA	Val	GCA	Ala	GAA	Glu	GGA	Gly
	GUG	Val	GCG	Ala	GAG	Glu	GGG	Gly

the first two letters of each codon are the primary deter-
minants of its specificity. The third position (i.e., the
nucleotide at the 3′-OH end of the codon) is of less importance; it is loose and tends to "wobble."

Another conspicuous feature of the genetic code is that
no punctuation or signal is required to indicate the end
of one codon and the beginning of the next. The reading
frame must therefore be correctly "set" at the beginning
of the readout of an mRNA molecule and then moved
sequentially from one triplet to the next, or all codons
will be out of register and lead to formation of a missense
protein with a garbled amino acid sequence.

The code words shown in Figure 23-12 have been shown
to be identical in man, E. coli, the tobacco plant, certain
amphibia, and the guinea pig, among other species. In all
probability the genetic code words for amino acids are
universal for all species.

Code Words for the Initiation and
Termination of Polypeptide Chains

We have already seen that all polypeptide chains in
E. coli (and probably all other prokaryotic cells) begin
with the N-formylmethionine residue, signalled by a
special fMet codon. Three of the 64 triplets (UAG, UAA,
and UGA) do not code for any known amino acids (Figure 23-12). They have been called "nonsense" codons.

However, they are now known to be signals for the termination of polypeptide chains. The UAG codon is often called *amber* whereas the UAA codon is called *ochre*. The latter is believed to act as the normal termination codon.

Regulation of Protein Synthesis

Living cells must possess accurately programmed mechanisms for regulating the relative amounts of different types of proteins that are synthesized. For example, the number of molecules of those enzymes catalyzing a mainstream metabolic pathway is much greater than the number of molecules of the enzymes catalyzing the biosynthesis of the coenzymes, which are needed in only trace amounts. Moreover, regulation of the rate of enzyme synthesis provides "coarse" control over metabolism, in contrast to allosteric regulation, which yields fine control. Regulation of protein synthesis is also important in the differentiation of cells, not only with respect to the relative numbers of each type of protein that are made, but also to the time and sequence of their appearance during development.

Most of our present knowledge of the regulation of protein synthesis has come from studies on the *induction* and *repression* of enzyme synthesis in bacteria. Certain enzymes in bacterial cells may vary dramatically in concentration in response to the nature of the nutrients in the culture medium. Such an enzyme, which is called an *adaptive*, or an *inducible enzyme*, is present only in traces when its substrate is lacking from the culture medium, but greatly increases in amount when its substrate is added to the medium. Such enzyme induction is most conspicuous when the inducing substrate is the only available source of carbon or nitrogen in the medium. The induced enzyme is then required by the cell to transform its substrate into a metabolite that can be directly utilized by the cell for energy and growth. The classic example of an inducible enzyme is *β-galactosidase*. When wild-type *E. coli* cells, which grow readily on glucose, are placed in a culture medium containing lactose as sole carbon source, they are at first unable to utilize lactose, but very soon they respond by synthesizing β-galactosidase in large amounts. Since this enzyme can hydrolyze lactose to D-galactose and D-glucose, the induced enzyme makes it possible for the cell to use glucose formed from the lactose. If the induced cells are now transferred to a medium containing glucose but no lactose, a condition

Figure 23-13 (at right)
Steps in repression and induction of β-galactosidase activity in E. coli. The regulator gene R is transcribed to form mRNA coding for synthesis of repressor molecules, which can bind specifically to operator region O and thus prevent transcription of the structural gene for β-galactosidase.

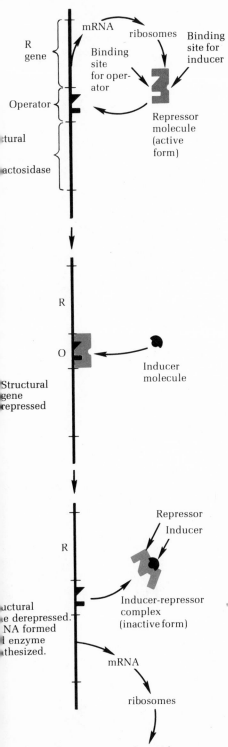

R
gene

mRNA
ribosomes

Binding
site
for oper-
ator

Binding
site for
inducer

Operator

Repressor
molecule
(active
form)

:tural

actosidase

R

O

Inducer
molecule

Structural
gene
repressed

R

Repressor

Inducer

Inducer-repressor
complex
(inactive form)

uctural
e derepressed.
NA formed
l enzyme
thesized.

mRNA

ribosomes

β-galactosidase

under which the induced β-galactosidase is no longer needed, the latter enzyme is no longer made by the cell. This effect is called *repression*. Enzyme induction and repression are reflections of the principle of cellular economy: Enzymes are made only when they are needed.

A whole group of enzymes catalyzing a sequence of consecutive metabolic reactions may be induced or repressed together. For example, when histidine is not present in the culture medium for *E. coli*, all the enzymes required to make histidine from its precursors are synthesized by the cell. But when preformed histidine is added to the culture medium the synthesis of the histidine-forming enzymes immediately stops.

Jacob and Monod have formulated a general hypothesis for the regulation of the synthesis of inducible enzymes (Figure 23-13). In their hypothesis enzyme repression is the basic process; enzyme induction is simply the release of repression. The gene coding for the amino acid sequence of a given enzyme is called a *structural gene*. This gene will normally be transcribed to yield the mRNA coding for this enzyme unless transcription is repressed by the action of a specific regulator gene R. Monod and Jacob postulated that the R gene codes for the amino acid sequence of a specific protein called the *repressor*. The repressor molecule was postulated to diffuse from the ribosomes, where it is formed, and to become physically bound to a specific segment of DNA near the structural gene of the enzyme whose synthesis it controls. When the repressor is bound, it prevents the transcription of the structural gene to form the corresponding mRNA and thus presents synthesis of the enzyme. That segment of DNA to which the repressor is bound is called the *operator*, or *O site*.

How can this hypothesis explain the phenomenon of enzyme induction? In the absence of inducer, Jacob and Monod postulated that the repressor molecule occurs in its free or active state, in which it can combine with operator and prevent transcription of the corresponding structural gene (Figure 23-13). But when the inducing substrate is present in the cell, it combines with the repressor protein at a specific complementary site to form an inactive *repressor–inducer complex*. This complex is unable to bind to the segment of DNA containing the structural gene. The structural gene thus becomes free for transcription and the enzyme is then synthesized. The interaction between the inducing agent and the repressor molecule is reversible, to account for the fact that when the inducing substrate is removed from the culture medium or is used up by enzymatic action, enzyme synthesis is again repressed by the binding of the repressor to the operator. The repressor molecule was thus conceived to be a protein having two specific binding sites, one for the inducer and one for the operator locus. When the inducer site is filled, the operator binding site is no longer functional.

The Operon and the Operator

Jacob and Monod extended their hypothesis for the regulation of protein synthesis to provide a mechanism for *coordinate repression*, in which a group of enzymes may be repressed by a single repressor, and *coordinate induction*, in which a group of enzymes can be induced by a single inducer. For example, when the presence of histidine in the medium represses histidine synthesis, it represses not only the formation of the last enzyme in the sequence forming histidine but several preceding enzymes as well. Such a set of metabolically related enzymes which can be induced or repressed as a group is coded by a group of structural genes called an *operon*. The formation of the entire set of enzymes synthesizing histidine is repressed by the presence of free histidine, which functions as a *corepressor*. Free histidine combines with the repressor molecule generated from the R gene of the histidine operon to form a *repressor-corepressor* complex, which then combines with the operator gene and thus prevents transcription of the structural genes coding for the sequence of enzymes responsible for histidine synthesis. When histidine is lacking in the medium, the repressor molecule alone cannot bind to the operator. As a consequence, all the histidine-forming enzymes are then made by the cell.

Specific repressor molecules have been isolated from bacterial cells. They have been proved to bind to specific sites on the chromosome and to prevent the synthesis of their corresponding enzymes. It is now believed that this type of control, called *transcriptional control*, is one of the most important mechanisms regulating protein synthesis.

Summary

Amino acids are first activated for protein synthesis by specific aminoacyl-tRNA synthetases in the cytoplasm; they catalyze the formation of the aminoacyl esters of homologous tRNAs with simultaneous cleavage of ATP to AMP and pyrophosphate.

There is more than one type of tRNA for each amino acid. The anticodon nucleotide triplet of tRNA is responsible for the specificity of interaction of the aminoacyl-tRNA with the complementary codon triplet of mRNA. The growth of polypeptide chains on ribosomes begins with the NH_2-terminal amino acid and proceeds by successive additions of new residues to the COOH-terminal end. In bacteria the initiating NH_2-terminal amino acid of all proteins is methionine, which enters as N-formylmethionyl-tRNA. After binding of fMet-tRNA and mRNA to the free 30S subunits of ribosomes to form the initiation complex, the 50S subunit is bound to form a complete 70S ribosome. In the subsequent elongation steps, GTP is required for binding of the incoming aminoacyl-tRNA to the aminoacyl binding site on the ribosome. In the peptidyl transferase reaction the tRNA of the previously bound residue is displaced by the amino group of the incoming aminoacyl tRNA. The elongated peptidyl-tRNA is then translocated from the aminoacyl site to the peptidyl site, a process requiring hydrolysis of GTP. Polyribosomes consist of mRNA molecules to which are attached several or many ribosomes, each independently reading the mRNA and forming protein. At least three high-energy phosphate bonds are required to generate each peptide bond.

The genetic code words (codons) for amino acids consist of specific triplets of successive nucleotides in DNA. The codons were deduced from experiments with synthetic messenger RNAs of known composition and sequence. The amino acid code is degenerate, that is, it has multiple code words for nearly all the amino acids. The third position in each codon is much less specific than the first and second. The genetic code words are probably universal in all species. The "nonsense" triplets UAA (ochre), UAG (amber), and UGA code for no amino acids but are signals for chain termination. N-formylmethionine is coded by AUG.

Protein synthesis is regulated primarily at the level of the transcription of DNA to yield mRNA. Transcription of a gene or a set of genes (an operon) can be repressed by binding of a repressor substance to an operator gene. It can be derepressed by the presence of certain regulatory metabolites, which may be a substrate of the enzyme coded by the repressed gene.

References

See References to Chapter 22.

Problems

1. Predict the amino acid sequences of peptides formed by ribosomes in response to the following messengers, assuming that the left-hand end has a free 5′-hydroxyl group. Indicate the amino-terminal ends.
   (a) GGUCAGUCGCUCCUGAUU
   (b) UUGGAUGCGCCAUAUUUUGCU
   (c) CACGACGCUUGUUGCUAU
   (d) AUGGACGAA

2. One strand of double helical DNA contains the following sequence, reading from 5′ → 3′:
   TTCGTCGACGATGATCATCGGCTTCTCGAG
   Write (a) the sequence of bases in the complementary strand of the DNA, (b) the sequence of bases in the mRNA transcribed from the first strand of DNA, and (c) the actual amino acid sequence coded.

3. Write a base sequence for a messenger RNA molecule coding for the synthesis of the A chain of bovine insulin.

4. Write the base sequence of the single strand DNA coding for the messenger RNA of Problem 1(b).

5. How many high-energy phosphate bonds are required for the synthesis of the single polypeptide chain of bovine ribonuclease from its constituent amino acids?

APPENDIXES

SOLUTIONS TO PROBLEMS

Chapter 1

1 (a) pH 12.0 (b) pH 3.0 (c) pH 4.0 (d) pH 12.48
  (e) pH 11.53
2 (a) 5.01×10^{-3} M (b) 1.59×10^{-3} M
  (c) 3.16×10^{-7} M (d) 1.59×10^{-7} M
  (e) 2.51×10^{-7} M
3 $K' = 2.57 \times 10^{-6}$ M
4 $pK' = 5.59$
5 Lactic acid; its pK' is closest to pH 4.0
6 [Proton acceptor]/[proton donor] = 1.38
7 To 1.71 volumes of 0.1 M lactic acid add 1.0
  volume of 0.1 N NaOH
8 pH = 3.26

Chapter 2

1 pH 2.34
2 1.12×10^{-6} M
3

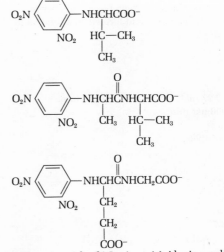

4

5 (a) Aspartic acid (b) Lysine (c) Alanine, valine,
  and threonine
6 Alanine (a) +; (b) O; (c) −
  Glutamic acid (a) +; (b) −; (c) −
  Lysine (a) +; (b) +; (c) −

Chapter 3

1 (a) Lys; Asp-Gly-Ala-Ala-Glu-Ser-Gly
  (b) Ala-Ala-His-Arg; Glu-Lys; Phe-Ile-Gly-Glu-
  Gly-Glu
  (c) Tyr-Cys-Lys; Ala-Arg; Arg; Gly
  (d) Phe-Ala-Glu-Ser-Ala-Gly-Lys

  (a) DNP-Lys; DNP-Asp
  (b) DNP-Ala; DNP-Glu; DNP-Phe
  (c) DNP-Tyr; DNP-Ala; DNP-Arg; DNP-Gly
  (d) DNP-Phe
2 (a) Val-Ala-Lys; Glu-Glu-Phe-Val-Met-Tyr-Cys-
  Glu-Trp-Met-Gly-Gly-Phe-Arg; Phe-Trp-Val-
  Lys; Ala-Gly-Ser-Phe-Gly
  (b) Val-Ala-Lys-Glu-Glu-Phe; Val-Met-Tyr; Cys-
  Glu-Trp; Met-Gly-Gly-Phe; Arg-Phe; Trp; Val-
  Lys-Ala-Gly-Ser-Phe; Gly
3

	pH 1.0	pH 6.5	pH 11.1
(a)	C	C	A
(b)	C	O	A
(c)	C	A	A
(d)	C	A	A
(e)	C	C	A

4 (a) Toward anode
  (b) At pH 5.0 toward cathode; at pH 7.0, toward
  anode

(c) At pH 5.0, toward cathode; at pH 9.5, stationary; at pH 11, toward anode

5   When an oligomeric protein is dissociated into its separate polypeptide chains by heating or treatment with urea, the amino-terminal residue of each chain will react with a molecule of 2,4-dinitrofluorobenzene. Therefore, the number of polypeptide chains is given by the number of moles of DNP-amino acid formed from one mole of oligomeric protein.

Chapter 4

1   $K_M = 2.27$ mM; $V_{max} = 0.51$ mg per min.
2   336,000 per min.
3   1.67 mM
4   Competitively

Chapter 5

1   8
2   (a) 65.5°
    (b) 52.7°
3   (a)   (b)

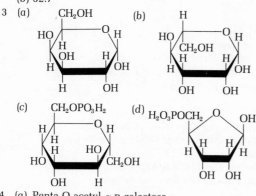

    (c)   (d)

4   (a) Penta-O-acetyl-α-D-galactose

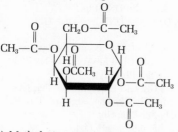

    (b) Methyl-2,3,4,6-tetra-O-methyl-α-L-mannopyranoside

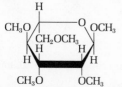

(c) D-Galactitol

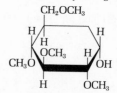

(d) 2,3,4,6-Tetra-O-methyl-α-D-glucopyranose

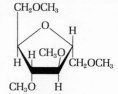

and 1,3,4,6-tetra-O-methyl-β-D-fructofuranose (this isomerized rapidly with the α-D form)

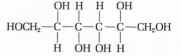

Chapter 6

1   PPP, OOO, SSS, PPO, PPS, OOP, OOS, SSP, SSO, POP, PSP, OPS, OSO, SPS, SOS, OPP, SPP, OSS, PSS, POO, SOO, OPO, POS, PSO, SPO, OSP, SOP. In those forms with different fatty acids in the 1 and 3 positions, the 2-carbon atom is asymmetric; two stereoisomers of such forms are then possible.

2   (a) Sodium stearate, 2 molecules of sodium palmitate, and glycerol
    (b) Sodium palmitate, sodium oleate, glycerol 3-phosphorylcholine
    (c) Sodium palmitate, sodium oleate, glycerol 3-phosphate, choline

3   (a) O
    (b) O
    (c) A

4   Phospholipase B: Palmitic acid, 2-linoleylglycerol 3-phosphorylcholine
    Phospholipase D: 1-palmitoyl 2-linoleylphosphatidic acid; choline

Chapter 7

1   (a)

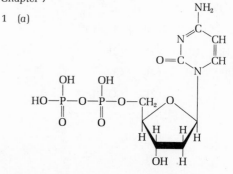

(b)

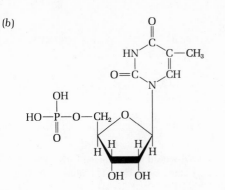

(e)

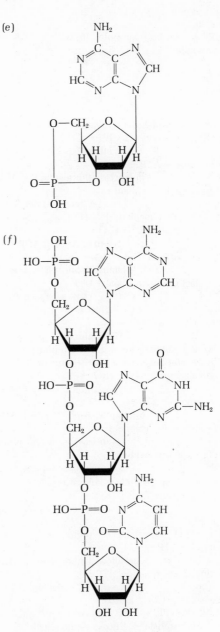

(f)

(c)

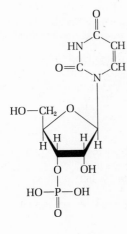

(d)

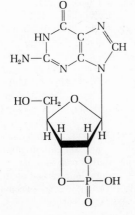

2 (a) Guanine and ribose 5-phosphate
   (b) U; C; (5′)GU(3′)
   (c) Phosphoric acid and cytidine
   (d) Cytidine, CMP, 2 GMP, 2 UMP, AMP
3 (a) 2dG5′p; 3dA5′p; 2dT5′p
   (b) 2dG3′p; 3dA3′p; dT3′p, dT; P_i

Chapter 10

1 (a) $K' = 263$ (b) $K' = 1.33 \times 10^5$ (c) $K' = 3.54$
2 (a) $\Delta G^{\circ\prime} = -1.14$ kcal per mole   (b) $\Delta G^{\circ\prime} = +11.1$ kcal per mole   (c) $\Delta G^{\circ\prime} = +16.5$ kcal per mole

423

3 $K' = 8.96$
4 (a) $\Delta G^{\circ\prime} = -3.0$ kcal per mole   (b) $\Delta G^{\circ\prime} = -3.5$
   kcal per mole

Chapter 11

1 (a) D-Mannose + $2P_i$ + $2NAD^+$ →
                2 phosphoenolpyruvate + 2NADH + $2H^+$
   (b) D-Galactose + $2P_i$ + 2ADP →
                            2 ethanol + $2CO_2$ + 2ATP
2 (a) D-Fructose + ATP →
                    D-fructose 6-phosphate + ADP
      D-fructose 6-phosphate →
                            D-glucose 6-phosphate
      D-Glucose 6-phosphate + H_2O → D-glucose + P_i
      Sum: D-fructose + ATP + H_2O →
                            D-glucose + ADP + P_i
   (b) D-Galactose + ATP →
                    D-galactose 1-phosphate + ADP
      D-Galactose 1-phosphate + UTP →
                            UDP-galactose + PP_i
      UDP-Galactose → UDP-glucose
      UDP-Glucose + PP_i →
                            UTP + glucose 1-phosphate
      Glucose 1-phosphate → glucose 6-phosphate
      Glucose 6-phosphate + H_2O → glucose + P_i
      Sum: D-galactose + ATP + H_2O →
                            D-glucose + ADP + P_i
   (c) D-mannose + ATP →
                    D-mannose 6-phosphate + ADP
      D-mannose 6-phosphate →
                            D-fructose 6-phosphate
      D-fructose 6-phosphate →
                            D-glucose 6-phosphate
      D-glucose 6-phosphate + H_2O → D-glucose + P_i
      Sum: D-mannose + ATP + H_2O →
                            D-glucose + ADP + P_i
3 85.7 percent
4 Lactose + H_2O → D-galactose + D-glucose
   D-galactose + ATP →
                    D-galactose 1-phosphate + ADP
   D-galactose 1-phosphate + UTP →
                            UDP-galactose + PP_i
   UDP-galactose → UDP-glucose
   UDP-glucose + PP_i → UTP + glucose 1-phosphate
   Glucose 1-phosphate → glucose 6-phosphate
   D-glucose + ATP → D-glucose 6-phosphate + ADP
   2 D-glucose 6-phosphate → 2 fructose 6-phosphate
   2 D-fructose 6-phosphate + 2ATP →
                    D-fructose 1,6-diphosphate + 2ADP
   2 D-fructose 1,6-diphosphate →
                        2 dihydroxyacetone phosphate +
                        2 glyceraldehyde 3-phosphate
   2 Dihydroxyacetone phosphate →
                        2 glyceraldehyde 3-phosphate
   Sum: Lactose + 4ATP + H_2O → 4 D-glyceraldehyde
   3-phosphate + 4ADP

Chapter 12

1 (a)   (b)

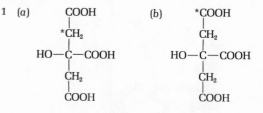

   (c) Citric acid would be unlabeled   (d) Same as
   in 1(a)   (e) Same as in 1(a)   (f) Same as in 1(b)
2 (a) Citrate + GDP + P_i + $2NAD^+$ →
      Succinate + $2CO_2$ + GTP + $2NADH^+$ + $2H^+$ + H_2O
   (b) Pyruvate + fumarate + $4NAD^+$ + P_i + GDP →
            succinate + $3CO_2$ + GTP + 4NADH + $4H^+$ + H_2O
3 Fructose 6-phosphate + $38P_i$ + 39ADP + $6O_2$ →
                            $6CO_2$ + 39ATP + $38H_2O$
4 2 Acetate + NAD^+ + 2ATP + $2H_2O$ →
            succinate + NADH + H^+ + 2AMP + $2PP_i$
   The carboxyl groups of succinate will contain the
   ^{14}C.
5 Glucose + ATP + $2NADP^+$ + H_2O → D-ribose
            5-phosphate + ADP + 2NADPH + $2H^+$ + CO_2

Chapter 13

1 (a) -26.8 kcal per mole   (b) -25.8 kcal per mole
   (c) -12.0 kcal per mole
2 (a) To right   (b) To left
3 39.6 percent
4 (a) 37ATP   (b) 12ATP   (c) 19ATP   (d) 72ATP

Chapter 14

1 (a) Stearic acid + 9CoA + $8O_2$ + $38P_i$ + 40ADP →
                        9 acetyl CoA + 39ATP + AMP
   (b) Myristoyl CoA + 6CoA + $6O_2$ + $30P_i$ +
      30 ADP → 7 acetyl CoA + 30ATP
   (c) D-β-hydroxybutyric acid + 2CoA + $\frac{1}{2}O_2$ +
      3ADP + P_i → 2 acetyl CoA + CoA + 2ATP + AMP
2 (a) Lauric acid + $17O_2$ + $95P_i$ + 97ADP →
                        96ATP + AMP + $12CO_2$
   (b) Butyric acid + $5O_2$ + $27P_i$ + 29ADP →
                        28ATP + $4CO_2$ + AMP
   (c) Acetoacetic acid + $4O_2$ + $22P_i$ + 24ADP →
                        23ATP + AMP + $4CO_2$
3 Label at C_9 (a) Carboxyl carbon atom
                    (b) 1-carboxyl carbon atom
   Label at C_{16} (a) Methyl carbon atom
                    (b) Methylene carbon atom
4 Glucose + $14P_i$ + 14ADP + $2O_2$ → acetoacetic
   acid + 14ATP + $2CO_2$ + $14H_2O$

Chapter 15

1 Phenylalanine + NADPH + H$^+$ + O$_2$ →

$\qquad$ tyrosine + NADP$^+$

Tyrosine + α-ketoglutarate →

$\qquad$ glutamate + p-hydroxyphenylpyruvate

p-Hydroxyphenylpyruvate + O$_2$ →

$\qquad$ homogentisic acid + CO$_2$

Homogentisic acid + O$_2$ → 4-maleylacetoacetate

4-Maleylacetoacetate → 4-fumarylacetoacetate

4-Fumarylacetoacetate + H$_2$O →

$\qquad$ fumarate + acetoacetate

Acetoacetate + ATP + H$_2$O + CoA →

$\qquad$ acetoacetyl CoA + AMP + PP$_i$

Acetoacetyl CoA + CoA → 2 acetyl CoA

24ATP from oxidation of 2 acetyl CoA plus 3ATP
from the oxidation of fumarate to oxaloacetate.

Sum: phenylalanine + α-ketoglutarate +
3O$_2$ + 2H$_2$O + NADPH + H$^+$ + ATP + 2CoA →
fumarate + glutamate + 2 acetyl CoA +
CO$_2$ + NADP$^+$ + AMP + PP$_i$

2 Valine + α-ketoglutarate →

$\qquad$ α-ketoisovalerate + glutamate

Glutamate + NAD$^+$ + H$_2$O →

$\qquad$ α-ketoglutarate + NH$_3$ + NADH + H$^+$

CO$_2$ + NH$_3$ + 2ATP + H$_2$O →

$\qquad$ carbamyl phosphate + P$_i$ + 2ADP

Carbamyl phosphate + ornithine → citrulline + P$_i$

Glutamate + oxaloacetate →

$\qquad$ α-ketoglutarate + aspartate

Citrulline + aspartate + ATP →

$\qquad$ argininosuccinate + AMP + 2P$_i$

Argininosuccinate → fumarate + arginine

Arginine + H$_2$O → urea + ornithine

Sum: valine + oxaloacetate + NAD$^+$ + CO$_2$ +
3ATP + glutamate + 3H$_2$O → α-ketoisovalerate +
α-ketoglutarate + fumarate + urea + NADH +
H$^+$ + 2ADP + AMP + 4P$_i$

3 (a) Both N atoms   (b) Both carboxyl carbon atoms

(c) H$_2$^{15}N—C—NHCH$_2$CH$_2$CH$_2$CHNH$_2$COOH

$\qquad \qquad \|$

$\qquad \quad ^{15}$NH

(d) H$_2$^{15}N—C—NHCH$_2$CH$_2$CH$_2$CHNH$_2$COOH

$\qquad \qquad \|$

$\qquad \qquad$ O

(e) Ornithine not labeled   (f) ^{15}N in amino
group, ^{14}C in both carboxyl groups.

Chapter 17

1 Carbon atoms 3 and 4

2 Carbon atoms 2 and 5

3 2 Alanine + 2 α-ketoglutarate →

$\qquad$ 2 pyruvate + 2 glutamate

2 Pyruvate + 2CO$_2$ + 2ATP + 2H$_2$O →

$\qquad$ 2 oxaloacetate + 2ADP + 2P$_i$

2 Oxaloacetate + 2NADH + 2H$^+$ →

$\qquad$ 2 malate + 2NAD$^+$

2 Malate + 2NAD$^+$ →

$\qquad$ 2 oxaloacetate + 2NADH + 2H$^+$

2 Oxaloacetate + 2GTP →

$\qquad$ 2 phosphoenolpyruvate + 2CO$_2$ + 2GDP

2 Phosphenolpyruvate + 2H$_2$O →

$\qquad$ 2 2-phosphoglycerate

2 2-Phosphoglycerate → 2 3-phosphoglycerate

2 3-Phosphoglycerate + 2ATP →

$\qquad$ 2 1,3-diphosphoglycerate + 2ADP

2 1,3-Diphosphoglycerate + 2NADH + 2H$^+$ +
H$_2$O → 2 glyceraldehyde 3-phosphate +

Glyceraldehyde 3-phosphate →

$\qquad$ dihydroxyacetone phosphate

Glyceraldehyde 3-phosphate + dihydroxyacetone
phosphate → fructose 1,6-diphosphate

Fructose 1,6-diphosphate + H$_2$O →

$\qquad$ fructose 6-phosphate + P$_i$

Fructose 6-phosphate → glucose 6-phosphate

Glucose 6-phosphate + H$_2$O → glucose + P$_i$

Sum: 2 Alanine + 2 α-ketoglutarate + 4ATP +
2GTP + 2NADH + 6H$_2$O + 2H$^+$ → glucose +
2 glutamate + 4ADP + 2GDP + 6P$_i$ + 2NAD$^+$

4 Six

5 D-Glucose + ATP + UTP + (glucose)$_n$ + H$_2$O →

$\qquad$ (glucose)$_{n+1}$ + ADP + UDP + PP$_i$

Chapter 18

1 9 Acetyl CoA + 16NADPH + 16H$^+$ + 8ATP +
8H$_2$O → stearic acid + 9CoA + 16NADP$^+$ +
8ADP + 8P$_i$

2 Formation of pyruvate from glucose

12 Glucose + 24P$_i$ + 24ADP + 24NAD$^+$ →
24 pyruvate + 24NADH + 24H$^+$ + 24ATP + 24H$_2$O

Conversion of pyruvate to acetyl CoA

24 pyruvate + 24CoA + 24NAD$^+$ → 24 acetyl
CoA + 24CO$_2$ + 24NADH + 24H$^+$

Formation of glycerol 3-phosphate from glucose

½ Glucose + ATP + NADH + H$^+$ →

$\qquad$ glycerol 3-phosphate + ADP + NAD$^+$

Synthesis of 3 molecules of palmitic acid

24 Acetyl CoA + 42NADPH + 42H$^+$ + 21ATP +
21H$_2$O → 3 palmitic acid + 24CoA + 42NADP$^+$ +
21ADP + 21P$_i$

Synthesis of tripalmitin

3 Palmitic acid + 3ATP + glycerol 3-phosphate +
7H$_2$O → tripalmitin + 3AMP + 7P$_i$

Sum: 12½ glucose + 2ADP + ATP + 47NAD$^+$ +
42NADPH + 4H$_2$O → tripalmitin + 24CO$_2$ +
3AMP + 4P$_i$ + 47NADH + 42NADP$^+$ + 5H$^+$

3 Lactic acid (2 carbons), D-fructose (4 carbons),
succinate (2 carbons), L-glutamate (2 carbons)

4 Oleic acid + ATP + CoA + H$_2$O →

$\qquad$ oleyl CoA + AMP + PP$_i$

Palmitic acid + ATP + CoA + H$_2$O →

$\qquad$ palmitoyl CoA + AMP + PP$_i$

2PP$_i$ + 2H$_2$O → 4P$_i$

Dihydroxyacetone phosphate + NADH + H$^+$ →
glycerol 3-phosphate + NAD$^+$

Oleyl CoA + palmitoyl CoA + glycerol
3-phosphate → phosphatidic acid + 2CoA

Phosphatidic acid + CTP + H$_2$O →
cytidine diphosphate diacylglycerol + PP$_i$

PP$_i$ + H$_2$O → 2P$_i$

Cytidine diphosphate diacylglycerol + serine →
CMP + phosphatidylserine

Phosphatidylserine →
phosphatidylethanolamine + CO$_2$

Sum: Oleic acid + palmitic acid + dihydroxyacetone phosphate + serine + 2ATP + CTP +
NADH + H$^+$ + 6H$_2$O → phosphatidylethanolamine + CO$_2$ + 2AMP + CMP + NAD$^+$ + 6P$_i$

5   From 2-^{14}-C-D-glucose carbon atoms 1, 3, 5, 7, 9,
11, 13, and 15 of palmitic acid will be labeled.
From β-^{14}C-L-alanine, carbon atoms 2, 4, 6, 8, 10,
12, 14, 16 will be labeled.

Chapter 19

1   ½ D-glucose + 2NAD$^+$ + glutamate + tetrahydrofolate → glycine + α-ketoglutarate + N^5,N^{10}-
methylenetetrahydrofolate + 2NADH + 2H$^+$

2   None

3   The α-carbon atom

4   α-Ketoglutarate + NH$_3$ + NADPH + 2NADH +
3H$^+$ → proline + NADP$^+$ + 2NAD$^+$ + H$_2$O
^{14}C label on α-carbon

Chapter 21

1   (3')TACGGCATACGTAAG(5')

2   21 percent adenine, 21 percent thymine, 29
percent guanine, and 29 percent cytosine

3   1.1×10^{-3} g

4   12 g; 2.32×10^9 miles

5   (a) 372   (b) 423   (c) 735

6   (a) 1265 Å   (b) 1438 Å   (c) 2499 Å

Chapter 22

1   (3')TCGAACGTTGCAACGTAATC(5')

2   5dTTP + 6dATP + 5dCTP + 4dGTP + 40H$_2$O →
DNA segment + 40P$_i$

3   (3')UCGAACGUUGCAACGUAAUC(5')

4   6ATP + 5UTP + 4GTP + 5CTP + 40H$_2$O →
mRNA + 40P$_i$

5   7000 nucleotide residues per sec per cell

Chapter 23

1   (a) (NH$_2$) Gly-Gln-Ser-Leu-Leu-Ile   (b) (NH$_2$) Leu-
Asp-Ala-Pro-Tyr-Phe-Ala   (c) (NH$_2$) His-Asp-Ala-
Cys-Cys-Tyr   (d) (NH$_2$) Met-Asp-Glu

2   (a) (3')AAGCAGCTGCTACTAGTAGCCGAAGAG
CTC(5')   (b) (3')AAGCAGCUGCUACUAGUAG
CCGAAGAGCUC(5')   (c) (NH$_2$) Leu-Glu-Lys-Pro-
Met-Ile-Ile-Val-Asp-Glu

3   Many answers are possible. One is
(5')GGUAAUGUUGAACAAUGUUGUGCUAGU
GUUUGUAGUUUAUAUCAACUUGAAAAUUAU
UGUAAU(3')

4   (3')AACCTACGCGGTATAAAACGA(5')

5   At least 372

LIST OF ABBREVIATIONS

ACTH	Adrenocorticotrophic hormone	G6P	Glucose 6-phosphate
AMP, ADP, ATP	Adenosine 5'-mono, di- and triphosphate	GSH, GSSG	Glutathione and its oxidized form
dAMP, dGMP, dADP, etc.	2'-Deoxyadenosine 5'-monophosphate, 2'-deoxyguanosine 5'-monophosphate, 2'-deoxyadenosine 5'-diphosphate, etc.	Hb, HbO_2, HbCO	Hemoglobin, oxyhemoglobin, carbon monoxide hemoglobin
		LDH	Lactate dehydrogenase
		MDH	Malate dehydrogenase
		Mb, MbO_2	Myoglobin; oxymyoglobin
ATPase	Adenosine triphosphatase	NAD^+, NADH, DPN^+, DPNH	Nicotinamide adenine dinucleotide (diphosphopyridine nucleotide) and its reduced form
CMP, CDP, CTP	Cytidine nucleotides		
CM-cellulose	Carboxymethyl cellulose	$NADP^+$, NADPH TPN^+, TPNH	Nicotinamide adenine dinucleotide phosphate (triphosphopyridine nucleotide) and its reduced form
CoASH, acyl-CoA, acyl-S-CoA	Coenzyme A and its acyl derivatives		
CoQ	Coenzyme Q; ubiquinone	P_i	Inorganic orthophosphate
DEAE-cellulose	Diethylaminoethyl cellulose	PAB or PABA	p-Aminobenzoic acid
DFP	Diisopropyl phosphofluoridate	PEP	Phosphoenolpyruvate
DNA	Deoxyribonucleic acid	3PG	3-Phosphoglycerate
DNase	Deoxyribonuclease	PGA	Pteroylglutamic acid (folic acid)
DNP	2,4-Dinitrophenol	PP_i	Inorganic pyrophosphate
EDTA	Ethylenediaminetetraacetic acid	PRPP	5-phosphoribosyl 1-pyrophosphate
Fd	Ferredoxin	RNA	Ribonucleic acid
FA	Fatty acid	mRNA	Messenger RNA
FAD, $FADH_2$	Flavin adenine dinucleotide and its reduced form	rRNA	Ribosomal RNA
		tRNA	Transfer RNA
FDNB (DNFB)	Fluorodinitrobezene	RNase	Ribonuclease
FDP	Fructose 1,6-diphosphate	TMP, TDP, TTP	Thymidine nucleotides
FH_2, FH_4	Dihydro- and tetrahydrofolic acid	TMV	Tobacco mosaic virus
FMN, $FMNH_2$	Flavin mononucleotide and its reduced form	TPP	Thiamine pyrophosphate
		UDP-gal	Uridine diphosphate galactose
GDH	Glutamate dehydrogenase	UDP-glucose	Uridine diphosphate glucose
GMP, GDP, GTP	Guanosine nucleotides	UMP, UDP, UTP	Uridine nucleotides
G3P	Glyceraldehyde 3-phosphate	UV	Ultraviolet

GLOSSARY

Absolute configuration: The specific configuration in space of four substituent groups around an asymmetric carbon atom.

Absorption: Transport of the products of digestion from the small intestine into the blood.

Acceptor control: The regulation of the rate of respiration by the availability of ADP as phosphate acceptor.

Actin: Muscle protein; component of thin myofilaments

Activation energy: Amount of energy in kcal required to bring all the molecules in one mole of a reacting substance to the transition state.

Active site: Region of enzyme surface which binds the substrate molecule and transforms it.

Active transport: Energy-requiring transport of a solute across a membrane in the direction of increasing concentration.

Activity: The true thermodynamic potential or activity of a substance, as distinguished from its molar concentration.

Activity coefficient: The factor by which the concentration of a solute must be multiplied to give its true thermodynamic activity.

Actomyosin: A molecular complex of actin and myosin; the basic contractile element in muscle.

ADP (adenosine diphosphate): A ribonucleoside 5′-diphosphate serving as phosphate group acceptor in the cell energy cycle.

Adipose tissue: Specialized connective tissue which functions to store large amounts of neutral fat.

Aerobes: Organisms that live in and utilize oxygen.

Aldose: A simple sugar in which the carbonyl carbon atom is at the end of the carbon chain.

Alkaloids: Nitrogen-containing organic compounds of plant origin, often basic, which have intense biological activity.

Allosteric enzymes: Enzymes whose catalytic activity is modulated by the binding of a specific metabolite at a site other than the catalytic site (also called *regulatory enzymes*).

Allosteric site: The specific "other" site on the surface of an allosteric enzyme molecule to which the effector or modulator molecule is bound.

Amino acid activation: The preparation of an amino acid for protein synthesis by ATP-dependent enzymatic esterification of its carboxyl group with a corresponding transfer RNA molecule.

Aminoacyl synthetase: An enzyme that catalyzes amino acid activation.

Aminotransferases: Enzymes that catalyze transfer of amino groups from one metabolite to another; also called *transaminases.*

Amphibolic pathway: A metabolic pathway used in both catabolism and anabolism.

Amphipathic compound: A compound whose molecule contains both polar and nonpolar zones.

Amphoteric compound: A compound capable of either donating or accepting protons, thus able to serve as either an acid or a base.

Anabolism: That phase of intermediary metabolism concerned with the energy-requiring biosynthesis of cell components from smaller precursor molecules.

Anaerobes: Organisms that can live without oxygen.

Anaplerotic reaction: Enzyme-catalyzed reaction that replenishes the supply of intermediates of the tricarboxylic acid cycle.

Angstrom (Å): A unit of length 10^{-8} cm) used to indicate molecular dimensions.

Anomers: Two stereoisomers of a given sugar differing only in the configuration about the carbonyl (anomeric) carbon atom.

Antibiotic: One of many different organic compounds formed and secreted by various species of microorganisms and plants that are toxic to other species and presumably function in a defensive role.

Anticodon: A specific sequence of three nucleotides in a transfer RNA molecule complementary to a codon in a messenger RNA molecule.

Asymmetric carbon atom: A carbon atom covalently bonded to four different groups, which may occupy two different tetrahedral configurations.

ATP (adenosine triphosphate): A ribonucleoside 5′-triphosphate functioning as phosphate-group donor in the cell energy cycle.

ATPase: An enzyme hydrolyzing ATP to yield ADP and phosphate, usually coupled to some process requiring energy.

ATP synthetase: An enzyme complex in the inner mitochondrial membrane with the function of forming ATP from ADP and phosphate during oxidative phosphorylation.

Autotrophs: Organisms that can build their own macromolecules from very simple nutrient molecules, such as carbon dioxide and ammonia.

Auxotrophic mutant: A mutant of a microorganism defective in the synthesis of a given biomolecule, which must thus be supplied for normal growth.

Avogadro's number: The number of molecules in a gram molecular weight of any compound (6.023×10^{23}).

Bacteriophage: A virus capable of replicating in a bacterial cell.

Bilayer: A double layer of oriented phospholipid molecules, in which the hydrocarbon tails face inward to form a continuous nonpolar phase.

Bond energy: The energy required to break a bond; it is not to be confused with the term "phosphate bond energy."

Buffer: A system capable of resisting changes in pH, consisting of a conjugate acid-base pair in which the ratio of proton donor to proton acceptor is near unity.

Building-block molecule: A molecule serving as a structural unit of a biological macromolecule, such as an amino acid, a sugar, or a fatty acid.

Calorie: A measure of energy; the amount of heat required to raise the temperature of 1.0 g of water from 14.5° to 15.5°C.

Catabolism: That phase of metabolism involved in energy-yielding degradation of nutrient molecules.

Catalytic site: That site on an enzyme molecule involved in the catalytic process.

Central dogma: The principle that genetic information flows from DNA to RNA to protein.

Chlorophyll: Green pigments involved in photosynthesis, consisting of magnesium-porphyrin complexes.

Chloroplasts: Chlorophyll-containing membrane-surrounded organelles in the cytoplasm of eukaryotic photosynthetic cells: they are the sites of conversion of light energy into chemical energy.

Chromatography: Process by which complex mixtures of molecules may be separated by many repeated partitioning processes between a flowing phase and a stationary phase.

Chromosome: A single large double-helical DNA molecule containing many genes and functioning to store and transmit genetic information.

Chylomicron: Large neutral lipid droplets in blood, stabilized by small amounts of protein and phospholipid.

Codon: A sequence of three adjacent nucleotides in a nucleic acid that codes for a specific amino acid.

Coenzyme: An organic cofactor required for the action of certain enzymes, often containing a vitamin as a building block.

Coenzyme A: Pantothenic acid-containing coenzyme serving as acyl group carrier in certain enzymatic reactions.

Cofactor: A small-molecular weight inorganic or organic substance required for the action of an enzyme.

Colligative properties: A group of properties of solutions depending on the number of solute particles per unit volume.

Common intermediate: A chemical compound that is common to two chemical reactions, as either a reactant or a product.

Competitive inhibition: Type of enzyme inhibition reversed by increasing the substrate concentration.

β-Configuration: An extended, zig-zag configuration of a polypeptide chain.

Conformation: The three-dimensional shape or form of a macromolecule.

Conjugate acid-base pair: A proton donor and its deprotonated species; an example is acetic acid–acetate ion.

Conjugated protein: A protein containing a metal or an organic prosthetic group in addition to a polypeptide chain.

Coupled reactions: Two chemical reactions which have a common intermediate and thus a means by which energy can be transferred from one to the other.

Cyclic AMP: Important "second messenger" within cells, whose formation by adenyl cyclase is stimulated by certain hormones.

Cyclic electron flow: Light-induced flow of electrons in green plant cells originating from and returning to the same chlorophyll molecule(s).

Cytochromes: Heme proteins serving as electron carriers in electron transport chains involved in respiration and photosynthesis.

Cytosol: The continuous aqueous phase of cytoplasm, with its dissolved solutes.

Dalton: The weight of a single hydrogen atom (1.67×10^{-24} g).

Dark reactions: The light-independent enzymatic reactions in photosynthetic cells concerned in synthesis of glucose from CO_2, ATP, and NADPH.

Deamination: The enzymatic removal of amino groups from amino acids.

Dehydrogenases: Enzymes catalyzing removal of pairs of hydrogen atoms from specific substrates.

Denaturation: Partial or complete unfolding of the specific native configuration of the polypeptide chain(s) of proteins.

Deoxyribonucleotides: Nucleotides containing 2-deoxy-D-ribose as pentose component.

Dextrorotatory isomer: Stereoisomer rotating the plane of plane-polarized light to the right.

Diabetes mellitus: Metabolic disease due to deficiency of the hormone insulin; it is characterized by failure of glucose to be transported from the blood into cells at normal glucose concentrations.

Differential centrifugation: Separation of cell organelles by their different rates of sedimentation in a centrifugal field.

Diffusion: The tendency for molecules to move in the direction of a lesser concentration, thus making the concentration uniform throughout the system.

Digestion: Enzymatic hydrolysis of major nutrients in stomach and small intestine to yield their building-block components.

Dipole: A molecule having both positive and negative charges.

Diprotic acid: Acid capable of dissociating two protons.

Disaccharides: Carbohydrate consisting of two monosaccharide units joined by a glycosidic linkage.

Dissociation constant: An equilibrium constant for the dissociation of a compound into its components, such as the dissociation of an acid to yield a proton and its anion.

Disulphide bridge: Covalent cross-link between two polypeptide chains formed by a cystine molecule.

DNA (deoxyribonucleic acid): Polynucleotide having specific sequence of deoxyribonucleotide units and serving as the carrier of genetic information in chromosomes.

DNA polymerase: An enzyme catalyzing synthesis of DNA from its deoxyribonucleoside 5'-triphosphate precursors.

Double helix: The natural coiled configuration of two antiparallel complementary DNA chains.

Effector (modulator): A metabolite which, when bound to the allosteric site of a regulatory enzyme, alters the maximum velocity or K_M of the enzyme.

Electron acceptor: Substance acting to receive electrons in an oxidation-reduction reaction.

Electron carriers: Specialized proteins such as flavoproteins and cytochromes which can gain and lose electrons reversibly and function to transfer electrons from organic nutrients to oxygen.

Electron donor: Donor of electrons in an oxido-reduction reaction.

Electron transport: The movement of electrons from substrates to oxygen catalyzed by the respiratory chain.

Electrophoresis: Transport of charged solutes in response to an electrical field, often used to separate mixtures of ions.

Eluate: The effluent from a chromatographic column.

Enantiomers: Isomers which are mirror-images of each other.

Endergonic reaction: A chemical reaction with a positive standard free energy change; an "uphill" reaction.

Endonuclease: An enzyme capable of hydrolyzing interior internucleotide bonds of a nucleic acid, at points other than the terminal bonds.

Endoplasmic reticulum: An extensive system of double membranes in the cytoplasm of eukaryotic cells; it is often coated with ribosomes.

End-product (feedback) inhibition: Inhibition of the first (regulatory) enzyme of a multienzyme sequence by the end-product of the sequence, which serves as an allosteric modulator.

Entropy: The randomness or disorder of a system.

Epimers: Two stereoisomers of a compound having two or more asymmetric carbon atoms which differ only in the configuration about a single carbon atom.

Equilibrium: The state of a system in which no further change is occurring and in which its free energy is at a minimum.

Equilibrium constant: A constant, characteristic for each chemical reaction, designating the specific concentrations or activities of all components of the system at equilibrium at a given temperature.

Escherichia coli: A common aerobic bacterium found in the small intestine of vertebrates.

Essential fatty acids: Group of unsaturated fatty acids of plants required in the diet of mammals.

Eukaryotic cells: Major class of cells having nuclear membranes, membrane-surrounded organelles, and multiple chromosomes; they divide by mitosis.

Excited state: An energy-rich state of an atom or molecule existing after an electron has been moved from its normal stable orbital to an outer orbital having a higher energy level, as a result of the absorption of light energy.

Exergonic reaction: A chemical reaction with a negative standard free energy change; a "downhill" reaction.

Exonuclease: An enzyme hydrolyzing only terminal nucleotides from a nucleic acid.

Facultative cells: Cells that can live either in the presence or absence of oxygen.

FAD (See flavin adenine dinucleotide).

Feedback inhibition (See end-product inhibition).

Fermentation: Energy-yielding anaerobic breakdown of fuel molecules such as glucose.

Fibrous proteins: Insoluble structural proteins in which the polypeptide chain is extended or coiled along one dimension.

First law of thermodynamics: In all processes the total energy of the universe remains constant.

Flavin adenine dinucleotide (FAD): Coenzyme of certain oxido-reduction enzymes; it contains riboflavin.

Flavin-linked dehydrogenases: Dehydrogenases requiring one of the riboflavin coenzymes FMN or FAD.

Flavin mononucleotide (FMN): Coenzyme of certain oxido-reduction enzymes; it contains riboflavin.

Flavoprotein: An enzyme containing a flavin nucleotide as prosthetic group.

Free energy: That component of the total energy of a system which can do work under conditions of constant temperature and pressure.

Furanose: A sugar containing the 5-membered furan ring.

Gene: A mutable segment of a chromosome coding for a single polypeptide chain or RNA molecule.

Genetic code: The set of triplet code words used in DNA to specify the various amino acids of proteins.

Genetic information: The hereditary information contained in a sequence of nucleotide bases in chromosomal DNA or RNA.

Genetic map: A diagram showing the sequence of specific genes along the chromosomal DNA molecule.

Globular proteins: Proteins in which the polypeptide chain is folded in three dimensions to yield a globular shape.

Gluconeogenesis: The biosynthesis of new carbohydrate from noncarbohydrate precursors.

Glycolysis: That form of fermentation in which glucose is broken down anaerobically into two molecules of lactic acid.

Glyoxylate cycle: A variation of the tricarboxylic acid cycle which makes possible net conversion of acetate into succinate, and eventually, new carbohydrate.

Glyoxysomes: Membranous vesicles containing certain enzymes of the glyoxylate cycle.

Golgi body: Complex membranous organelles of eukaryotic cells functioning in the formation of new membranes, particularly plasma membrane.

Half-life: The time required for disappearance or decay of one-half of a given component.

Heat of vaporization: The number of calories required to convert one gram of a liquid into the vapor state at the same temperature.

α-Helix: A coiled, helical configuration of the polypeptide chain with maximal intrachain hydrogen bonding; found in α-keratins.

Heme: The iron-porphyrin prosthetic group of heme proteins.

Heme protein: Protein containing a heme as prosthetic group.

Hemoglobin: A heme protein of the red blood cell functioning in O_2 transport and containing four polypeptide chains and four heme groups.

Henderson-Hasselbalch equation: Equation relating the pH, the pK', and the ratio of proton acceptor to proton donor species (pH = pK' + log [proton acceptor]/[proton donor]).

Heterotrophic cells: Cells that require complex nutrient molecules such as glucose, amino acids, etc., to yield energy and to provide building blocks for synthesis of macromolecules.

Hexose: A simple sugar having a linear backbone chain of six carbon atoms.

High-energy bond: A chemical bond that yields a large decrease in free energy upon hydrolysis under standard conditions.

High-energy phosphate compound: A phosphorylated compound yielding a large decrease in standard free energy on hydrolysis.

Hill reaction: Oxygen evolution and photoreduction of an artificial electron acceptor by a chloroplast preparation in the absence of carbon dioxide.

Homologous proteins: Proteins having identical functions in different species, such as the hemoglobins.

Hormone: A chemical substance synthesized in trace amounts in one organ acting as a messenger to modulate the functions of another tissue or organ.

Hydrogen bond: A weak electrostatic attraction between one electronegative atom and a hydrogen atom covalently linked to a second electronegative atom.

Hydrolysis: The cleavage of a molecule into two or more smaller molecules by reaction with water.

Hydronium ion: The hydrated hydrogen ion (H_3O^+).

Hydrophilic: "Water-loving"; refers to molecules or groups that associate with H_2O.

Hydrophobic: "Water-hating"; refers to molecules or groups that are only poorly soluble in water.

Hydrophobic interactions: The association of nonpolar groups with each other in aqueous systems because of the tendency of the surrounding water molecules to seek their most stable configuration.

Hyperchromic effect: The large increase in light absorption at 260 nm occurring as a double-helical DNA is melted.

Informational molecules: Those containing information in the form of specific sequences of different building blocks; they include proteins and nucleic acids.

Intermediary metabolism: The enzyme-catalyzed reactions in cells that extract chemical energy from nutrient molecules and utilize it to assemble macromolecules required in cell growth.

In vitro: (Latin: in glass); refers to experiments done on isolated cells, tissues, or cell-free extracts in (glass) reaction vessels.

In vivo: (Latin: in life); refers to experiments done on intact living organisms.

Ion-exchange resin: A polymeric resin containing fixed charged groups, used in chromatography to separate ionic compounds.

Ion product of water (K_w): The product of the H^+ and OH^- concentrations of water ($K_w = 1 \times 10^{-14}$ at 25°).

Irreversible process: A process in which the entropy of the universe increases.

Isoelectric pH: The pH at which a solute has no net electrical charge.

Isoprene: The hydrocarbon 2-methyl-butadiene-1,3, which serves as a recurring structural unit of the terpenes.

Isothermal process: A process occurring at constant temperature.

Isotopes: Forms of elements used as tracers; they differ in atomic weight and thus in mass; some are unstable and emit radioactivity.

Isozymes (isoenzymes): Multiple forms of an enzyme which differ in their substrate affinity or maximum activity.

Ketone bodies: Name applied to acetoacetic acid and D-β-hydroxy-butyric acid of the blood.

Ketose: A simple monosaccharide having its carbonyl group at other than a terminal position.

Law of mass action: The rate of any given chemical reaction is proportional to the product of the active masses (activities) of the reactants.

Levorotatory isomer: That isomer of an optically active compound rotating the plane of plane-polarized light to the left.

Light reactions: Those reactions of photosynthesis which require light and cannot occur in the dark.

Lineweaver-Burk equation: An algebraic transformation of the Michaelis-Menten equation.

Lipoic acid: A vitamin for some microorganisms. It serves as an intermediate carrier of hydrogen atoms and acyl groups in α-ketoacid dehydrogenases.

Lipoproteins: Conjugated proteins contain a lipid or group of lipids.

Lithosphere: The inorganic (mineral) portion of the earth's surface.

Low-energy phosphate compound: A phosphorylated compound yielding a relatively low standard free energy of hydrolysis.

Lysosome: A membrane-surrounded organelle in the cytoplasm of eukaryotic cells in which are segregated many hydrolytic enzymes.

Macromolecules: Molecules having molecular weights in the range of a few thousand to many millions.

Messenger RNA: Class of RNA molecules, complementary to one strand of cell DNA, which serve to carry the genetic message from chromosome to ribosomes.

Metabolic turnover: The constant, steady-state metabolic replacement of cell components.

Metabolism: The enzyme catalyzed transformations of organic nutrient molecules in living cells.

Metabolite: A chemical intermediate in the enzyme-catalyzed reactions of metabolism.

Metalloenzyme: An enzyme having a metal ion as its prosthetic group.

Micelle: An association of a number of amphipathic molecules in water into a structure in which their nonpolar portions are in the interior and the polar portions on the exterior, exposed to water.

Michaelis constant, K_M: The substrate concentration at which an enzyme shows one-half its maximum velocity.

Michaelis-Menten equation: An equation relating the velocity and the substrate concentration of an enzyme.

Microbodies: Cytoplasmic membrane-surrounded vesicles containing certain oxidative enzymes and catalase.

Microsomes: Vesicles derived from endoplasmic reticulum following disruption and differential centrifugation of eukaryotic cells.

Mitochondria: Membrane-surrounded organelles in the cytoplasm of aerobic cells which contain the enzyme systems required in the tricarboxylic acid cycle, electron transport, and oxidative phosphorylation.

Mitosis: Replication of chromosomes in eukaryotic cells.

Mixed function oxidases: Enzymes catalyzing simultaneous oxidation of two substrates, one of which is often NADPH or NADH.

Molal solution: One mole of a solute dissolved in 1000 grams of water.

Molar solution: One mole dissolved in water and made up to 1000 ml.

Mole: One gram-molecular weight of a compound.

Monolayer: A single layer of oriented lipid molecules.

Monoprotic acid: An acid capable of losing only one proton.

Monosaccharide: Carbohydrate consisting of a single simple sugar unit.

Mucopolysaccharide: Acidic polysaccharides found in mucous secretions and in intercellular space of higher animals.

Mucoproteins: A conjugated protein containing an acid mucopolysaccharide.

Multienzyme system: Sequence of enzymes participating in a given metabolic pathway.

Mutagenic agent: A chemical agent capable of producing a genetic mutation.

Mutarotation: Change in specific rotation of a pyranose or furanose sugar or glycoside accompanying equilibration of its α- and β-forms.

Mutation: A heritable change in a chromosome.

Myofibrils: Unitary set of thick and thin filaments of muscle fibers.

Myosin: Muscle protein; component of the thick filaments of the contractile system.

NAD, NADP (*See* nicotinamide adenine dinucleotide and nicotinamide adenine dinucleotide phosphate).

Neurospora crassa: A common mold widely used in genetic analysis of metabolic pathways.

Neutral fats: Fatty acid esters of the three hydroxyl groups of glycerol; also called triacylglycerols.

Nicotinamide adenine dinucleotide (NAD) and nicotinamide adenine dinucleotide phosphate (NADP): Nicotinamide-containing coenzymes functioning as carriers of protons and electrons in certain enzymatic oxidation-reduction reactions.

Ninhydrin reaction: A color reaction given by amino acids and peptides on heating with ninhydrin; it is widely used for their detection and estimation.

Nitrogen cycle: Cycling of various forms of nitrogen through the plant, animal, and microbial worlds.

Noncompetitive inhibition: Type of enzyme inhibition not reversed by increasing the substrate concentration.

Noncyclic electron flow: Light-induced flow of electrons from water to NADP$^+$ in oxygen-evolving photosynthesis.

Nonheme-iron proteins: Electron-carrying proteins containing iron atoms but no porphyrin groups.

Nonpolar groups: Hydrophobic groups, usually hydrocarbon in nature.

Nonsense codons: Codons that do not specify amino acids; they indicate termination of polypeptide chains.

Nuclease: Enzyme capable of hydrolyzing internucleotide linkages of a nucleic acid.

Nucleic acids: Polynucleotides; long chains of mononucleotides linked by successive 3',5'-phosphodiester bonds.

Nucleolus: Round, granular structure found in nucleus of eukaryotic cells. Involved in rRNA synthesis and ribosome formation.

Nucleophilic group: An electron-rich group with a strong tendency to donate electrons to an electron-deficient nucleus.

Nucleoside: A compound consisting of a purine or pyrimidine base covalently linked to a pentose.

Nucleoside diphosphate sugar: A coenzyme like carrier of a sugar molecule functioning in enzymatic synthesis of polysaccharides and sugar derivatives.

Nucleoside diphosphokinase: An enzyme catalyzing transfer of the terminal phosphate of a nucleoside 5'-triphosphate to a nucleoside 5'-monophosphate.

Nucleotide: A nucleoside phosphorylated at one of its pentose hydroxyl groups.

Oligomeric protein: A protein having two or more polypeptide chains.

Oligosaccharide: Several monosaccharide groups joined by glycosidic bonds.

Open system: A system which exchanges matter and energy with its surroundings.

Optical activity: The capacity of some substances to rotate the plane of plane-polarized light.

Optimum pH: The pH at which an enzyme shows maximum catalytic activity.

Organelles: Membrane-surrounded structures found in eukaryotic cells; they contain enzymes for specialized cell functions.

Orthophosphate cleavage: The enzymatic cleavage of ATP to yield ADP and phosphate, usually coupled to an energy-requiring process or reaction.

Osmosis: Bulk flow of water through a semipermeable membrane into an aqueous phase containing a solute present in a higher concentration.

Osmotic pressure: Pressure generated by osmotic flow of water through a membrane into an aqueous phase containing a solute in higher concentration.

Oxidation: The loss of electrons from a compound.

β-Oxidation: Oxidative degradation of fatty acids into acetyl CoA by successive oxidations at the β-carbon atom.

Oxidation-reduction reaction: Reaction in which electrons are transferred from a donor to an acceptor molecule

Oxidative phosphorylation: The enzymatic phosphorylation of ADP to ATP coupled to electron transport from substrate to molecular oxygen.

Oxidizing agent (oxidant): The acceptor of electrons in an oxidation-reduction reaction.

Partition coefficient: A constant which expresses the ratio in which a given solute will be partitioned or distributed between two given immiscible liquids at equilibrium.

Pentose: A simple sugar whose backbone contains five carbon atoms.

Peptide: Two or more amino acids covalently joined by peptide bonds.

Peptide bond: A covalent bond between two amino acids in which the α-amino group of one is covalently bonded to the α-carboxyl group of the other through the removal of a molecule of water.

pH: The negative logarithm of the hydrogen ion concentration of an aqueous solution.

Phosphate bond energy: The decrease in free energy as one mole of a phosphorylated compound undergoes hydrolysis to equilibrium at pH 7.0 and 25°, in a $1.0M$ solution.

Phosphodiester: A molecule that contains two alcohols esterified to one molecule of phosphoric acid, which thus serves as a bridge between them.

Phosphogluconate pathway: An oxidative pathway beginning with glucose 6-phosphate and leading to formation of NADPH, pentoses, and other products via 6-phosphogluconate.

Phospholipids: Lipids containing one or more phosphate groups.

Photon: Ultimate unit of light energy.

Photoreduction: Light-induced reduction of an electron acceptor in photosynthetic cells.

Photosynthesis: The enzymatic conversion of light energy into chemical energy and use of the latter to form carbohydrates and oxygen from CO_2 and H_2O in green plant cells.

Photosynthetic phosphorylation (photophosphorylation): The enzymatic formation of ATP from ADP coupled to light-dependent transport of electrons from excited chlorophyll in photosynthetic organisms.

Photosystem: A functional set of light-absorbing pigments and its immediate electron acceptors in photosynthetic cells.

pK: The negative logarithm of an equilibrium constant.

Plasma proteins: The proteins present in blood plasma.

Pleated sheet: Side-by-side hydrogen-bonded arrangement of parallel polypeptide chains in the extended β-configuration.

Polar group: A hydrophilic or water-loving group.

Polynucleotide: A covalently linked sequence of nucleotides in which the 3′ position of the pentose of one nucleotide is linked through a phosphate group to the 5′ position of the pentose of the next.

Polypeptide: A long chain of amino acids linked by peptide bonds.

Polyribosome: A complex of a messenger RNA molecule and two or more ribosomes.

Polysaccharides: Linear or branched macromolecules composed of many monosaccharide units linked by glycosidic bonds.

Polysomes (See polyribosome).

Porphyrins: Complex ring-like nitrogenous compounds containing four pyrrole rings usually complexed with a central metal atom.

Primary structure of proteins: The covalent backbone structure of a protein, including its amino acid sequence and its inter- and intra-chain disulfide bridges.

Prokaryotes: Simple unicellular organisms (bacteria and blue-green algae) with no nuclear membrane, no membrane-bound organelles, and a single chromosome.

Prosthetic group: A metal ion or an organic group, other than an amino acid, which is bound to a protein and serves as its active group.

Proteolytic enzyme: An enzyme catalyzing hydrolysis of proteins or peptides.

Proton acceptor: An anionic compound capable of accepting a proton from a proton donor.

Proton donor: An acid, i.e., the donor of a proton in an acid-base reaction.

Purine: A basic nitrogenous compound found in nucleotides and nucleic acids, containing fused pyrimidine and imidazole rings.

Puromycin: An antibiotic that inhibits polypeptide synthesis by competing with aminoacyl tRNAs for incorporation into the polypeptide chain.

Pyranose: A simple sugar containing the pyrane ring.

Pyridine-linked dehydrogenases: Dehydrogenases requiring as coenzyme either one of the pyridine coenzymes NAD and NADP.

Pyridine nucleotide: A nucleotide-like coenzyme containing the pyridine derivative nicotinamide.

Pyridoxal phosphate: A coenzyme containing the vitamin pyridoxol and functioning in reactions involving amino group transfer.

Pyrimidine: A nitrogenous heterocyclic base which is a component of a nucleotide or nucleic acid.

Pyrophosphatase: An enzyme hydrolyzing inorganic pyrophosphate to yield two molecules of (ortho)phosphate.

Pyrophosphate cleavage: Enzymatic cleavage of ATP to yield AMP and pyrophosphate, usually coupled to synthesis of some other bond.

Pyrophosphorylases: Enzymes catalyzing the formation of nucleoside diphosphate sugars and pyrophosphate from a sugar phosphate and a nucleoside 5′-triphosphate.

Quantum: Ultimate unit of light energy.

Quaternary structure: The three-dimensional structure of an oligomeric protein, particularly the manner in which the chains fit together.

R group: The distinctive side-chain of an α-amino acid.

Racemate: An equimolar mixture of the D- and L-stereoisomers of an optically active compound.

Radioactive isotope: An isotopic form of an element with an unstable nucleus that stabilizes itself by emitting ionizing radiation.

Redox couple: An electron donor and its corresponding oxidized form.

Reducing agent (reductant): An electron donor in an oxidation-reduction reaction.

Reducing equivalent: General term for an electron or an equivalent hydrogen atom in an oxidation-reduction reaction.

Reduction: The gain of electrons by a compound.

Regulatory (allosteric) enzyme: Enzyme having a regulatory function through its capacity to undergo a change in catalytic activity on binding a specific modulating metabolite.

Replication: The synthesis of a daughter DNA molecule complementary to a parental DNA.

Respiration: The oxidative breakdown and release of energy from fuel molecules by reaction with oxygen in aerobic cells.

Respiratory chain: The electron transport chain, a sequence of electron-carrying proteins which transfer electrons from substrates to molecular oxygen.

Respiratory chain phosphorylation: Oxidative phosphorylation; phosphorylation of ADP coupled to electron transport between substrate and oxygen.

Reversible process: A process which proceeds with no change in entropy.

Ribonuclease: An (endo)nuclease capable of hydrolyzing certain internucleotide linkages of RNA.

Ribonucleotides: Nucleotides containing D-ribose as their pentose component.

Ribosomal RNA (rRNA): A class of RNA molecules serving as components of ribosomes.

Ribosomes: Small particles in cells, about 200 Å in diameter, made up of RNA and protein, which serve as the site of protein synthesis.

RNA (ribonucleic acid): A polyribonucleotide linked by successive 3',5'-phosphodiester linkages.

RNA polymerase: An enzyme that catalyzes the formation of RNA from ribonucleoside triphosphates, using a strand of DNA (or RNA) as a template.

Saponification: Alkaline hydrolysis of neutral fats (triacylglycerols) to yield fatty acids as soaps.

Sarcomere: Functional and structural unit of muscle contractile system.

Saturated fatty acid: Fatty acid containing a fully saturated alkyl chain.

Secondary structure: The configuration of the backbone of a polypeptide chain in its extended configuration along one axis, as in fibrous proteins.

Second law of thermodynamics: The entropy of the universe always tends to increase.

Sedimentation coefficient: A physical constant specifying the rate of sedimentation of a particle in a centrifugal field under specified conditions.

Simple protein: A protein yielding only amino acids on hydrolysis.

Specific activity: The number of micromoles of substrate transformed by an enzyme preparation per min per mg protein at 25°C.

Specific rotation: The rotation in degrees of the plane of plane-polarized light (D-line of sodium) of an optically active compound at 25°C at a specified concentration and light path.

Spontaneous process: A process accompanied by an increase in entropy.

Standard free energy change: The gain or loss of free energy in calories as one mole of reactants in the standard state is converted into one mole of products, under standard conditions of temperature, pressure, and concentration.

Standard reduction potential: The electromotive force exhibited at an electrode by equal concentrations of a reducing agent and its oxidized form at 25°C, a measure of the relative tendency of the reducing agent to lose electrons.

Standard state: The most stable form of a pure substance at 1.0 atmosphere pressure and 25°C (298°K). For reactions occurring in solution, the standard state of a solute is a 1.0 molal solution.

Steady state: A nonequilibrium state of a system through which matter is flowing and in which all components remain in constant concentration.

Stereoisomers: Isomers which are nonsuperimposable mirror images of each other.

Steroids: Class of lipids containing cyclopentanophenanthrene ring system.

Structural gene: A gene coding for the structure of a protein.

Substrate: The specific compound acted upon by an enzyme.

Substrate-level phosphorylation: Phosphorylation of ADP or some other nucleoside 5'-diphosphate coupled to a one-step oxidation of an organic substrate, prior to electron transport from the first electron acceptor to oxygen.

System: An isolated collection of matter. All other matter in the universe apart from the system is called the surroundings.

Template: A macromolecular mold or pattern for the synthesis of another macromolecule.

Terpene: An organic hydrocarbon or hydrocarbon derivative constructed from recurring isoprene units.

Tertiary structure (of a protein): The three-dimensional configuration of the polypeptide chain of a globular protein in its native folded state.

Tetrahydrofolic acid: The reduced, active coenzyme form of the vitamin folic acid.

Thioester: An ester of a carboxylic acid with a thiol or mercaptan.

Third law: The entropy or randomness of a perfect crystal is zero at absolute zero.

Titration curve: A plot of the pH versus the equivalents of base added during titration of an acid.

Tocopherols: Forms of vitamin E.

Toxins: Proteins elaborated by some organisms which are toxic to some other species.

Transaminases: Enzymes catalyzing transfer of amino groups; also called aminotransferases.

Transamination: The enzymatic transfer of an amino group from an amino acid to a keto acid.

Transcription: The enzymatic process whereby the genetic information contained in DNA is used to specify a complementary sequence of bases in an RNA chain.

Transfer RNAs (tRNAs): A class of RNA molecules (mol wt 25,000–30,000) which combines covalently with a specific amino acid; the resulting aminoacyl-tRNA then hydrogen-bonds to a mRNA nucleotide triplet or codon.

Transition state: An activated form

of a molecule capable of undergoing a chemical reaction.

Translation: The process in which the genetic information present in an mRNA molecule directs the sequence of amino acids during protein synthesis.

Triacylglycerol: A neutral fat; ester of glycerol with three molecules of fatty acid.

Tricarboxylic acid cycle: Cyclic series of enzymatic reactions for the oxidation of acetyl residues to CO_2; a central pathway of respiration.

Turnover number: The number of times an enzyme molecule transforms a substrate molecule per min under conditions giving maximal activity.

Uncoupling agent: A substance which can uncouple phosphorylation of ADP from electron transport; an example is 2,4-dinitrophenol.

Unsaturated fatty acid: A fatty acid containing one or more double bonds.

Urea cycle: A metabolic pathway in the liver responsible for the synthesis of urea from amino groups and CO_2.

V_{max}**:** The maximum velocity of a given enzymatic reaction.

Viruses: Self-replicating, infectious nucleic acid-protein complexes, which require intact host cells for

their replication and which contain a chromosome of either DNA or RNA.

Vitamins: Trace organic substances, required in the diet of some species, most of which function as components of certain coenzymes.

X-ray crystallography: Use of x-ray scattering by crystals to determine the three-dimensional structure of molecules.

Zwitterion: A dipolar ion, one with spatially separated positive and negative charges.

Zymogen: An inactive precursor of an enzyme, such as pepsinogen.

ACKNOWLEDGMENTS

Page 17, Figure 1-2 (right), reprinted from Linus Pauling, *The Nature of the Chemical Bond*. Copyright 1939 and 1940 by Cornell University. Third edition © 1960 by Cornell Used by permission of Cornell University Press.

Page 51, Figure 3-1, E. Margoliash.

Page 63, Figure 3-7, from E. Margoliash in B. Chance and R. Estabrook (eds.), *Hemes and Hemoproteins*, Academic Press, New York, 1966, p. 373.

Page 64, Figure 3-8, (left) from Linus Pauling, *The Nature of the Chemical Bond*. Copyright 1939 and 1940 by Cornell University. Third edition © 1960 by Cornell University. Used by permission of Cornell University Press. (right) From Linus Pauling, *Nature*, **171**:59 (1953).

Page 65, Figure 3-9 (top), redrawn from L. Pauling and R. B. Cory, *Proc. Nat. Acad. Sci.*, **37**:729 (1951).

Page 67, Figure 3-11, redrawn from R. E. Dickerson in H. Neurath (ed.), *The Proteins*, **II**, Academic Press, 1964, p. 634.

Page 102, Figure 5-17, D. Fawcett.

Page 104, Figure 5-19, R. D. Preston.

Page 135, Figure 7-12 (left), L. D. Simon; (right) T. F. Anderson.

Page 169, Figure 9-8, electron micrograph from G. Decker.

Page 171, Figure 9-9, electron micrograph from G. Decker.

Page 222, Figure 12-4, redrawn from L. J. Reed.

Page 251, Figure 13-15, K. R. Porter.

Page 282, Figure 16-1 (right), M. C. Ledbetter.

Page 282, Figure 16-3, redrawn from F. T. Haxo and L. R. Blinks, *J. Gen. Physiol.*, **33**:408 (1950).

Page 347, Table 20-3, data from W. S. Hoffman, *The Biochemistry of Clinical Medicine*, 4th ed., 1970, page 142. Year Book Medical Publishers, Inc., Chicago.

Page 350, Figure 20-6 (top), redrawn from Sylvia Colard Keene in W. Bloom and D. W. Fawcett, *Textbook of Histology*, W. B. Saunders Co., Philadelphia, 1968.

Page 376, Figure 21-1 (left), W. Büchi; (right) M. H. F. Wilkins, Medical Research Council, Biophysics Unit, King's College, London.

Page 377, Figure 21-2, redrawn from L. Pauling and R. B. Corey, *Arch. Biochem. Biophys.*, **65**:164 (1956), Academic Press, New York.

Page 380, Figure 21-5, redrawn from P. Doty in D. J. Bell and J. K. Grant (eds.), "The Structure and Biosynthesis of Macromolecules," *Biochem. Soc. Symposia*, **21**:8 (1962), Cambridge Press.

Page 381, Figure 21-6, redrawn from P. Doty in D. J. Bell and J. K. Grant (eds.), "The Structure and Biosynthesis of Macromolecules," *Biochem. Soc. Symposia*, **21**:8 (1962), Cambridge Press.

Page 382, Figure 21-7, data from H. R. Mahler, B. Kline, and B. D. Mehrotra, *J. Mol. Biol.*, **9**:801 (1964).

Page 383, Figure 21-8, (left) A. Kornberg; (right) L. MacHattie and C. A. Thomas, Jr.

Page 384, Figure 21-9, J. Cairns.

Page 406, Figure 23-1, K. R. Porter.

Page 408, Figure 23-4, redrawn from W. Fuller and A. Hodgson, *Nature*, **215**:817 (1967).

eukaryotes, 168, 279
DNA, 129
ribosomes, 133
evolution, 170, 194
excited state, 282
exergonic reactions, 181
exonucleases, 132

F₁, 250
FAD (*see* flavin adenine dinucleotide)
farnesyl pyrophosphoric acid, 318
fats, 110
caloric value, 255
release, 360
storage, 110, 360
(*see also* triacylglycerols)
fatty acids, 107
activation, 189, 256, 257
chemical properties, 109
elongation, 316
essential, 316
esters, 115
in carbohydrate synthesis, 301
in triacylglycerols, 110
melting points, 109
oxidation, 163, 172, 242, 255
saturated, 107
synthesis, 161, 163, 172, 230, 311
unsaturated, 107
unsaturated, synthesis, 316
transport, 54
fatty acid activating enzyme, 257, 317
fatty acid: CoA ligase, 257
fatty acid synthetase, 311
fatty acid thiokinase, 257
fatty acyl adenylate, 258
fatty acyl carnitine, 258
fatty acyl CoA
formation, 257
in lipid synthesis, 316
fatty acyl CoA dehydrogenase, 259
feedback inhibition, 87, 165
in amino acid synthesis, 326
fermentation, 73, 166, **193**
alcoholic, 141, 193, **206**
lactic, 194
ferredoxin
in nitrogen fixation, 333
in photosynthesis, 290
ferritin, 52, 54
fibrinogen, 54, 343
fibroin, 3-dimensional structure, 63
fibrous proteins, 53, 63
first law of thermodynamics, 178
Fischer, E., 94
Fiske, C., 177
flavin adenine dinucleotide, 127, 138,
141, **142**, 241
in fatty acid oxidation, 258
in pyruvate dehydrogenase complex,
221
in succinate dehydrogenase, 225
flavin-linked dehydrogenases, 241
flavin mononucleotide, 127, 138, 141,
142, 241
in amino acid oxidation, 267
reduction, 242
flavin nucleotides, 141
flavoenzymes, 141, 242
flickering clusters, 16

fluorescence, 282
fluoride, 196, 204
1-fluoro-2,4-dinitrobenzene, 41
FMN (*see* flavin mononucleotide)
folic acid, 138, **145**, 148, 325
formic acid, pK', 26
N-formylmethionine, 409, 417
N-formylmethionyl-tRNA, 409
N¹⁰-formyltetrahydrofolate, 409
free energy, 4, 163, **180**, 295
free energy change, 180
standard, 180
freezing point depression, 19
α-D-fructofuranose, 97
β-D-fructofuranose, 96, 97
fructokinase, 210
fructose, **94**, 338, 341, 344
glycolysis, 209
in polysaccharides, 103
in sucrose, 101
mutarotation, 97
optical activity, 94
phosphorylation, 197, 210
specific rotation, 94
fructose diphosphatase, 299, 300
fructose 1,6-diphosphate, 196, **199**, 299
fructose 1-phosphate, 210
fructose 1-phosphate aldolase, 210
fructose 6-phosphate
in glucose synthesis, 299
in transaldolase reaction, 232
in transketolase reaction, 232
standard free energy of hydrolysis, 184
α-D-fructose 6-phosphate, 100, **199**
fumarase, 225
molecular weight, 226
optimum pH, 78
fumarate
from amino acid degradation, 269, 272
from urea cycle, 274
in glucose synthesis, 301
in tricarboxylic acid cycle, 225
fumaric acid 218, 225
(*see also* fumarate)
4-fumarylacetoacetic acid, 270
fungi, 170
furan, 96
furanose, 96
furfurals, 97

ΔG° (*see* standard free energy change)
G-actin, 349
G factor, 411
D-galactosamine, 99, 100
galactose
formation from glucose, 307
in glycolysis, 209
in lactose, 101
phosphorylation, 210
D-galactose, **93**, 95, 344
galactose 1-phosphate, 210, 211
β-galactosidase, 79, 418
gangliosides, 113
gas constant, 341
gastric juice, 25
gastrointestinal tract, 337, 338
GDP (*see* guanosine diphosphate)
genes, 370, 383
derepression, 165
regulatory, 419

structural, 419
transcription, 401
genetic code, 415–417
genetic information, 7
genetic maps, 370, 385
genetic mutations, 166, 269, 369, **385**
genetic recombination, 395
genetics, 369
geranyl pyrophosphoric acid, 318
germ cells, 374
glass electrode, 25
gliadin, 54
globular actin, 349
globular proteins, 53, 66, 68
α-globulins, 343
β-globulins, 343
γ-globulin, 52, 343
isoelectric pH, 61
molecular weight, 53
glucagon, 356, 359
glucans, 102
α(1 → 4)glucan phosphorylases, 207
glucokinase, 199
gluconeogenesis, **297**, 345, 348
gluconeogenesis, regulation, 300
D-gluconic acid, 100
α-D-glucopyranose, 95, 97
β-D-glucopyranose, 95, 97
D-glucosamine, 99, 100, 197
glucosamine 6-phosphate, regulation of
glutamine synthetase, 327
glucose
acetylation, 97
anomers, 96
as energy source, 163
crystals, 95
fermentation, 193
furfural formation, 97
glycolysis, 193
in disaccharides, 101
in fatty acid synthesis, 311
in polysaccharides, 102
in starch hydrolysis, 102
mutarotation, 97
optical activity, 94
oxidation, 194, 216, 247
phosphorylation, 197
reduction, 98
ring structure, 95
specific rotation, 94
synthesis, 162, 172, 297
D-glucose, 91, 93, 95, 158, 338, 341
β-D-glucose, 95
L-glucose, 95, 341
glucose 1,6-diphosphate, 209
glucose phenylosazone, 98
glucose 6-phosphatase, **299**, 344
glucose 1-phosphate
in glycogen synthesis, 306
standard free energy of hydrolysis, 184
α-D-glucose 1-phosphate, **100**, 207, 357
glucose 6-phosphate, **100**, 188, 197, 199,
208, 344
from photosynthesis, 303
in carbohydrate synthesis, 298
in phosphogluconate pathway, 230
in regulation of glycogen synthesis, 307
in regulation of hexokinase, 197
in starch synthesis, 307
oxidation, 233
standard free energy of hydrolysis, 184,